Molecular Mechanisms of Programmed Cell Death

Molecular Mechanisms of Programmed Cell Death

Edited by

Yufang Shi

University of Medicine and Dentistry of New Jersey
Robert Wood Johnson Medical School
Piscataway, New Jersey

John A. Cidlowski

National Institutes of Health
Research Triangle Park, North Carolina

David Scott

American Red Cross
Rockville, Maryland

Jia-Rui Wu

Chinese Academy of Sciences
Shanghai, China

and

Yun-Bo Shi

National Institutes of Health
Bethesda, Maryland

Kluwer Academic / Plenum Publishers
New York, Boston, Dortrecht, London, Moscow

Library of Congress Cataloging-in-Publication Data

International Symposium on Programmed Cell Death (2nd : 2002 : Shanghai, China)
Molecular mechanisms of programmed cell death / edited by Yufang Shi . . . [et al.].
p. cm.
Includes bibliographical references and index.
ISBN 0-306-47778-5
1. Apoptosis—Congresses. 2. Immune response—Regulation—Congresses. 3. Cancer cells—Growth—Regulation—Congresses. I. Shi, Yufang, 1960– II. Title.
[DNLM: 1. Apoptosis—Congresses. 2. Models, Immunological—Congresses. 3. Signal Transduction—Congresses. QH 671 I615m 2004]
QH671.I585 2002
571.9′36–dc22

2003060725

This volume is based on proceedings of the Second International Symposium of Programmed Cell Death held September 1–3, 2002, in Shanghai, China.

ISBN: 0-306-47778-5

233 Spring Street, New York, New York 10013

http://www.wkap.nl/

10 9 8 7 6 5 4 3 2 1

Sponsors

National Institutes of Child Health and Development, National Institutes of Health

National Institutes of Environmental Health Sciences, National Institutes of Health

American Red Cross

University of Medicine and Dentistry of New Jersey-Robert Wood Johnson Medical School

Shanghai Institutes for Biological Sciences, Chinese Academy of Science

Preface

The 2002 Nobel Prize in Physiology or Medicine was awarded to Sydney Brenner, H. Robert Horvitz, and John E. Sulston for their seminal discoveries concerning "genetic regulation of organ development and programmed cell death." This clearly marked the prime importance of understanding the molecular mechanisms controlling cell death.

The 1st International Symposium on Programmed Cell Death was held in the Shanghai Science Center of the Chinese Academy of Sciences on September 8–12, 1996. A number of key issues in apoptosis were discussed at the meeting, and progress in major areas of apoptosis research was summarized by expert participants at the meeting and published by Plenum Publishing Corporation as a book entitled *Programmed Cell Death.* In the last six years, we have witnessed a real explosion in our knowledge on how cells undergo apoptosis, thereby participating in various developmental and pathophysiological processes. At this ever-exciting time, we organized the 2nd International Symposium on Programmed Cell Death.

The chapters in the present volume include contributions from invited speakers. Given the explosive growth and progress in the apoptosis field, it is clear that no single meeting will be able to cover all important areas. This volume emphasizes key areas such as cell volume changes, the role of Bcl-2 family proteins, signaling of the TNF family molecules, extracellular matrix, and the role of apoptosis in the regulation of the immune system and cancer. In each case, the contributors have emphasized the areas that are still open for further exploration. In addition, potential applications for understanding and treatment of diseases are discussed. Some diagrammatic representations are provided, which will be invaluable for summarizing the wealth of information. We expect that this volume will help those in basic research in this fascinating area, as well as those actively involved in drug discovery.

We are fortunate to have had financial support from National Institute of Child Health and Development of the National Institutes of Health, National Institute of Environmental Health Sciences of the National Institutes of Health, the American Red Cross, University of Medicine and Dentistry of New Jersey-Robert Wood Johnson Medical School, and Shanghai Institutes for Biological Sciences of Chinese Academy of Sciences. We are also deeply indebted to Mr. Richard Wernoski, Ms. Bo Zhou, and Ailan Chang for their support in putting the meeting together. Their efforts not only ensured the success of the meeting but also made it exciting and memorable.

YUFANG SHI
JOHN CIDLOWSKI
DAVID SCOTT
JIA-RUI WU
YUN-BO SHI

Contents

Chapter 1

Akt and Bcl-x_L Are Independent Regulators of the Mitochondrial Cell Death Pathways

DAVID R. PLAS, JEFFREY C. RATHMELL, JAMES E. THOMPSON, AND CRAIG B. THOMPSON*

ABSTRACT: In vivo, hematopoietic cells require continuous signals from their microenvironment to prevent activation of the endogenous programmed cell death machinery. Cell survival is therefore limited by the availability of ligands for the receptors that can influence cell survival. Following loss of receptor engagement, IL-3-dependent hematopoietic cells undergo a rapid decline in cellular metabolism, characterized by reductions in surface expression of the glucose transporter GLUT-1, mitochondrial potential, and cellular ATP. Two distinct classes of oncogenes can prevent cell death in response to declines in glucose uptake and metabolism following growth factor withdrawal: pro-survival Bcl-2 proteins, such as Bcl-x_L, or an activated form of Akt. However, Bcl-x_L and Akt appear to promote survival by distinct mechanisms. Expression of activated Akt leads to maintenance of glucose transporter expression, glycolytic activity, mitochondrial potential, and cell size, while Bcl-x_L-expressing cells deprived of growth factor survive in a more vegetative state characterized by small cells with reduced mitochondrial potential and glycolytic activity. Akt-mediated survival is dependent on promoting glycolysis and maintaining a physiologic mitochondrial potential. In contrast, Bcl-x_L maintains mitochondrial integrity in the face of a reduced mitochondrial membrane potential in growth factor-deprived cells. Thus, Akt and Bcl-x_L suppress mitochondrial-initiated apoptosis by distinct mechanisms.

Introduction

It is poorly understood how multicellular organisms maintain relatively constant numbers of cells throughout adult life. It has long been hypothesized that one critical mechanism underlying the control of cell numbers is the observation that the majority, if not all, cells in multicellular organisms lack the autonomous ability to replicate. Thus, metazoan cells have become dependent on extracellular signals for both initiating and progressing through the cell cycle. As such, cells are incapable of accumulating in a cell autonomous way, establishing a mechanism whereby the accumulation of excess cells during adult life is prevented by limiting the availability of necessary growth factors (Hanahan and Weinberg, 2000). Recently, we and others have hypothesized that a simple extension of this model could

*Departments of Medicine and Cancer Biology, Abramson Family Cancer Research Institute, University of Pennsylvania, Philadelphia, PA 19104.

also serve to explain the regulation of cell death in multicellular organisms. This proposal suggests that all cells within a multicellular organism are also dependent on extracellular survival signals to prevent the induction of cell death. When cells are deprived of necessary survival factors for a sustained period of time, an endogenous cell suicide pathway commonly referred to as apoptosis, or programmed cell death, is activated (Rathmell and Thompson, 2002).

A number of laboratories have provided evidence for the dependency of cell survival on extracellular signals using lineage-specific survival factors (Marrack et al., 2000). Recent evidence suggests that apoptotic death has a number of common features, independent of the cell's lineage. A central event in the induction of apoptosis in response to numerous apoptotic stimuli is the formation of a caspase-9-activating complex comprised of cytochrome c and Apaf-1 and ATP or dATP (Zou et al., 1997). In intact cells, cytochrome c is sequestered in the intermembrane space of mitochondria, where it functions as a component of the electron transport chain. Apoptotic stimuli induce events which eventually trigger the translocation across the outer mitochondrial membrane into the cytosol where it oligomerizes with Apaf-1. As a result of the cleavage of the proenzyme caspase-9 to its active form, a caspase cascade is activated that is responsible for the morphologic features associated with apoptosis (Shi, 2002). Studies to characterize how cytochrome c translocation is accomplished suggest a disruption in the integrity of the outer mitochondrial membrane that releases not only cytochrome c but also a number of other proapoptotic molecules normally resident in the inter-membrane space, such as AIF or Smac (Du et al., 2000; Susin et al., 1999; Verhagen et al., 2000).

Much attention has now been focused on the molecular basis for the loss in the mitochondrial outer membrane integrity that results in redistribution of cytochrome c and other proapoptotic factors. This step in many cell types appears to be a point of irreversible commitment to cell death. Two hypotheses concerning the role of mitochondria in programmed cell death have been developed. In the prevailing model, mitochondria are viewed primarily as a storage site for various proapoptotic proteins. In this view, mitochondrial permeability is triggered as a result of events in the cytosol, which stimulate apoptotic control proteins in the cytosol to directly induce mitochondria to release cytochrome c (Huang and Strasser, 2000). In the extreme, this model proposes that mitochondrial physiology plays no central role in the regulation of apoptosis. An alternative model suggests that an impairment in mitochondrial function causes the loss in the integrity of the outer mitochondrial membrane (Vander Heiden and Thompson, 1999). In this model, loss of the integrity of the outer mitochondrial membrane is viewed as an irreversible loss of the ability of mitochondria to maintain organelle physiology.

A family of proteins localized to the outer mitochondrial membrane can regulate the ability of mitochondria to release cytochrome c and other pro-apoptotic molecules. The prototype of this family, Bcl-2, was first demonstrated to regulate the induction of apoptosis in leukemia cells overexpressing Bcl-2 as a result of a chromosomal translocation (Tsujimoto et al., 1985). Following the identification of cytochrome c as a critical mitochondrial constituent that is required for the activation of caspase 9, it was shown that the ability of Bcl-2 and related proteins to block programmed cell death is explained by an ability to prevent cytochrome c exit from mitochondria (Kluck et al., 1997; Yang et al., 1997). This finding has implications for models explaining the mechanism of cytochrome c release from mitochondria. If mitochondria act to simply release cytochrome c when acted upon by events in the cytosol, then Bcl-2 family proteins must regulate these cytosolic events. Alternatively, if a failure in mitochondrial function is tied to the release of cytochrome c, then Bcl-2 family proteins would be expected to actively support mitochondrial physiology.

To examine the role of mitochondrial physiology in the induction of programmed cell death, we have for several years been studying growth factor-dependent cell lines derived from hematopoietic lineages, characterizing changes in mitochondrial bioenergetics following removal of growth factors from cultures. It was originally hypothesized that decreases in signal transduction through growth factor receptors might result in an increased level of cellular ATP and a decline in ADP, since growth factor signal transduction depends on phosphorylation reactions to engage kinase cascades, activate new transcription, and stimulate translation. All three of these processes are energy-dependent. Despite this expectation, we have found in multiple cell lines that withdrawal of growth factor or serum leads to a reproducible and continuous decline in the ATP/ADP ratio (Vander Heiden et al., 1999). Following withdrawal of growth factors, the fall in the ATP/ADP ratio can be accounted for by a decline in mitochondrial substrates with which to maintain electron transport chain activity and the mitochondrial membrane potential (Vander Heiden et al., 2001). Careful analysis of NADH compartments in cells suggests that the majority of the decline in NADH available to maintain electron transport comes from loss of the NADH produced through glycolysis, and demonstrated that a common feature of growth factor signal transduction is to maintain glucose uptake and glycolytic metabolism (Harris et al., manuscript submitted). Anti-apoptotic Bcl-2 proteins such as Bcl-2 or Bcl-x_L act prior to apoptosis to dampen the decline in the ATP/ADP ratio, suggesting that they act to sustain the ability of mitochondria to maintain ATP production in the face of growth factor withdrawal (Vander Heiden et al., 1999).

Experiments on a number of cellular systems suggest that one common feature of cytokines involved in cell survival is their ability to maintain cellular glucose uptake. In studies characterizing IL-3 withdrawal-induced death in the IL-3-dependent cell line, FL5.12, we find that following IL-3 withdrawal, there is a rapid loss of the expression of three enzymes involved in the proximal steps of glucose uptake and glycolytic commitment (Rathmell et al., 2000). There is a rapid decline in both the mRNA and protein levels of GLUT-1, the major glucose transporter of hematopoietic cells, hexokinase, and phosphofructokinase-2 (Vander Heiden et al., 2001). As a result, within 12 hours of growth factor withdrawal, the cell experiences a 10-fold decline in its ability to take up glucose from its extracellular environment and cannot generate a sufficient supply of NADH to maintain electron transport at levels that would sustain cellular ATP levels. This has led us to a relatively simple model that suggests a common feature of growth factor survival-mediated signal transduction is the stimulation of nutrient uptake necessary for the production of NADH required to maintain mitochondrial bioenergetics (Figure 1). When growth factors are withdrawn, the loss of the ability of cells to autonomously take up sufficient nutrients to maintain the NADH supply leads to a progressive loss of remaining intracellular stores of metabolites that can be utilized to produce NADH. When NADH levels fall beyond a given threshold, mitochondria are no longer able to generate a sufficient mitochondrial membrane potential to maintain ion homeostasis. This results in a disruption in their physiology, the non-specific rupture of the outer mitochondrial membrane, and the release of cytochrome c into the cytosol, where it can initiate formation of an apoptosome.

Surprisingly, we have found that Bcl-2 or Bcl-x_L overexpression, while preventing cytochrome c release, has no effect on the loss of glucose uptake and glycolysis in response to growth factor withdrawal (Rathmell et al., 2000). As a result, Bcl-x_L can promote cell

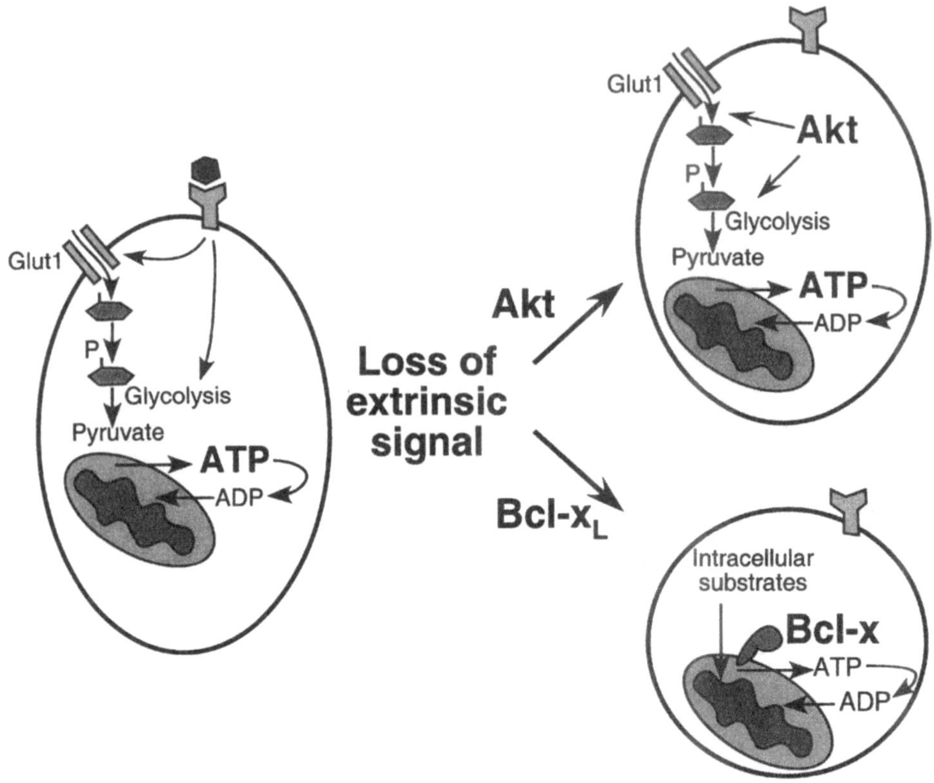

Figure 1. Bcl-x$_L$ and Akt mediate cell survival via distinct mechanisms. Signals emanating from growth factor receptors support both glucose uptake and glycolysis, providing substrates required for maintaining mitochondrial homeostasis. When growth factors are removed from the medium, activated Akt can promote cell survival by supporting continued glucose uptake and metabolism (top), while Bcl-x$_L$ can prevent the disruption of mitochondrial homeostasis, allowing cells to adapt to a diminished supply of energy driven by the consumption of intracellular nutrients (bottom).

survival under conditions of not only growth factor withdrawal but also glucose withdrawal (Figure 2). Nevertheless, the changes in mitochondrial physiology that immediately precede cytochrome c release are not detected when Bcl-x$_L$ is expressed, suggesting that in the presence of Bcl-x$_L$, the decline in glycolysis is not causing a disruption in mitochondrial physiology. Thus, when antiapoptotic Bcl-2 proteins are overexpressed, cells may have sufficient time to alter their physiology to adapt to long-term maintenance of their mitochondrial integrity through the induction of autophagy, in which efficient utilization of intracellular substrates through lysosomal degradation is utilized to maintain a supply of NADH and thus mitochondrial integrity over a prolonged period of time (Vander Heiden et al., 1997). Under this model, the expression of the proapoptotic Bcl-2 family members, such as Bax and Bak, prevents cells from undergoing such an adaptation and leads to the induction of apoptosis (Figure 2A). Based on these models, it appears that the relative balance of pro- and anti-apoptotic Bcl-2 family members determines the sensitivity of cells to undergo cell death in response to growth factor withdrawal. Sensitivity of cells to apoptosis can be further modulated by transcriptionally-activated or post-translationally-modified BH3-containing proteins, which act as regulators of the functions of Bcl-2 proteins (Huang

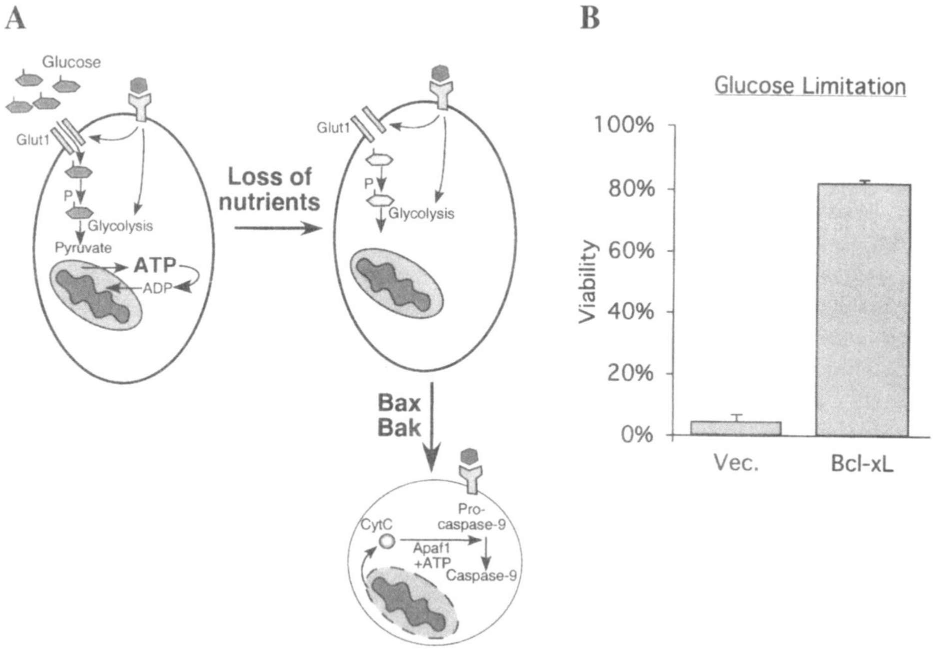

Figure 2. Control of apoptosis under nutrient-limited conditions. A. When nutrients such as glucose become limiting, cells fail to provide mitochondria with the nutrients required to maintain mitochondrial homeostasis, even though growth factor may still be present. Cytochrome c is released in a Bax- and Bak-dependent manner, resulting in the activation of Caspase 9 and downstream caspases. B. Bcl-x_L prevents cell death when glucose is limiting. Vector control or Bcl-x_L-expressing FL5.12 cells were cultured in medium containing a limiting concentration of glucose (0.02 mM). Cell viability was measured by propidium iodide exclusion after 48 hours of culture.

and Strasser, 2000). Consistent with this, we find that cells deficient in both Bax and Bak are profoundly resistant to apoptosis following growth factor withdrawal (Figure 2A) (Cheng et al., 2001; Lindsten et al., 2000; Zong et al., 2001). This model has important implications for how long cells can survive in the absence of growth factor withdrawal.

Based on the model presented in Figure 1, either a loss of the proapoptotic functions of Bax and Bak, or an overexpression of Bcl-2 or Bcl-x_L, will allow cells to acclimate to a more efficient, highly coupled form of mitochondrial maintenance. However, this occurs as a result of the progressive consumption of cellular constituents, since intracellular organelles and contents must be oxidized to provide a continuous supply of NADH. As a result, cells progressively atrophy in the absence of extracellular signal transduction, and although they remain alive, they exist in a state which requires prolonged recovery before they can reinitiate cell proliferation following growth factor readdition (Casey J. Fox and C.B.T, manuscript submitted). This is best seen when cells overexpressing Bcl-x_L are followed over a long period of time of growth factor withdrawal. Under these conditions, Bcl-x_L-expressing cells maintain nearly uniform survival for approximately one week. Thereafter there is a continual loss of cell viability, with few, if any, recoverable cells three weeks after growth factor withdrawal. Throughout this entire time course, cells undergo progressive atrophy, which causes a progressively longer delay before cells can reenter S phase following readdition of growth factors. Thus, Bcl-x_L-dependent cell survival is accompanied by a cost: the progressive atrophy of the cell and loss of its capabilities to carry out effector functions or cell proliferation.

These observations suggest that growth factor signal transduction must have components that directly stimulate glucose uptake and glycolytic commitment. The insulin receptor signaling pathway represents a paradigm for growth factor signal transduction that directly modulates glucose metabolism (Hajduch et al., 2001). In both muscle and fat cells, the ligand-induced activation of the tyrosine kinase activity of the insulin receptor leads to the phosphorylation of IRS-1, which in turn acts as an adaptor in the activation of PI3 kinase. The products of PI3 kinase lead to the membrane recruitment and activation of the serine/threonine kinase Akt, which can activate the three proximal steps of glycolysis, including glucose transporter expression and surface translocation, hexokinase-2 expression, and PFK-1 activity through allosteric regulation of PFK-2 (J.C.R. and C.B.T. manuscript submitted).

Like many other survival cytokines involved in maintaining cell survival, IL-3 stimulates Akt activity, suggesting that growth factor receptors modulate cellular bioenergetics in hematopoietic cells, as has been described for the insulin receptor in fat and muscle cells (Songyang et al., 1997). Therefore, the bioenergetic effects of overexpression of a constitutively active form of Akt on the survival of FL5.12 cells in the absence of growth factor were examined (Plas et al., 2001). Cells overexpressing constitutively active myristoylated Akt were produced, and their ability to survive following growth factor withdrawal was examined over a three week time period. Akt provided potent ability of cells to survive over this period of time such that approximately 30% of cells were capable of surviving and undergoing rapid recovery following readdition of IL-3, even after three weeks of growth factor withdrawal. However, a decidedly different pattern of cell survival kinetics was observed. A large number of Akt cells underwent cell death in the first week following growth factor withdrawal. However, after the first week, cells apparently had undergone an accommodation that allowed them to survive for a prolonged period of time without significant further diminution in their survival. This occurred in the absence of proliferation and was observed in multiple clones. Furthermore, unlike Bcl-x_L-transfected cells, Akt-transfected cells (although they withdrew from the cell cycle) maintained greater cell size in G1 throughout the period of time of growth factor withdrawal. This correlated with the ability of Akt to maintain GLUT-1 expression, glycolytic activity, and mitochondrial membrane potential (Plas et al., 2001). Furthermore, although constitutive Akt activation provides prolonged protection from apoptosis induced by growth factor withdrawal, it fails to protect from glucose withdrawal (Figure 3). Thus, it appears that following growth factor withdrawal, the activity of Akt on the proximal steps of glycolysis allows cells to maintain a higher level of cell autonomous nutrient uptake, thereby preventing disruption of mitochondrial homeostasis and the release of cytochrome c.

The ability of Akt to maintain glucose transport expression on the cell surface was examined by confocal microscopy. In addition to maintaining the overall expression of GLUT-1 on cells, Akt selectively maintained its expression at the cell surface, accounting for its ability to sustain glucose uptake and glycolysis. This activity of Akt to maintain nutrient transporters was not confined only to the uptake of glucose. A number of amino acids can also contribute to the production of NADH under conditions of nutrient limitation, and cell culture media contains high levels of both essential and nonessential amino acids. In growth factor-dependent cells, removal of growth factor induces a rapid internalization of amino acid transporters as visualized by the antibody 4F2, which is directed against the light chain of the common amino acid transporter. This intracellular sequestration and

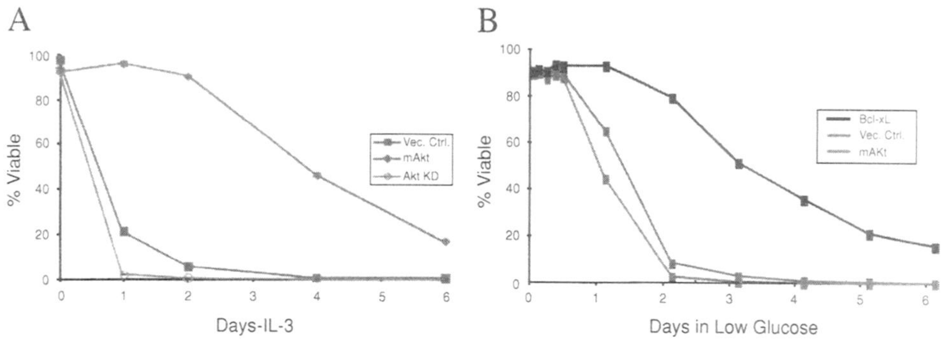

Figure 3. Akt prevents death in response to growth factor withdrawal, but not glucose withdrawal. A. Cells expressing activated Akt (mAkt) or kinase-deficient Akt (Akt KD) were cultured in the absence of growth factor for up to 6 days. Cell viability in mAkt, Akt KD, and vector control cells was measured by propidium iodide exclusion at the indicated time points. B. Vector control, mAkt-, or Bcl-x_L-expressing cells were cultured in limiting concentrations of glucose. Cell viability was measured as described in A.

degradation within the lysosome of amino acid transporters was prevented in Akt-expressing cells (Edinger and Thompson, 2002). Thus, Akt promotes not only the uptake of glucose as previously described in insulin responsive tissues, but also maintains amino acid uptake in hematopoietic cells. This increased nutrient uptake induced by Akt is potentially sufficient to account for its ability to maintain macromolecular synthesis and cell size in the face of growth factor withdrawal on a cell autonomous basis.

Based on these studies, it appears that the regulation of mitochondrial bioenergetics contributes to the susceptibility of cells to induction of programmed cell death, and that mitochondrial bioenergetics are directly regulated by growth factor receptor modulation of the ability of cells to take up nutrients. In the absence of extracellular signals to direct nutrient uptake, cells lack the autonomous ability to take up sufficient nutrients to maintain themselves. How long cells can survive in the absence of extracellular nutrients is regulated either directly or indirectly by the activities of the Bcl-2 family members. Alternatively, enzymes that control nutrient uptake downstream of growth factor receptors, such as Akt, can be directly activated in a cell-autonomous fashion and contribute to oncogenic transformation by supporting uptake of extrinsic nutrients. Thus, it appears that Akt and Bcl-x_L regulate metabolism by distinct mechanisms, and this regulation can directly contribute to the apoptotic sensitivity of cells (Figure 1). Investigation of other genes involved in cell survival is likely to reveal an intricate connection between the sustained cellular metabolism and the ability of cells to suppress the induction of programmed cell death.

References

Cheng, E. H., Wei, M. C., Weiler, S., Flavell, R. A., Mak, T. W., Lindsten, T., and Korsmeyer, S. J. (2001). BCL-2, BCL-X(L) sequester BH3 domain-only molecules preventing BAX- and BAK-mediated mitochondrial apoptosis. Mol Cell *8*, 705–711.

Du, C., Fang, M., Li, Y., Li, L., and Wang, X. (2000). Smac, a mitochondrial protein that promotes cytochrome c-dependent caspase activation by eliminating IAP inhibition. Cell *102*, 33–42.

Edinger, A. L., and Thompson, C. B. (2002). Akt Maintains Cell Size and Survival by Increasing mTOR-dependent Nutrient Uptake. Mol Biol Cell *13*, 2276–2288.

Hajduch, E., Litherland, G. J., and Hundal, H. S. (2001). Protein kinase B (PKB/Akt)—a key regulator of glucose transport? FEBS Lett *492*, 199–203.

Hanahan, D., and Weinberg, R. A. (2000). The hallmarks of cancer. Cell *100*, 57–70.

Huang, D. C., and Strasser, A. (2000). BH3-Only proteins-essential initiators of apoptotic cell death. Cell *103*, 839–842.

Kluck, R. M., Bossy-Wetzel, E., Green, D. R., and Newmeyer, D. D. (1997). The release of cytochrome c from mitochondria: a primary site for Bcl-2 regulation of apoptosis. Science *275*, 1132–1136.

Lindsten, T., Ross, A. J., King, A., Zong, W. X., Rathmell, J. C., Shiels, H. A., Ulrich, E., Waymire, K. G., Mahar, P., Frauwirth, K., *et al.* (2000). The combined functions of proapoptotic Bcl-2 family members bak and bax are essential for normal development of multiple tissues. Mol Cell *6*, 1389–1399.

Marrack, P., Bender, J., Hildeman, D., Jordan, M., Mitchell, T., Murakami, M., Sakamoto, A., Schaefer, B. C., Swanson, B., and Kappler, J. (2000). Homeostasis of alpha beta TCR+ T cells. Nat Immunol *1*, 107–111.

Plas, D. R., Talapatra, S., Edinger, A. L., Rathmell, J. C., and Thompson, C. B. (2001). Akt and Bcl-xL Promote Growth Factor-independent Survival through Distinct Effects on Mitochondrial Physiology. J Biol Chem *276*, 12041–12048.

Rathmell, J. C., and Thompson, C. B. (2002). Pathways of apoptosis in lymphocyte development, homeostasis, and disease. Cell *109 Suppl*, S97–107.

Rathmell, J. C., Vander Heiden, M. G., Harris, M. H., Frauwirth, K. A., and Thompson, C. B. (2000). In the absence of extrinsic signals, nutrient utilization by lymphocytes is insufficient to maintain either cell size or viability. Mol Cell *6*, 683–692.

Shi, Y. (2002). Mechanisms of caspase activation and inhibition during apoptosis. Mol Cell *9*, 459–470.

Songyang, Z., Baltimore, D., Cantley, L. C., Kaplan, D. R., and Franke, T. F. (1997). Interleukin 3-dependent survival by the Akt protein kinase. Proc Natl Acad Sci USA *94*, 11345–11350.

Susin, S. A., Lorenzo, H. K., Zamzami, N., Marzo, I., Snow, B. E., Brothers, G. M., Mangion, J., Jacotot, E., Costantini, P., Loeffler, M., *et al.* (1999). Molecular characterization of mitochondrial apoptosis-inducing factor. Nature *397*, 441–446.

Tsujimoto, Y., Cossman, J., Jaffe, E., and Croce, C. M. (1985). Involvement of the bcl-2 gene in human follicular lymphoma. Science *228*, 1440–1443.

Vander Heiden, M. G., Chandel, N. S., Schumacker, P. T., and Thompson, C. B. (1999). Bcl-xL prevents cell death following growth factor withdrawal by facilitating mitochondrial ATP/ADP exchange. Mol Cell *3*, 159–167.

Vander Heiden, M. G., Chandel, N. S., Williamson, E. K., Schumacker, P. T., and Thompson, C. B. (1997). Bcl-xL regulates the membrane potential and volume homeostasis of mitochondria. Cell *91*, 627–637.

Vander Heiden, M. G., Plas, D. R., Rathmell, J. C., Fox, C. J., Harris, M. H., and Thompson, C. B. (2001). Growth Factors Can Influence Cell Growth and Survival through Effects on Glucose Metabolism. Mol Cell Biol *21*, 5899–5912.

Vander Heiden, M. G., and Thompson, C. B. (1999). Bcl-2 proteins: regulators of apoptosis or of mitochondrial homeostasis? Nat Cell Biol *1*, E209–216.

Verhagen, A. M., Ekert, P. G., Pakusch, M., Silke, J., Connolly, L. M., Reid, G. E., Moritz, R. L., Simpson, R. J., and Vaux, D. L. (2000). Identification of DIABLO, a mammalian protein that promotes apoptosis by binding to and antagonizing IAP proteins. Cell *102*, 43–53.

Yang, J., Liu, X., Bhalla, K., Kim, C. N., Ibrado, A. M., Cai, J., Peng, T. I., Jones, D. P., and Wang, X. (1997). Prevention of apoptosis by Bcl-2: release of cytochrome c from mitochondria blocked. Science *275*, 1129–1132.

Zong, W. X., Lindsten, T., Ross, A. J., MacGregor, G. R., and Thompson, C. B. (2001). BH3-only proteins that bind pro-survival Bcl-2 family members fail to induce apoptosis in the absence of Bax and Bak. Genes Dev *15*, 1481–1486.

Zou, H., Henzel, W. J., Liu, X., Lutschg, A., and Wang, X. (1997). Apaf-1, a human protein homologous to C. elegans CED-4, participates in cytochrome c-dependent activation of caspase-3. Cell *90*, 405–413.

Chapter 2

Thyroid Hormone-Induced Apoptosis during Amphibian Metamorphosis

TOSIKAZU AMANO, LIEZHEN FU, ATSUKO ISHIZUYA-OKA, AND YUN-BO SHI*

ABSTRACT: Anuran metamorphosis involves thyroid hormone (TH)-induced, systematic transformations of individual organs. The vast majority of the larval tissues are removed during this process. Among them is the complete degeneration of the tail and gills and reduction of small intestine by about 90% (lengthwise). Various morphological and cellular studies have shown that the removal of larval organs/tissues is through programmed cell death or apoptosis. Recent cloning and characterization of TH-regulated genes revealed that a group of genes encoding matrix metalloproteinases (MMPs) are activated by TH during metamorphosis in various organs. The activation of MMPs, which are extracellular or membrane-associated enzymes capable of degrading extracellular matrix (ECM) proteins, are in agreement with the previously observed remodeling/degradation of the ECM during metamorphosis. In vivo and in vitro studies have provided evidence to support that ECM remodeling by MMPs plays an important role in regulating apoptosis and cell migration during tissue remodeling.

Key Words: *Xenopus laevis*; thyroid hormone receptor; extracellular matrix (ECM); matrix metalloproteinase (MMP).

Abbreviations: TH, thyroid hormone; TR, TH Receptor; RXR, retinoid X receptor or 9-cis-retinoic acid receptor; MMP, matrix metalloproteinase, ECM, extracellular matrix; ST1, stromelysin-1; ST3, stromelysin-3; Col3, collagenase-3; Col4, collagenase-4, GLA, gelatinase-A; GLB, gelatinase-B.

Introduction

Anuran development is a biphasic process. Their embyogenesis leads to the formation of free-living aquatic tadpoles. After a finite period of growth, the tadpoles then undergo metamorphosis. This postembryonic process involves systematic transformations of essentially every organ/tissue, leading to the formation of adult organs/tissues that have

*Laboratory of Gene Regulation and Development, National Institute of Child Health and Human Development, National Institutes of Health, Bethesda, MD 20892-5431, U.S.A. and Department of Histology and Neurobiology, Dokkyo University School of Medicine, Mibu, Tochigi 321-0293, Japan. *Correspondence:* Yun-Bo Shi, Laboratory of Gene Regulation and Development, Building 18T, Rm. 106, NICHD, NIH, Bethesda MD, 20892, USA, 301-402-1004, 301-402-1323 fax; Shi@helix.nih.gov; Or, Atsuko Ishizuya-Oka, Department of Histology and Neurobiology, Dokkyo University School of Medicine, Mibu, Tochigi 321-0293, Japan, 81-282-87-2124, +81-282-86-1463 fax; i-oka@dokkyomed.ac.jp similar structure and function of their counterparts in mammals. While it has been difficult to study postembryonic development in mammals due to the difficulty to access and manipulate the fetus, the external developmental of amphibian embryos and subsequent metamorphosis has long been used as a model for vertebrate development. More importantly, the metamorphic transformations of all organs/tissues are under the control of a single hormone, the thyroid hormone (TH) (Dodd and Dodd, 1976; Shi, 1999). Furthermore, TH appears to regulate metamorphosis organ autonomously. Thus it is easy to manipulate this process by simply altering the TH levels in tadpoles or in cultures of individual organs. This has allowed many morphological to cellular studies of this postembryonic process and led to the demonstration that cell death is responsible for the degeneration of most, if not all, larval tissues during metamorphosis (Dodd and Dodd, 1976; Shi, 1999).

TH functions by binding to thyroid hormone receptor (TR). TR is a sequence specific DNA-binding transcription factor that belongs to the nuclear receptor super-family. Numerous studies suggest that TR forms a heterodimer with retinoid X receptor (RXR, or 9-cis retinoic acid receptor) to regulate gene expression (Mangelsdorf et al., 1995; Tsai and O'Malley, 1994; Yen, 2001). Thus, TH is believed to induce a gene regulation cascade in individual tissues/organs to effect their transformations. Many laboratories have attempted to study the molecular mechanisms governing tissue resorption and cell death through isolation and characterization of TH-regulated genes during metamorphosis (Shi, 1999). Our own studies on the TH-induced matrix metalloproteinase (MMP) genes have provided evidence for a role of MMPs in regulating apoptosis and tissue morphogenesis during intestinal remodeling.

Cell Death during Metamorphosis

Different tissues/organs undergo distinct changes at different developmental stages (Dodd and Dodd, 1976; Shi, 1999). Some organs, such as the limbs, develop *de novo* from undifferentiated cells preserved in the tadpoles. The tail and gills represent a second type of transformation, i.e., complete resorption. The vast majority of the organs are, however, present in both the tadpole and frog. They undergo partial yet drastic remodeling to adapt to their roles in the frog, which have different diets and living habitats. All three types of transformations involve cell death, e.g., the removal of interdigital cells of the developing limbs, the degeneration of the larval epithelium during remodeling of the intestine, and resorption of all tail tissues (Fig. 1).

Electron microscopic analyses of Kerr et al. (Kerr et al., 1974) were the first to show that tail resorption during anuran metamorphosis of the dwarf tree frog *Litoria glauerti* occurs through apoptosis. Similar observations were subsequently made for *Rana japonica* (Kinoshita et al., 1985) and *Xenopus laevis* (Nishikawa and Hayashi, 1995).

Our own studies have shown that the apoptosis is responsible for the removal of larval intentinal epithelium during *Xenopus laevis* metamorphosis. The herbivorous tadpole intestine is a simple tubular organ consisting of predominantly a single layer of larval epithelium. There is little connective tissue, except in the single epithelial fold, the typholosole, and the muscle layers are thin. During metamorphosis, the larval or tadpole epithelium undergoes degeneration and is replaced by the adult epithelium through the proliferation of adult epithelial stem cells, whose origin remains unclear, and their subsequent differentiation. Concurrently, the connective tissue and muscles develop extensively (Fig. 1) (Shi and Ishizuya-Oka, 1996). Electron microscopic studies showed that the dying larval

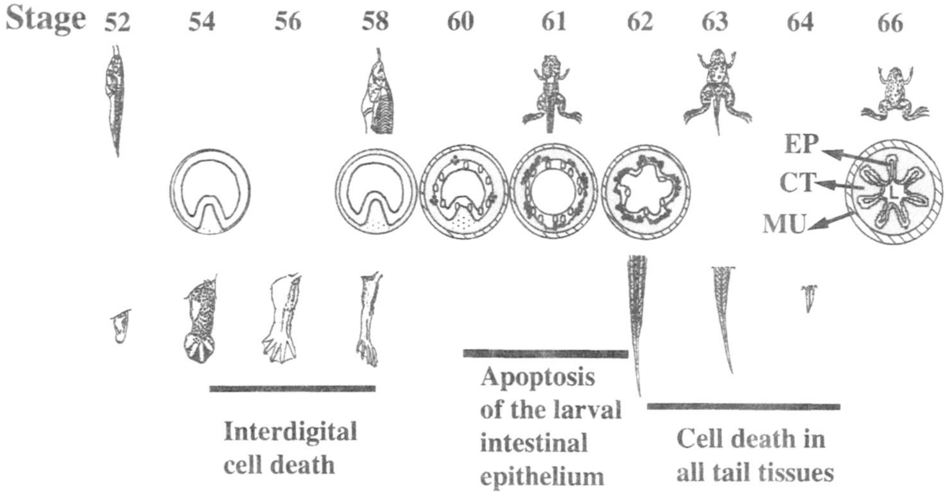

Figure 1. Stage-dependent apoptosis in different organ transformations during *Xenopus laevis* metamorphosis. The developmental stages are based on Nieuwkoop and Faber (Nieuwkoop and Faber, 1956). The tails at stages 62–66 are drawn to the same scale to show the resorption (no tail remains by stage 66), while the tadpoles, intestinal cross-sections (middle) and the hindlimbs at different stages are not in the same scale in order to highlight the morphological differences. Tadpole small intestine has a single epithelial fold, where connective tissue (CT) is abundant, while a frog has a multiply folded intestinal epithelium (EP), with elaborate connective tissue and muscle (MU). Dots: proliferating adult intestinal epithelial cells. Open circles: apoptotic primary intestinal epithelial cells. L: intestinal lumen.

epithelial cells have apoptotic morphology and that the apoptotic bodies are removed at least in part through phagocytosis (Ishizuya-Oka and Ueda, 1996; Ishizuya-Oka et al., 1997; Shi and Ishizuya-Oka, 1996).

The organ autonomous nature of TH regulation of metamorphosis allows studies of metamorphosis of individual organs/tissues in vitro. Thus, intestinal fragments of premetamorphic *Xenopus laevis* tadpoles can be isolated and cultured in vitro (Ishizuya-Oka and Shimozawa, 1991). The addition of TH to the culture medium can induce precocious remodeling just like in intact animals (Fig. 2), i.e., larval epithelial cell death and adult epithelial development. Furthermore, the connective tissue and the epithelium of the intestine can be separated and cultured individually (Fig. 2) (Ishizuya-Oka and Shimozawa, 1992; Ishizuya-Oka and Shimozawa, 1994). The addition of TH leads to larval epithelial cell death. However, the development of the adult epithelium occurs only in the presence of the connective tissue and conversely, connective tissue development also requires the epithelium. These results indicate that larval epithelial cell death is cell autonomous, at least in vitro, while cell-cell interactions are important for adult tissue development.

To investigate the nature of the intestinal epithelial cell death, we have isolated the larval epithelial and fibroblastic cells from the intestine of *Xenopus laevis* tadpoles and cultured them *in vitro* on plastic dishes in the presence or absence of TH (Su et al., 1997a; Su et al., 1997b). Our results indicate that the larval epithelial cells are induced to die by physiological concentrations of TH. Microscopic examination of the dying cells revealed the existence of apoptotic bodies typical of mammalian apoptotic cell death. Furthermore, it also had another hallmark of apoptosis, i.e., the formation of a nucleosomal sized nuclear DNA ladder (Fig. 3A). In contrast to the epithelial cells, the fibroblastic cells were refractory to TH-induced death under the same conditions (Fig. 3B), indicating that the in vitro responses

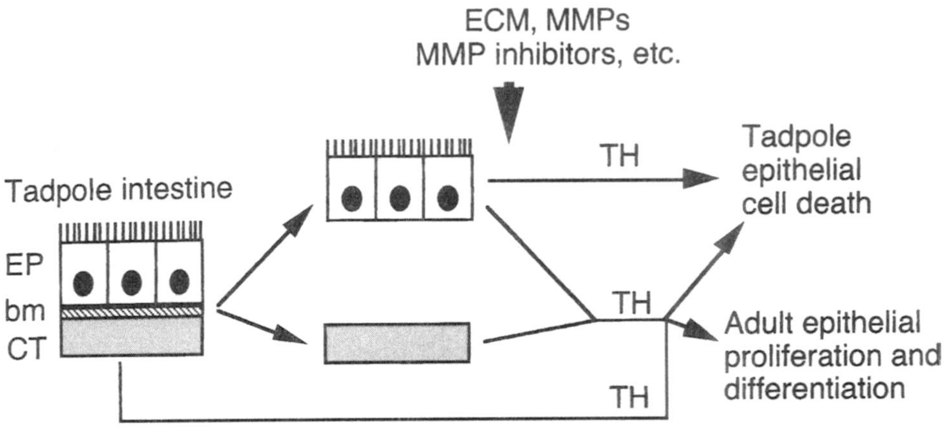

Figure 2. Schematic diagram for intestinal transformation in organ cultures. The tadpole intestine consists of three major tissue types: the inner (facing the lumen) epithelium (EP), connective tissue (CT), and outer muscle layers (not shown). EP is separated from CT by a special ECM, the basal lamina or basement membrane (bm). The most dramatic changes occur in the epithelium during metamorphosis. In response to TH, the larval epithelial cells undergo apoptosis and adult epithelial cells, whose origin is yet unknown, proliferate and differentiate. These changes can be reproduced in vitro in the presence of TH by culturing intestinal fragments (bottom) or in recombinant organ cultures of EP and CT (middle). If EP is cultured alone, only cell death occurs (top). In addition, dissociated cells from EP and CT can be isolated for primary cell culture studies (not shown, see Fig. 3). The effects of ECM on tissue transformation can be studied by culturing the tissues on ECM or by adding MMPs and MMP inhibitors, etc. to the organ culture medium (arrowhead).

mimics the natural metamorphosis. Interestingly, when the larval intestinal epithelial cells were cultured on plastic dishes coated with various components of the ECM, such as fibronectin and laminin, etc., the TH-induced cell death was inhibited (Su et al., 1997b). However, the ECM components failed to influence TH-stimulated cellular DNA synthesis or the downregulation of two epithelial specific genes. Thus, the ECM specifically affects cell death, suggesting that the TH-induced epithelial apoptosis is not entirely cell autonomous *in vivo* but depends on cell-cell and/or cell-ECM interactions.

Similar studies have been carried out on tail resportion. First, isolated epidermal cells from the tail of *Rana catesbeiana* tadpoles undergo cell death when cultured in the presence of TH (Nishikawa and Yoshizato, 1986), again suggesting that the TH-dependent epidermal cell death is cell autonomous, at least when cultured in isolation *in vitro*. Second, a cell line derived from *Xenopus laevis* tail muscles was found to respond to TH by undergoing apoptosis (Yaoita and Nakajima, 1997). On the other hand, the effects of ECM have not been examined on these in vitro cell cultures. However, it has been shown that the removal of tail epidermis prevents TH-induced tail resorption in organ cultures (Niki et al., 1982). In addition, adult-type non-T leukocytes have been implicated to participate in the removal of tail cells (Izutsu et al., 1996). Thus, cell-cell interactions are also likely important for apoptosis during tail resorption as well.

Roles of MMPs and ECM Remodeling during Apoptotic Tissue Remodeling

TH is believed to induce a series of gene regulation steps during metamorphosis. The first step is the transcriptional regulation of direct or early TH response genes by TRs.

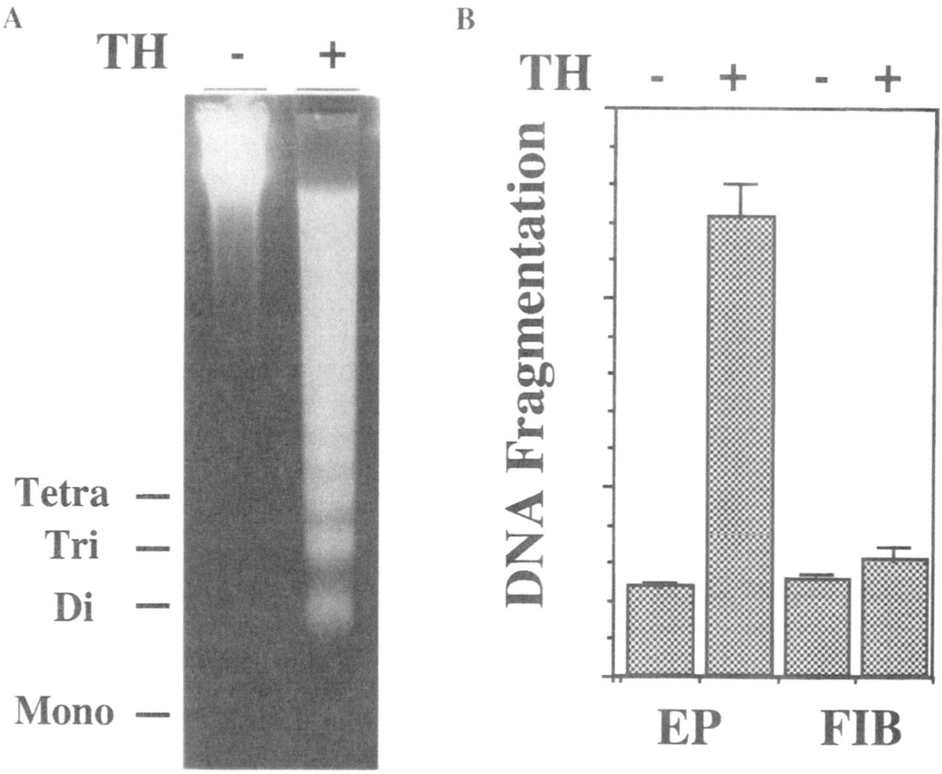

Figure 3. Primary cell culture studies indicate that TH-induces apoptosis of the intestinal epithelial cells (EP) but not the fibroblasts (FIB). (A). TH induces the formation of a nucleosomal DNA ladder in the epithelial cells. Dissociated tadpole intestinal epithelial cells were cultured in the presence or absence of TH for 1 day. The nuclear DNA was isolated and separated on an agarose gel. Note the formation of a nucleosomal DNA ladder with the mono- to tetra-nucleosomal-sized DNA bands indicated. (B). Quantitative DNA fragmentation assay indicates that EP but not FIB undergo TH-induced cell death in vitro. Primary cell cultures were treated with TH for 3 days and DNA fragmentation was assayed. See (Su et al., 1997a; Su et al., 1997b) for more details.

The products of these genes in turn affect the expression of downstream, or late TH response genes. Alternatively, they may directly participate in tissue transformation. Thus, to understand the molecular pathways of metamorphosis, a key step is to identify the TH response genes. Many such genes have been isolated over the years in many different laboratories by using various means (Shi, 1999).

Among these genes are those encoding MMPs. MMPs are Zn^{2+}-dependent proteinases that are extracellular or bound to plasma membrane (Alexander and Werb, 1991; Birkedal-Hansen et al., 1993; McCawley and Matrisian, 2001; Nagase, 1998; Pei, 1999). They can cleave specific proteinaceous components of the extracelluar matrix (ECM) and non-ECM proteins. Numerous studies in mammals have suggested that MMPs are involved in diverse developmental and pathological processes by modifying or degrading the ECM, thus influencing cell behavior (Murphy and Gavrilovic, 1999; Sang, 1998; Stetler-Stevenson, 1996; Vu and Werb, 2000).

So far, six MMP genes have been cloned and found to be upregulated during anuran metamorphosis. The *Rana catesbeiana* collagenase-1 (Col1) (Oofusa et al., 1994) and *Xenopus laevis* stromelysin-3 (ST3) (Shi and Brown, 1993; Wang and Brown, 1993) are direct TH response genes while *Xenopus* Col3, Col4, gelatinase-A (GLA) genes are indirectly regulated by TH as their upregulation requires more than 2 days of TH treatment (Patterton et al., 1995; Stolow et al., 1996). More recently, a fragment of *Xenopus* gelatinase-B (GLB) was cloned and used to show that GLB was upregulated during tail resoprtion (Jung et al., 2002), although it is unclear how it is regulated by TH. Finally, using the human ST1 cDNA as a probe, we showed that the putative *Xenopus* ST1 gene is also indirectly upregulated by TH (Patterton et al., 1995). All of the MMPs are upregulated during tail resorption. In contrast, during intestinal remodeling, only ST3 and GLA are strongly upregulated while the expression of Col3, Col4 and ST1 changes little (no data is reported on the expression of Col1 and GLB during intestinal metamorphosis.

Through Northen blot and in situ hybridization analyses, we and others have shown that *Xenopus* Col3, Col4, GLB, and ST3 are expressed in tissues where apoptosis occurs during metamorphosis (Shi and Ishizuya-Oka, 2001) (Berry et al., 1998a; Berry et al., 1998b; Damjanovski et al., 1999; Ishizuya-Oka et al., 2000; Ishizuya-Oka et al., 1996; Jung et al., 2002; Patterton et al., 1995). In particular, we have demonstrated a tight spatial and temporal correlation of ST3 mRNA and protein with epithelial apoptosis during intestinal metamorphosis (Ishizuya-Oka et al., 2000; Patterton et al., 1995). In addition, ST3 expression is also correlated with the remodeling of the ECM (the basal lamina or basement membrane) that separates the epithelium and the connective tissue.

The basal lamina is composed of laminin, entactin, collagens, and proteglycans, etc. (Hay, 1991; Timpl and Brown, 1996). In premetamorphic *Xenopus laevis* tadpoles, the intestinal basal lamina is thin. It becomes much thicker and mutlply folded during metamorphosis along with massive epithelial apoptosis (Ishizuya-Oka and Shimozawa, 1987; Murata and Merker, 1991). Toward the end of metamorphosis, with the progress of intestinal morphogenesis as the adult epithelial cells differentiate, the basal lamina becomes thin and flat again (Shi and Ishizuya-Oka, 1996).

The expression of ST3 is restricted to the fibroblastic cells underneath the thick basal lamina during metamorphosis. Thus, it is quite likely that ST3 directly or indirectly participate in the modification of the basal lamina, which in turn affects cell behavior and fate during metamorphosis. Such a mechanism is also consistent with the in vitro observation that ECM inhibits TH-induced epithelial cell death and the cell-cell interactions are important for adult tissue development (see above).

To directly demonstrate the involvement of ST3 in cell death and tissue morphogenesis, we have made use of the ability to induce intestinal remodeling in organ cultures in vitro. As ST3 is a secreted protein, we reasoned that we might be able to inhibit its function by adding function-blocking antibodies to the organ culture medium (Fig. 2). Thus, we generated a polyclonal antibody against the catalytic domain of *Xenopus* ST3. As expected, this antibody was able to block the catalytic function of ST3 in vitro (Fig. 4A) (Ishizuya-Oka et al., 2000). The addition of this antibody to intestinal organ culture medium led to dose-dependent inhibition of larval epithelial cell death (Fig. 4B) as well as the remodeling (thickening and folding) of the basal lamina. In addition, it also inhibited the invagination of the adult epithelial cells into the connective tissue (Ishizuya-Oka et al., 2000), a process critical for the adult epithelial morphogenesis. Similarly, we also found that a synthetic, broad-spectrum MMP inhibitor inhibited TH-induced apoptosis in the intestinal organ cultures (Ishizuya-Oka et al., 2000). These results suggest that during metamorphosis, TH induces the expression of ST3 in the fibroblastic cells in the connective tissue (Fig. 5). The MMP activity of ST3 then directly or indirectly participates in the remodeling of the

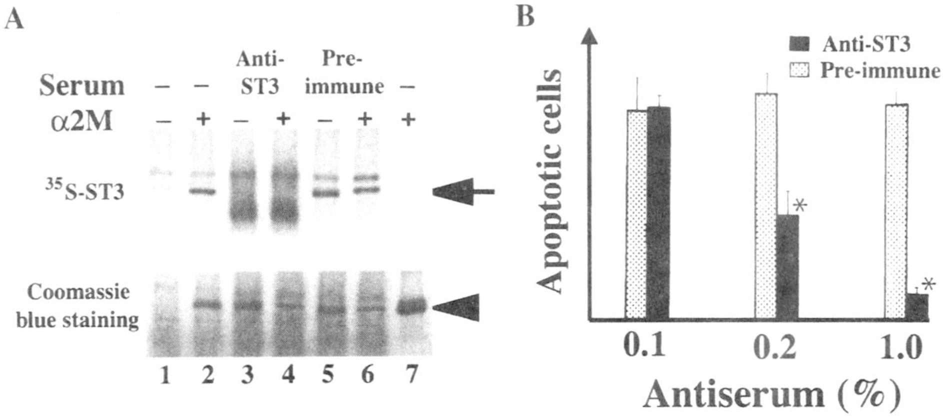

Figure 4. An anti-ST3 antibody inhibits TH-induced epithelial apoptosis. (A). a_2-macroglobulin (a_2M) capture assay shows that the anti-ST3 antibody blocks ST3 function. In vitro synthesized, ^{35}S-labeled ST3 catalytic domain was incubated with a_2M in the presence or absence of anti-ST3 antiserum or preimmune serum. The resulting mixture was electrophoresed on a native polyacrylamide gel. The gel was stained with coomassie blue (lower panel) and then dried and autoradiographed (top panel). The arrowhead and arrow indicate the position of a_2M and a_yM-ST3 complex, respectively (under the gel conditions, free ST3 failed to enter the gel while a_2M and a_2M-ST3 complex migrated at the same position). Note that the preimmune serum had no effect on a_2M-ST3 complex formation. In fact, ST3 also formed a complex with endogenous a_2M in the serum. In contrast, anti-ST3 antibody inhibited the formation of the a_yM-ST3 complex (with both exogenously added and endogenous serum a_2M) (the faster migrating band in the presence of anti-ST3 serum is the antibody-ST3 complex). Lane 7 contained only a_2M but at a higher level to facilitate its identification by coomassie blue staining. (B). Intestinal explants were cultured for 3 days in the presence of TH and indicated concentrations of anti-ST3 serum or preimmune serum. The apoptosis in the epithelium was detected with the TUNEL method. Labeling indices of apoptotic epithelial cells in the explants were measured after 3 days of treatment in the presence of increasing concentrations of preimmnune (dotted bars) or anti-ST3 antiserum (solid bars). *, $P<0.01$. See (Ishizuya-Oka et al., 2000) for more details.

basal lamina ECM. The consequence of this ECM remodeling is the facilitation of larval epithelial cell death and adult tissue morphogenesis (the invagination of the adult epithelial cells into the connective tissue (Fig. 5). In addition, the remodeled basal lamina may also allow extensive contacts between the proliferating adult epithelium and the fibroblasts and migration of the macrophages of the connective tissue into the epithelium, where they engulf the apoptotic bodies (Fig. 5).

Conclusion and Prospects

Programmed cell death or apoptosis is an essential aspect anuran metamorphosis. It is involved in the transformations of most, if not all, organs, from de novo development to complete resorption. Molecular and cellular studies have provided strong evidence for a role of TH-induced MMP genes in ECM remodeling and apoptosis. How MMPs function remains to be determined. They may affect tissue remodeling by directly modifying the ECM or by cleaving non-ECM proteins to indirectly influence ECM remodeling and/or cell behavior. This latter case is particularly likely for ST3, which has only weak activities toward ECM proteins but can effectively cleave a 1-proteinase inhibitor (a 1-antitrypsin), a

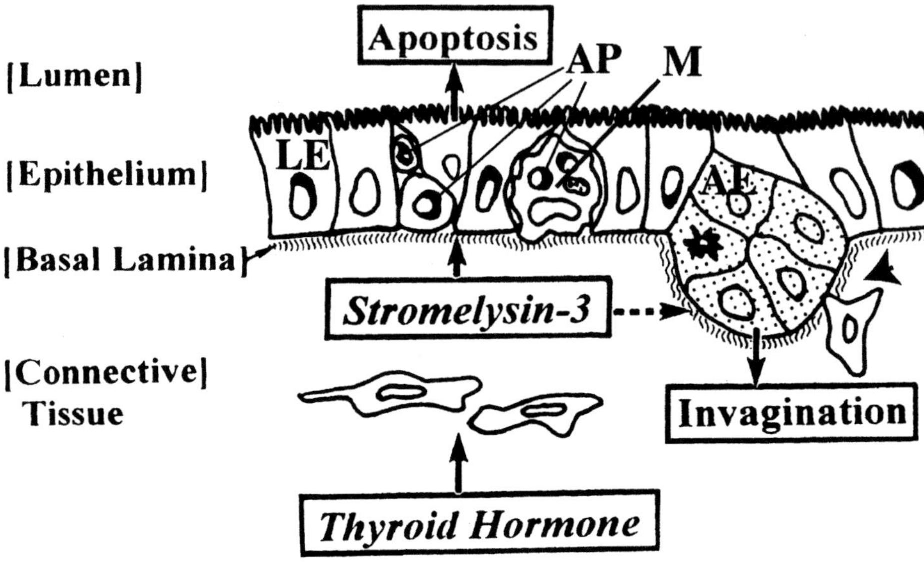

Figure 5. Schematic diagram showing the participation of ST3 in intestinal remodeling. TH induces the expression of ST3 in the fibroblasts of the connective tissue. The secreted ST3 directly or indirectly participates in the modification of the basal lamina, leading to increased thickness. This modified basal lamina in turn facilitates the death of the larval epithelial cells (LE) through apoptosis, and the invagination of the adult epithelial cells (AE) into the connective tissue. The modified basal lamina appears to be more permeable than pre- or post-metamorphic basal lamina as reflected by the frequently observed contacts between AE and fibroblasts in the connective tissue (arrowhead) and the migration of the macrophages (M) from the connective tissue into the epithelium, where they engulf the apoptotic bodies (AP).

non-ECM derived serine proteinase inhibitor (Murphy et al., 1993; Pei et al., 1994; Uria and Werb, 1998). The identification of its in vivo substrates will provide important clues on its mechanisms of action. In addition, analyses of the function of ST3 in developing tadpoles will be an important complement to the existing in vitro organ culture studies. This is made possible by development of transgenesis method for *Xenopus laevis* (Kroll and Amaya, 1996). Our preliminary transgenic studies using the ubiquitous promoter CMV to drive the expression of *Xenopus* Col4 or ST3 show that precocious overexpression of these MMPs leads to embryonic defects and lethality (Damjanovski et al., 2001). Future studies by using tissue specific and/or inducible promoter may allow the study of MMP function during metamorphosis in vivo.

The activation of MMP genes and the remodeling of the ECM argue that the nature of the ECM is important for cell behavior during development. Thus, another important area for study cell-cell and cell-ECM interactions during metamorphosis is to determine the nature of the ECM remodeling and how it in turn signals the cells. The downstream events of interests are how the ECM remodeling regulates the expression and/or activities of cell death genes, such as the caspases and bcl-2 superfamily of cell death regulators (Cryns and Yuan, 1998; Rao and White, 1997; Woo et al., 2000; Yang and M., 2000; Zhang and Xu, 2000). Some of these genes are cloned in amphibian but other are not. The cloning and characterization of these genes should provide insights on whether different cell death genes are affected by MMP-induced ECM remodeling in the apoptosis of different larval tissues.

References

Alexander, C. M., and Werb, Z. (1991). Extracellular matrix degradation. In Cell Biology of Extracellular Matrix, E. D. Hay, ed. (New York, Plenum Press), pp. 255–302.

Berry, D. L., Rose, C. S., Remo, B. F., and Brown, D. D. (1998a). The expression pattern of thyroid hormone response genes in remodeling tadpole tissues defines distinct growth and resorption gene expression programs., Dev Biol *203*, 24–35.

Berry, D. L., Schwartzman, R. A., and Brown, D. D. (1998b). The expression pattern of thyroid hormone response genes in the tadpole tail identifies multiple resorption programs., Develop Biol *203*, 12–23.

Birkedal-Hansen, H., Moore, W. G. I., Bodden, M. K., Windsor, L. T., Birkedal-Hansen, B., DeCarlo, A., and Engler, J. A. (1993). Matrix metalloproteinases: a review., Crit Rev in Oral Biol and Med *4*, 197–250.

Cryns, V., and Yuan, J. (1998). Proteases to die for [published erratum appears in Genes Dev 1999 Feb 1;13(3):371]., Genes Dev *12*, 1551–70.

Damjanovski, S., Amano, T., Li, Q., Pei, D., and Shi, Y.-B. (2001). Overexpression of Matrix Metalloproteinases Leads to Lethality in Transgenic Xenopus Laevis: Implications for Tissue-Dependent Functions of Matrix Metalloproteinases during Late Embryonic development., Dev Dynamics *221*, 37–47.

Damjanovski, S., Ishizuya-Oka, A., and Shi, Y. B. (1999). Spatial and temporal regulation of collagenases-3, -4, and stromelysin-3 implicates distinct functions in apoptosis and tissue remodeling during frog metamorphosis., Cell Res *9*, 91–105.

Dodd, M. H. I., and Dodd, J. M. (1976). The biology of metamorphosis. In Physiology of the amphibia, B. Lofts, ed. (New York, Academic Press), pp. 467–599.

Hay, E. D. (1991). Cell Biology of Extracellular Matrix, 2nd edn (New York, Plenum Press).

Ishizuya-Oka, A., Li, Q., Amano, T., Damjanovski, S., Ueda, S., and Shi, Y.-B. (2000). Requirement for matrix metalloproteinase stromelysin-3 in cell migration and apoptosis during tissue remodeling in Xenopus laevis., J Cell Biol *150*, 1177–88.

Ishizuya-Oka, A., and Shimozawa, A. (1987). Ultrastructural changes in the intestinal connective tissue of Xenopus laevis during metamorphosis., J Morphol *193*, 13–22.

Ishizuya-Oka, A., and Shimozawa, A. (1991). Induction of metamorphosis by thyroid hormone in anuran small intestine cultured organotypically in vitro., In Vitro Cell Dev Biol *27A*, 853–7.

Ishizuya-Oka, A., and Shimozawa, A. (1992). Connective tissue is involved in adult epithelial development of the small intestine during anuran metamorphosis in vitro., Roux's Arch Dev Biol *201*, 322–329.

Ishizuya-Oka, A., and Shimozawa, A. (1994). Inductive action of epithelium on differentiation of intestinal connective tissue of Xenopus laevis tadpoles during metamorphosis in vitro., Cell Tissue Res *277*, 427–36.

Ishizuya-Oka, A., and Ueda, S. (1996). Apoptosis and cell proliferation in the Xenopus small intestine during metamorphosis., Cell Tissue Res *286*, 467–76.

Ishizuya-Oka, A., Ueda, S., Damjanovski, S., Li, Q., Liang, V. C., and Shi, Y.-B. (1997). Anteroposterior gradient of epithelial transformation during amphibian intestinal remodeling: immunohistochemical detection of intestinal fatty acid-binding protein., Dev Biol *192*, 149–61.

Ishizuya-Oka, A., Ueda, S., and Shi, Y.-B. (1996). Transient expression of stromelysin-3 mRNA in the amphibian small intestine during metamorphosis., Cell Tissue Res *283*, 325–9.

Izutsu, Y., Yoshizato, K., and Tochinai, S. (1996). Adult-type splenocytes of Xenopus induce apoptosis of histocompatible larval tail cells in vitro., Differentiation *60*, 277–86.

Jung, J.-C., Leco, K. J., Edwards, D. R., and Fini, M. E. (2002). Matrix metalloproteinase mediate the dismantling of mesenchymal structures in the tadpole tail during thyroid hormone-induced tail resorption., dev dyn *223*, 402–413.

Kerr, J. F. R., Harmon, B., and Searle, J. (1974). An electron-microscope study of cell eletion in the anuran tadpole tail during spontaneous metamorphosis with special reference to apoptosis of striated muscle fibres., J Cell Sci *14*, 571–585.

Kinoshita, T., Sasaki, F., and Watanabe, K. (1985). Autolysis and heterolysis of the epidermal cells in anuran tadpole tail regression., J of Morphology *185*, 269–275.

Kroll, K. L., and Amaya, E. (1996). Transgenic Xenopus embryos from sperm nuclear transplantations reveal FGF signaling requirements during gastrulation., Development *122*, 3173–83.

Mangelsdorf, D. J., Thummel, C., Beato, M., Herrlich, P., Schutz, G., Umesono, K., Blumberg, B., Kastner, P., Mark, M., Chambon, P., and et al. (1995). The nuclear receptor superfamily: the second decade., Cell *83*, 835–9.

McCawley, L. J., and Matrisian, L. M. (2001). Matrix metalloproteinases: they're not just for matrix anymore!, Current Opinion in Cell Biology *13*, 534–540.

Murata, E., and Merker, H. J. (1991). Morphologic changes of the basal lamina in the small intestine of Xenopus laevis during metamorphosis., Acta Anat *140*, 60–9.

Murphy, G., and Gavrilovic, J. (1999). Proteolysis and cell migration: creating a path?, Current Opinion in Cell Biology *11*, 614–621.

Murphy, G., Segain, J.-P., O'Shea, M., Cockett, M., Ioannou, C., Lefebvre, O., Chambon, P., and Basset, P. (1993). The 28-kDa N-terminal domain of mouse stromelysin-3-has the general properties of a weak metalloproteinase., J Biol Chem *268*, 15435–15441.

Nagase, H. (1998). Cell surface activation of progelatinase A (proMMP-2) and cell migration., Cell Res *8*, 179–86.

Nieuwkoop, P. D., and Faber, J. (1956). Normal table of Xenopus laevis., 1st. edn (Amsterdam, North Holland Publishing).

Niki, K., Namiki, H., Kikuyama, S., and Yoshizato, K. (1982). Epidermal tissue requirement for tadpole tail regression induced by thyroid hormone., Dev Biol *94*, 116–20.

Nishikawa, A., and Hayashi, H. (1995). Spatial, temporal and hormonal regulation of programmed muscle cell death during metamorphosis of the frog Xenopus laevis., Differentiation *59*, 207–14.

Nishikawa, A., and Yoshizato, K. (1986). Hormonal regulation of growth and life span of bullfrog tadpole tail epidermal cells cultured in vitro., J Exp Zool *237*, 221–30.

Oofusa, K., Yomori, S., and Yoshizato, K. (1994). Regionally and hormonally regulated expression of genes of collagen and collagenase in the anuran larval skin., Int J Dev Biol *38*, 345–50.

Patterton, D., Hayes, W. P., and Shi, Y. B. (1995). Transcriptional activation of the matrix metalloproteinase gene stromelysin-3 coincides with thyroid hormone-induced cell death during frog metamorphosis., Dev Biol *167*, 252–62.

Pei, D. (1999). Leukolysin/MMP25/MT6-MMP: a novel matrix metalloproteinase specifically expressed in the leukocyte lineage., Cell Res *9*, 291–303.

Pei, D., Majmudar, G., and Weiss, S. J. (1994). Hydrolytic inactivation of a breast carcinoma cell-derived serpin by human stromelysin-3., J Biol Chem *269*, 25849–55.

Rao, L., and White, E. (1997). Bcl-2 and the ICE family of apoptotic regulators: making a connection., Curr Opin Genet Dev *7*, 52–8.

Sang, Q. X. (1998). Complex role of matrix metalloproteinases in angiogenesis., Cell Res *8*, 171–7.

Shi, Y.-B. (1999). Amphibian Metamorphosis: From morphology to molecular biology. (New York, John Wiley & Sons, Inc.).

Shi, Y.-B., and Brown, D. D. (1993). The earliest changes in gene expression in tadpole intestine induced by thyroid hormone., J Biol Chem *268*, 20312–20317.

Shi, Y.-B., and Ishizuya-Oka, A. (1996). Biphasic intestinal development in amphibians: Embryogensis and remodeling during metamorphosis., Current Topics in Develop Biol *32*, 205–235.

Shi, Y.-B., and Ishizuya-Oka, A. (2001). Thyroid hormone regulation of apoptotic tissue remodeling: Implications from molecular analysis of amphibian metamorphosis., Progress in Nucleic Acid Research and Molecular Biology *65*, 53–100.

Stetler-Stevenson, W. G. (1996). Dynamics of matrix turnover during pathologic remodeling of the extracellular matrix., American Journal of Pathology *148*, 1345–1350.

Stolow, M. A., Bauzon, D. D., Li, J., Sedgwick, T., Liang, V. C., Sang, Q. A., and Shi, Y. B. (1996). Identification and characterization of a novel collagenase in Xenopus laevis: possible roles during frog development., Mol Biol Cell *7*, 1471–83.

Su, Y., Shi, Y., and Shi, Y.-B. (1997a). Cyclosporin A But not FK506 Inhibits Thyroid Hormone-Induced Apoptosis in Xenopus Tadpole Intestinal Epithelium., FASEB J *11*, 559–565.

Su, Y., Shi, Y., Stolow, M., and Shi, Y.-B. (1997b). Thyroid hormone induces apoptosis in primary cell cultures of tadpole intestine: cell type specificity and effects of extracellular matrix., J Cell Biol *139*, 1533–1543.

Timpl, R., and Brown, J. C. (1996). Supramolecular assembly of basement membranes., BioEssays *18*, 123–132.

Tsai, M. J., and O'Malley, B. W. (1994). Molecular mechanisms of action of steroid/thyroid receptor superfamily members., Ann Rev Biochem *63*, 451–486.

Uria, J. A., and Werb, Z. (1998). Matrix metalloproteinases and their expression in mammary gland., Cell Res *8*, 187–94.

Vu, T. H., and Werb, Z. (2000). Matrix metalloproteinases: effectors of development and normal physiology., Genes & Dev *14*, 2123–33.

Wang, Z., and Brown, D. D. (1993). Thyroid hormone-induced gene expression program for amphibian tail resorption., J Biol Chem *268*, 16270–16278.

Woo, M., Hakem, R., and Mak, T. W. (2000). Executionary pathway for apoptosis: lessons from mutant mice., Cell Research *10*, 267–278.

Yang, Y. L., and M., L. X. (2000). The IAP family: endogenous caspase inhibitors with multiple biological activities., Cell Research *10*, 169–177.

Yaoita, Y., and Nakajima, K. (1997). Inductoin of Apoptosis and CPP32 Expression by Thyroid Hormone in a Myoblastic Cell Line Dervided from Tadpole Tail., J Biol Chem *272*, 5122–5127.

Yen, P. M. (2001). Physiological and molecular basis of thyroid hormone action., Physiol Rev *81*, 1097–142.

Zhang, J. H., and Xu, M. (2000). DNA fragmentation in apoptosis., Cell Research *10*, 205–211.

Chapter 3

The Endoplasmic Reticulum Stress Response in Health and Disease

MICHAEL BOYCE AND JUNYING YUAN*

ABSTRACT: The ER stress response is a signaling pathway that transmits information about the homeostasis of the ER lumen (or its disruption) to the nucleus and cytoplasm. Metazoan cells can respond to ER stress by upregulating protein folding, export and degradation machinery, by downregulating global translation, and by apoptosis. The ER stress response has been well characterized in *S. cerevisiae*, but the mammalian ER stress pathways are significantly more complex and only partially understood. Here we review the major components of both the yeast and mammalian ER stress response pathways and highlight their role in both normal cellular physiology and selected pathological conditions.

Introduction

Approximately one-third of all newly synthesized proteins in a eukaryotic cell pass through the endoplasmic reticulum (ER) for subcellular targeting or secretion outside the cell. Therefore, proper protein folding and sorting by the ER are critical for maintaining normal cellular homeostasis. To monitor and regulate protein trafficking in the ER, all eukaryotic cells possess signaling pathways known collectively as the ER stress response (ESR) [1–3]. The ESR is activated by perturbations in normal ER function, such as the accumulation of unfolded, misfolded or excessive protein, or changes in the redox or ionic conditions of the ER lumen [1, 2]. Once activated, the ESR results in the transcriptional upregulation of genes involved in protein folding and export (as a means of attempting to correct the burden in the ER), increased removal of proteins from the ER lumen, and, in metazoans, a global attenuation of translation [1, 2, 4]. In metazoan cells, persistent or intense ER stress can also result in programmed cell death, or apoptosis [5].

The UPR Pathways in Yeast

The foundation of our knowledge of the ESR is based primarily on the genetic analysis of the budding yeast *S. cerevisiae* [1, 2, 6]. In yeast, ER stress is monitored by Ire1p, a type I transmembrane protein that resides in the ER membrane with its N-terminal domain in the ER lumen and C-terminal serine-threonine kinase and endonuclease domains on the

*Department of Cell Biology, Harvard Medical School Boston, MA 02115. *Correspondence:* jyuan@hms.harvard.edu cytoplasmic or nuclear side of the ER membrane [7, 8]. Under normal conditions, the ER lumenal domain of one Ire1p molecule interacts with Kar2p, an ER lumenal chaperone of the Hsp70 family, which assists in folding nascent proteins translated by ER-associated ribosomes [9]. Under conditions of ER stress, such as when large amounts of unfolded proteins accumulate in the ER lumen, Kar2p is thought to be titrated away from Ire1p by unfolded protein [9]. The removal of Kar2p permits the oligomerization of Ire1p through its ER lumenal domain and the subsequent activation of Ire1p by trans-autophosphorylation via the C-terminal kinase domain [7–9].

Activated Ire1p is responsible for signaling to the nucleus to initiate the transcriptional branch of the ESR, often referred to as the unfolded protein response (UPR). To do this, the endonuclease domain of activated Ire1p splices the mRNA of the HAC1 gene, which encodes a bZIP transcription factor that drives expression of UPR-responsive genes in yeast [10, 11]. The HAC1 mRNA message is constitutively transcribed and present in the cytoplasm under resting conditions but contains an intron which has been shown to block translation through a poorly understood mechanism [11, 12]. When Ire1p is activated by conditions of ER stress, the translation attenuator intron is excised from the HAC1 message by the Ire1p endonuclease domain in a novel reaction which is independent of the spliceosome but dependent on tRNA ligase [12, 13]. This splicing event removes the 252-nucleotide intron present near the 3′ end of HAC1 transcript and replaces the last 10 codons of the HAC1 ORF with an exon encoding 18 amino acids, but does not affect the N-terminal 220-amino acid region of Hac1p that contains the DNA binding domain [11, 12, 14]. Interestingly, the C-terminal 18-amino-acid segment functions as a potent activation domain, significantly enhancing Hac1p-mediated transactivation of target genes [14]. Therefore, the splicing event not only allows the generation of productive mRNA but also connects the HAC1 DNA-binding domain to its activation domain in order to initiate the UPR efficiently.

Recent DNA microarray analyses have shown that genes for components of all stages of the secretory pathway are upregulated by Hac1p, including chaperones to assist protein folding in the ER lumen, enzymes involved in phospholipid synthesis and ER membrane biogenesis and components of the protein export pathway [15, 16]. These targets of Hac1p presumably help to alleviate the burden of unfolded protein in the ER by increasing the capacity and efficiency of protein folding as well as the clearance of protein from the ER lumen.

In addition to upregulating components of the secretory pathway, ER stress also activates a novel proteolytic pathway known as ER-associated degradation (ERAD) [2, 17]. In ERAD, misfolded proteins in the ER membrane or lumen are specifically retrotranslocated, or "dislocated," into the cytoplasm, where they are targeted for ubiquitin-mediated degradation by the proteasome [17]. The specific components that execute ERAD are not fully understood, but the dislocation step may occur through the Sec61 complex, which mediates the entry of nascent polypeptides into the ER lumen from ER-associated ribosomes under normal conditions [18]. In yeast, ERAD also requires a number of ER-resident proteins, including Der1p, Der3p/Hrd1p, and Hrd3p [18–21].

The protein degradation pathway of ERAD and the transcriptional activation pathway of the UPR are interconnected in yeast. It has been shown that *ire1*Δ strains cannot activate ERAD in response to ER stress, but that this defect can be overcome by forced expression of mature Hac1p, implicating the core signaling machinery of the UPR in the induction of ERAD [15, 16]. Peter Walter and colleagues used DNA microarray analysis of yeast under conditions of ER stress to demonstrate an intimate connection between ERAD and the UPR [15]. On the one hand, genes critical for ERAD, such as DER1 and HRD3, are strongly induced by treatment with the N-linked glycosylation inhibitor tunicamycin in wild type yeast but not the *ire1*Δ strain, indicating that the UPR can enhance the rate of ERAD [15]. Indeed, overexpression of Hac1p alone is sufficient to activate ERAD, confirming the crosstalk between the pathways. On the other hand, in strains lacking DER1, HRD1, HRD3 or PER100, and therefore defective in ERAD, the UPR is constitutively activated, indicating a feedback control of ERAD over the UPR [15]. In this instance, it is thought that the misfolded proteins present in the ER under normal growth conditions cannot be removed due to the lack of efficient ERAD in these mutants, which leads the cells to mobilize additional chaperone proteins via the UPR to cope with the situation.

In addition to demonstrating a close regulatory relationship between the UPR and ERAD, the DNA microarray analysis by Travers et al. revealed that the UPR may control the expression of genes involved in secretion or the biogenesis of secretory organelles, including genes involved in protein translocation, glycosylation, vesicular transport, cell wall biosynthesis, vacuolar protein targeting and phospholipid metabolism. Remarkably, an estimated 5% of all yeast ORFs are induced in response to ER stress [15]. These data suggest that the accumulation of unfolded proteins in the ER may alter the cell physiology in a global fashion.

The UPR Pathways in Mammalian Cells

In mammalian cells, many of the important features of the yeast ESR are preserved but with additional complexities [1–4]. Mammalian cells have two known homolog of Ire1p (termed Ire1a and Ire1b) and a homolog of Kar2p (termed BiP or Grp78), which are believed to carry out functions similar to those of the corresponding yeast proteins. Ire1a is ubiquitously expressed, while Ire1b gs only detected in the intestines [22, 23]. Overexpression of mammalian Ire1a activated the UPR in the absence of ER stress signal [22]. However, while Ire1p is absolutely required in yeast for initiation of the UPR, cells derived from the double knock-out mouse for Ire1a and Ire1b show no defect in the UPR, suggesting that other redundant pathways must exist in mammals [4].

The mammalian Ire1 protein has been shown to activate JNK and its associated stress kinase pathway [24, 25]. Homologous stress kinase pathways in yeast are involved in the pheromone mating pathway and other stresses but no role has been reported for them in the yeast UPR. Treating cells with the ER stress inducers such as tunicamycin, thapsigargin (an inhibitor of the ER Ca^{+2}-dependent ATPase) or dithiothreitol (a reducing agent), as well as the overexpression of Ire1a, induces the activation of JNK [25], although the magnitude of activation is less than that by canonical JNK activators such as ultraviolet light treatment or overexpression of constitutively active MEKK1. Interestingly, while ER stress-induced the activation of JNK in wild type mouse embryo fibroblasts (MEFs), the same stimuli caused the reduction of JNK activation in Ire1a $^{-/-}$ MEFs [25]. The mechanism of this reduction is still unknown but this is a striking observation, since Ire1a $^{-/-}$ MEFs are otherwise normal in their UPR. By searching for interacting proteins of the cytosolic domain of Ire1b, Urano et al. found that Ire1 may interact with TRAF2, an adaptor protein involved in mediating cellular signaling of proinflammatory cytokines such as TNFa and IL1 [25]. The interaction of Ire1 and TRAF2 may result in the formation of Ire1-TRAF2-ASK1 ternary complex which in turn could activate JNK [25, 26]. Since JNKs regulate gene expression through the phosphorylation and activation of transcription factors such as cJun and ATF2, these data raise the possibility that cJun and ATF2 may participate in the mammalian UPR.

The functional importance of JNK activation in the ER stress pathway has not yet been fully examined, but $ASK1^{-/-}$ cells are partially resistant to ER stress-induced cell death [27], suggesting that the JNK pathway may promote cell death under conditions of intense or prolonged ER stress.

Xbp-1 (X-box-binding protein-1), a basic region leucine zipper protein in the CREB/ATF (cyclic AMP response element binding protein/activating transcription factor) family of transcription factors, has recently been found to encode a mammalian functional homologue of yeast HAC1 [28, 29]. Xbp-1 is ubiquitously expressed in adult tissues but preferentially expressed in fetal exocrine glands, osteoblasts, chondroblasts and liver [30]. Like its homologue HAC1 in yeast, newly synthesized XBP-1 transcript has to be processed by activated Ire1 to form mature mRNA [28, 29]. When activated under conditions of ER stress, Ire1 excises a 26-nucleotide intron from the immature XBP-1 transcript to result in a frame shift at amino acid 165, and the replacement of C-terminal 97 amino acids in ORF 1 to the C-terminal 212 amino acids of ORF2 [28, 31]. Similar to yeast Hac1p, only the mature Xbp-1 protein produced after splicing can efficiently activate the transcription of UPR target genes in mammalian cells.

Unlike the case of yeast HAC1, mammalian Xbp-1 mRNA is not constitutively expressed [31]. The expression of Xbp-1 is regulated at the transcriptional level by ATF6, which itself is under UPR control [31]. ATF6 is a bZIP transcription factor with significant sequence identity to Hac1p in its N-terminal basic region [32, 33]. Although the overall structure of ATF6 does not resemble Hac1p and ATF6 protein levels are not regulated by mRNA processing, evidence suggests that ATF6 nevertheless plays an important role in the mammalian UPR, as overexpression of ATF6 constitutively activates many targets of the mammalian UPR [33]. ATF6 protein is a type II transmembrane protein in the ER membrane with its C-terminus in the ER lumen and the N-terminus in the cytoplasm or nucleus [32]. Under conditions of ER stress, ATF6 is proteolytically processed by the site-1 and site-2 proteases, the same enzymes which cleave membrane-bound sterol response element binding proteins during conditions of cholesterol starvation [34]. When ATF6 is cleaved, its N-terminal bZIP domain is detached from the membrane and enters the nucleus to activate transcription of XBP-1 and other UPR target genes [32, 34]. It will be interesting to learn in the future how the accessibility or activity of the site-1 and site-2 proteases to ATF6 are regulated by conditions of ER stress.

Another critical feature of mammalian ER stress signaling, and one that is apparently absent from yeast, is the prominent role of global translation attenuation [4]. This is achieved principally through the phosphorylation of the translation initiation factor eIF2a on Ser51. When eIF2a is phosphorylated, it binds to its guanine nucleotide exchange factor (GEF), eIF2B, with greatly enhanced affinity [35, 36]. This interaction sequesters eIF2B, preventing it from mediating the GDP/GTP exchange reaction that eIF2a must undergo prior to assembling the 43S translation initiation complex to begin a new round of translation initiation. Therefore, phosphorylated eIF2a downregulates global cap-dependent translation by inhibiting its own GEF.

In mammalian cells, four eIF2a kinases have been found in mammals: PKR, GCN2, HRI and PERK. PKR is activated by double stranded RNA in response to viral infection and is involved in the interferon-mediated anti-viral defense [37]. In yeast, Gcn2p is a dedicated eIF2a kinase that is specifically activated upon amino acid starvation but has no reported role in the yeast ER stress response [37]. Like its counterpart in yeast, mammalian GCN2 is also activated by amino acid starvation [38, 39]. HRI is specifically expressed in erythroid cells and inhibited by heme, where it blocks the synthesis of new hemoglobin polypeptide chains when insufficient heme exists to assemble holo-hemoglobin [40]. Finally, PERK phosphorylates eIF2a in response to ER stress [41].

Like Ire1, PERK is a type I transmembrane protein that resides in the ER membrane [41]. The ER lumenal domain of PERK shares homology with Ire1a and Ireb and binds Bip, leading to speculation that the mechanism of PERK activation by ER stress may resemble the activation of Ire1 itself [9, 42]. The large N-terminal domains of Ire1 and PERK reside in the ER lumen and have been proposed to dimerize in a ligand-independent manner under ER stress condition [42]. Because the lumenal domains of Ire1 and PERK have been shown to interact with Bip under resting conditions [9, 42, 43], it has been proposed that excess unfolded proteins under conditions of ER stress compete with the lumenal domains of Ire1 and PERK for binding to Bip. ER stress therefore would promote the reversible dissociation of BiP from the lumenal domains of PERK and Ire1, allowing Ire1 and PERK to homodimerize and activate themselves by *trans*-autophosphorylation. Displacement of BiP from Ire1 or PERK correlates with the formation of activated PERK and Ire1, and overexpression of BiP attenuates their activation [9]. These findings are consistent with a model in which BiP represses signaling through PERK and Ire1, and protein misfolding relieves this repression by releasing BiP from the PERK and Ire1 lumenal domains.

As in yeast, the phosphorylation of mammalian eIF2a attenuates the translation of most mRNAs while selectively promoting the translation of a minority of messages. In the case of ER stress, some components of the UPR signaling are induced at the translational level when eIF2a is phosphorylated by PERK, including the transcription factor ATF4 and its downstream target CHOP [38, 44]. Although ATF4 mRNA is constitutively expressed, it is normally associated with low molecular weight polyribosomes and monoribosomes, suggesting that it is not efficiently translated under resting condition [38, 44]. ATF4 mRNA quickly shifts to heavier ribosomal fractions and is selectively translated in a PERK-dependent fashion upon the induction of ER stress [38]. The "bypass scanning" of the upstream open reading frames (uORFs) in the 5′ leader of the ATF4 mRNA is thought to be the mechanism that allows the preferential translation of ATF4 mRNA while most protein translation is halted by eIF2a phosphorylation under ER stress conditions [38]. A similar mechanism is used in yeast to control the translation of GCN4 mRNA by Gcn2p-mediated eIF2a phosphorylation under conditions of amino acid starvation [35, 45]. Thus, although the phosphorylation of eIF2a prevents the efficient formation of a translation initiation complex which recognizes the m^7G-cap at the 5′ of most cellular mRNAs, it does not prevent the initiation through the internal ribosome entry sites formed by the small uORFs that are present on a subpopulation of transcripts such as ATF4 or GCN4 [35, 38, 45–47].

While the full spectrum of genes induced in mammalian cells under conditions of ER stress has not been revealed, elevated eIF2a phosphorylation is known to control the expression of genes other than ATF4. Consistent with this observation, an increased level of ATF4 is not sufficient to activate the UPR [38]. Furthermore, mice deficient for ATF4 were not reported to display defects associated with the UPR [48], though this possibility has not been examined in detail yet. One direct transcriptional target of ATF4 is CHOP, a member of the bZIP family of transcription factor, which contains an ER stress response element (ESRE) in its promoter [49–51]. CHOP forms heterodimers with members of the C/EBP and Fos-Jun families of transcription factors [52, 53] and induces apoptosis [50, 54]. $CHOP^{-/-}$ MEFs are partially resistant to ER stress-induced cell death, though $CHOP^{-/-}$ mice are not resistant to lethal doses of tunicamycin [50]. In addition, recent findings have indicated that $PERK^{-/-}$ and $ATF4^{-/-}$ cells are sensitized to oxidative stress caused by endogenously generated peroxides [116]. These data indicate that the translational control components of the ESR are needed for maintaining cellular redof homeostasis under conditions of amino acid starvation or ER stress.

Protein translation under normal and ER stress conditions is also controlled by protein dephosphorylation. Specifically, the phosphorylation status of eIF2a is regulated by a complex containing protein phosphatase 1 (PP1). In an expression screen for genetic elements that suppress the induction by ER stress of GFP driven by the CHOP promoter, David Ron and colleagues found that GADD34 (also known as MyD116) functions as a cofactor for PP1, binding the phosphatase catalytic subunit and directing it to dephosphorylate eIF2a [55]. GADD34 was originally discovered as a gene induced by cellular stresses such DNA damage [56, 57]. Interestingly, like CHOP, the expression of GADD34 itself is induced by eIF2a phosphorylation and GADD34 induction depends on the presence of GCN2 or PERK under conditions of amino acid starvation or ER stress, respectively [55]. Therefore, the regulation of eIF2a by GADD34/PP1 forms a feedback loop to control translation: under conditions of ER stress, PERK phosphorylates eIF2a □hich induces the expression of GADD34, which in turn eventually mediates the dephosphorylation of eIF2a by PP1, returning protein translation to its resting state after the UPR has succeeded in correcting the ER stress.

Finally, when ER stress is intense or sustained, mammalian cells initiate apoptosis. The upstream signaling and downstream transcriptional components of the UPR are believed to participate in ER stress-induced apoptosis, as the UPR activates the pro-survival NF-kB transcription pathway [58] as well as pro-apoptotic signaling molecules such as JNK [25], CHOP [49] and GSK3b [59]. Recently, our lab identified caspase-12 [60], a new member of the caspase family that is associated with the ER membrane and is activated by apoptotic stimuli that induce ER stress but not by other mechanisms (such as serum deprivation, inhibition of protein kinase C or activation of the Fas or TNF receptors). Subsequent work in our lab and by other groups has suggested that caspase-12 may be activated by the calcium-activated protease calpain [61], by caspase-7 [62] or by autoproteolysis mediated by direct interaction with Ire1a and the adaptor protein TRAF2 [63]. In addition, the ER stress-activated caspase-12 pathway may in turn activate the downstream components of the mitochondrial pathway. Morishima et al. found that caspase-12 can cleave and activate caspase-9 [64], which is consistent with the finding that ER stress induced the activation of caspase-9 independent of Apaf-1 [65]. However, the specific modes of activation and downstream substrates of caspase-12 remain to be worked out in detail. Recent findings have also shown that the Bcl-2 family of proteins is important for regulating apoptosis induced by ER stress. Specifically, the presence of the multidomain Bcl-2 family members Bax and Bak at the ER membrane was shown to be crucial for maintaining the correct resting level of ER lumenal Ca^{+2} [117]. Accordingly, in the absence of Bax and Bak, the release of Ca^{+2} from the ER during apoptosis is attenuated, preventing the proper initiation of cell death in response to several stress stimuli, including treatment with arachidonic acid or oxidative stress [117].

UPR in Development and Tissue Homeostasis

In yeast, Ire1p and Hac1p form a linear ER-to-nucleus signaling pathway that is essential for the UPR [1]. In contrast, the role of the homologous mammalian pathway mediated by Ire1a/b and Xbp-1 in the UPR is less clear. Significant evidence suggests that the mammalian Ire1/Xbp-1 pathway may have important functions distinct from their role in the

UPR. Ire1a $^{-/-}$ mouse embryos die around day 10.5 of gestation due to unknown causes, but Ire1a $^{-/-}$ cells exhibit no defect in the UPR [24, 66]. These data suggest that other genes play a partially redundant role with Ire1a in regulating the mammalian UPR. On the other hand, mice deficient for Ire1 normally expressed only in the intestines, are born without obvious abnormalities, though they display increased sensitivity to colitis [23].

As noted above, Xbp-1 is ubiquitously expressed in adult tissues but preferentially in fetal exocrine glands, osteoblasts, chondroblasts and liver [30]. Xbp-1 has been shown to be essential for fetal hepatocyte growth [67]. Mice deficient for Xbp-1 display hypoplastic fetal livers with reduced cell proliferation and increased apoptosis, and show reduced hematopoiesis that results in severe anemia and death [67]. However, it is not known whether the role of Xbp-1 in the UPR or in some other pathway is critical for liver development. Xbp-1 is also known to play an important role in plasma cell differentiation [68]. Xbp-1$^{-/-}$ lymphocytes transplanted into RAG-2 chimeric mice displayed a severe defect in the generation of plasma cells [68]. It has not been reported whether Xbp-1$^{-/-}$ cells demonstrate a defect in the UPR, but it is tempting to speculate that cell types such as hepatocytes and plasma cells, which are specialized for protein secretion and have large ERs, may depend on Xbp-1 and the UPR to maintain normal cell growth and survival during development. Interestingly, a recent study found that a polymorphism in the Xbp-1 promoter, which led to reduced Xbp-1 signaling and a weakened ER stress response, may be a genetic risk factor in the development of bipolar disorder [118]. The authors also showed that the clinical mood stabilizer valproate may help to offset this defect in ESR signaling by inducing ATF6, suggesting that pharmacological intervention to bolster the ESR may be of therapeutic value in the treatment of bipolar disorder and perhaps other diseases [118].

Although immature HAC1 mRNA may be the only target of yeast Ire1p, mammalian Ire proteins may have additional nucleic acid targets. Iwawaki et al. reported that mammalian Ire1b may cleave the 28S ribosomal RNA in response to ER stress [69], which might provide a PERK-independent means of downregulating translation during ER stress conditions. Since the role of Xbp-1 in mammalian ER stress has not been demonstrated, it is possible that mammalian Ire proteins may have additional targets in mediating ER stress response.

In contrast to the possibly redundant role of Ire1 and Xbp-1 in regulating the mammalian UPR, PERK and eIF2a phosphorylation clearly play a prominent role in the mammalian ER stress response. PERK is highly expressed in the pancreatic acini that produce and secrete digestive enzymes, and the islets of Langerhans, which produce and secrete the polypeptide hormones insulin and glucagon [70–72]. Phosphorylation of PERK, an indicator of its activation, is detected in pancreas, lung, liver, spleen and thymus [72]. Phosphorylation of eIF2a at Ser51, a target of PERK, is prominent in normal pancreas, spleen and thymus. The phosphorylation of eIF2a is lost in the PERK$^{-/-}$ pancreas and thymus but not in spleen [72]. Consistent with the role of PERK in the homeostatic regulation of the ER burden in pancreatic cells, compensatory activation of Ire1a is detected in the pancreas of PERK$^{-/-}$ mice [72].

PERK$^{-/-}$ mice survive to birth, at which point they develop progressive disturbances in glucose metabolism, pancreatic b-cell insufficiency, and hypoinsulinemia [72]. Although the neonatal pancreas of PERK$^{-/-}$ mice is largely normal, the number of insulin-producing cells decreases over time with a perhaps compensatory increase in the number of glucagon-positive cells [72]. These results underline the critical role of PERK in the normal tissue homeostasis in the pancreas. Again, because it is the professional secretory cells in the PERK$^{-/-}$ pancreas that are most affected, it may be that PERK-mediated translational control in response to ER stress is necessary in order not to exceed the ability of ER to process and export proteins such as insulin under normal physiological conditions.

Analyses of mice with a homozygous "knock-in" Ser → Ala mutation at the eIF2a phosphorylation site (Ser51Ala) provided one of the most dramatic examples of phosphorylation in regulating cellular function [73]. The rate of protein translation in homozygous A/AeIF2a knockin MEFs is significantly higher than that of wild type MEFs under unstressed conditions, confirming a role for eIF2a phosphorylation in homeostatic regulation of protein translation [73]. When exposed to ER stress, A/AeIF2a knockin MEFs fail to downregulate protein translation normally and are considerably more vulnerable to ER stress-induced death than are wild type cells [73]. Consistent with the role of eIF2a phosphorylation in regulating gene expression under ER stress, the induction of Bip, CHOP and approximately 1/3 of other ER stress-inducible genes is significantly attenuated in A/AeIF2a knockin MEFs [73].

Although $PERK^{-/-}$ and A/AeIF2a knockin mice are both predicted to reduce or eliminate Ser51 phosphorylation in response to signals from the ER, the phenotype of A/AeIF2a knockin mice is earlier and more severe than that of $PERK^{-/-}$ mice [72, 73]. A/AeIF2a knockin mice are born phenotypically indistinguishable from their wild type or heterozygous littermates but most of die within 18 hr after birth due to hypoglycemia [73]. The induction of genes involved in gluconeogenic enzymes, such as phosphoenol pyruvate carboxykinase (PEPCK) in the liver are significantly reduced [73]. Glycogen synthase expression is induced to promote the storage of maternal glucose as glycogen in the liver for survival during the early neonatal period. The reduced ability to store glycogen may contribute to the hypoglycemia in A/AeIF2a knockin neonatal mice. In addition, the insulin levels in the A/AeIF2a knockin embryonic and neonatal pancreas are significantly lower than that of wild type [73]. The severe defects in the insulin metabolism of both $PERK^{-/-}$ and A/AeIF2a knockin mice suggest that ER lumenal protein folding and the regulation of eIF2a phosphorylation are important homeostatic controls in glucose metabolism. The fact that A/AeIF2a knockin mice exhibit an earlier and more severe phenotype than that of $PERK^{-/-}$ mice indicates that additional eIF2a kinases may play a role in normal fetal and neonatal development. However, none of the single knockouts of the other three eIF2a kinases, GCN2 [74], PKR [75, 76] or HRI [77], exhibits any abnormality in glucose metabolism, suggesting that two or more of these enzymes in combination may be responsible for liver glycogen metabolism.

Many human diseases, such as a 1-antitrypsin deficiency and cystic fibrosis, are associated with the accumulation of unfolded proteins in the ER and subsequent disruption of ER function [78]. The loss of the ability to control eIF2a phosphorylation in humans also has devastating consequences. Wolcott-Rallison syndrome (WRS) is a rare, autosomal recessive disorder characterized by permanent neonatal or early infancy insulin-dependent diabetes, with epiphyseal dysplasia, osteoporosis and growth retardation occurring at a later age [79–81]. Other frequent multisystemic manifestations of WRS include hepatic and renal dysfunction, mental retardation and cardiovascular abnormalities [79–81]. Recently, loss-of-function mutations in the gene encoding the human PERK, known as the eukaryotic translation initiation factor 2-alpha kinase 3 gene (EIF2AK3), have been identified as the underlying genetic defect of WRS in the families analyzed [82]. These results led to the suggestion that defects in the signaling process of UPR may be relevant to additional forms of human diabetes mellitus [82].

UPR and Neurodegeneration

The induction of the ESR by acute ischemic brain injury has also been well documented. Protein synthesis is immediately inhibited in ischemic neurons upon reperfusion and then gradually resumes in regions more resistant to ischemic damage, but remains suppressed in more vulnerable neurons [83, 84]. Phosphorylation of eIF2a and inhibition of translational initiation have also been found in reperfused brains [85–87]. Immunoreactivity towards eIF2a (P) was very low in control brains, but became very intense in the perinuclear cytoplasm of affected neurons in reperfused brains after a brief ischemic period in both global and focal ischemia models [88, 89]. Since the inhibition of protein synthesis occurs immediately upon reperfusion and resolves gradually as neurons recover, it has generally been assumed that inhibition of protein synthesis is detrimental to the recovery of neurons from ischemia [90]. However, since the induction of the ESR also induces the expression of protective proteins such as Bip, and because the phosphorylation of eIF2a is clearly protective from other forms of stress, it remains possible that eIF2a contributes positively to the recovery of neurons or at least is a "mixed blessing", an interesting hypothesis to be tested in the future.

The mechanism of eIF2a phosphorylation in reperfusion is controversial. PP1-dependent phosphatase activity has been reported as either unchanged [91] or modestly reduced in reperfused brains [92]. PKR-, HRI- and GCN2-knockout mice showed no difference in reperfusion induced eIF2a phosphorylation [91, 93]. However, the phosphorylation of PERK, an indication of its activation, was detected in ischemic brain upon reperfusion [93]. Unfortunately, the perinatal lethality of the PERK$^{-/-}$ mice precludes testing them in a brain ischemia/reperfusion experiment directly.

The disturbance of ER calcium homeostasis, a common phenomenon in ischemic brain injury, may be the major inducer of ER stress in reperfused neurons. This model is supported by the fact that blockers of ER calcium release, such as dantrolene or TMB-8, can protect cells from damage induced by ischemia, or excitotoxicity [94–96]. Additionally, excitotoxic glutamate may mediate the release of ER calcium stores through the ryanodine receptor (RyaR)-sensitive pool. Interestingly, the RyaR antagonist dantrolene protects cells from damage in various pathological conditions [94, 95, 97, 98]. It will be interesting to determine whether it is the loss of calcium from the ER or the sudden spike in cytoplasmic calcium (or a combination) that is the major factor in neuronal death following ischemia.

ER stress has also been proposed as a contributor to chronic neurodegenerative conditions. Expression of mutant presenilin-1 (PS-1) found in patients with familial Alzheimer's disease has been reported to decrease the expression of chaperone proteins such as Bip and sensitize cells to ER stress-induced cell death [99, 100], although this finding has been contested [101]. Siman et al. reported that in PS-1 P264L knockin cells, the UPR is unaltered [102]. In the original report, the levels of Bip and Grp94 were found to be significantly lower in some of the familial Alzheimer's patients and overexpression of Bip in cells expressing mutant PS-1 can reduce the cell death induced by ER stress [99]. PS-1 was found to interact with Ire1a and expression of mutant PS-1 reduced the activation of Ire1a [99]. Since Ire1a $^{-/-}$ cells exhibit a normal UPR [25], a defect in the Ire1 pathway alone may not explain a possible defect in Bip induction. It remains possible, however, that the expression of mutant PS-1 may affect the ESR indirectly by affecting cellular regulation of calcium homeostasis.

UPR in Viral Infection

As intracellular parasites, viruses co-opt the host cell machinery to produce and process large amounts of their own proteins. Accordingly, viral infection might be predicted to place a large burden on organelles, such as the ER, that are sensitive to protein overload. In particular, some classes like the flaviviruses (which include important human pathogens such as the hepatitis C, dengue, yellow fever and West Nile viruses) use the ER as the primary site for polyprotein processing, envelope glycoprotein biogenesis, and virion formation. Such ER-tropic viruses may therefore be expected to perturb normal ER homeostasis and engage the host cell ER stress pathways.

Indeed, numerous studies have demonstrated that viral glycoproteins in the ER interact specifically with resident host chaperones such as Bip, Grp94, calnexin and calreticulin. For example, protein products of vesicular stomatitis [103], Sindbis [104], rabies [105], hepatitis C [106] and measles viruses [107] have all been shown to exist in specific complexes with ER lumenal chaperones. In some (but not all) cases, it has been demonstrated that the interaction between host chaperones and viral proteins is essential for the efficient processing of the viral proteins and the assembly of complete virions. Even in cases where the interaction is dispensable for virus assembly, large amounts of viral glycoproteins in the ER may compete with cellular proteins for ER chaperone binding, leading to ER stress and the activation of ESR components like Ire1 and PERK.

In fact, activation of the ESR by viral infection has recently been demonstrated in several instances. For example, Japanese encephalitis virus infection has been shown to cause the hypertrophy of the ER membrane and to induce the protein levels of such canonical UPR targets as Bip, calnexin, Grp94 and CHOP [108]. Similarly, Tardif et al. showed that hepatitis C virus (HCV) infection induces the cleavage and activation of ATF6, with concomitant upregulation of Bip [109]. The authors further showed that the induction of Bip by HCV infection was dependent on the ER stress response elements in the Bip promoter, thereby demonstrating the specificity of Bip induction via ER stress in this case [109]. Bolt and colleagues have also used DNA microarrays and other methods to show that measles virus (MV) infection results in the upregulation of myriad UPR targets, including Bip, calreticulin, calnexin, Grp94, CHOP, ATF4, ERp57 and TFII-I, strongly implying the activation of both the UPR and PERK [107, 110]. Interestingly, MV infection has long been known to cause an inhibition of lymphoproliferation, wherein lymphocytes arrest in the G_0 or G_1 phase of the cell cycle [111, 112]. Because of this observation, and because lymphocytes are relatively sensitive to ER stress, which can activate PERK and result in cell cycle arrest [113], Bolt and colleagues have speculated that ER stress may contribute to the lymphoproliferative impairment during MV infection [107, 110].

Some recent studies have also directly linked the apoptosis caused by viral infection to the induction of ER stress. Using a cytopathic strain of the flavivirus bovine viral diarrhea virus (BVDV), Jordan et al. showed that BVDV infection in culture resulted in the activation of PERK, the phosphorylation of eIF2a, the upregulation of pro-apoptotic molecules such as CHOP and caspase-12, and the downregulation of the anti-apoptotic Bcl-2 [114]. These observations are consistent with the model that ER stress induced by BVDV infection is the primary pro-apoptotic stimulus that kills infected cells [114]. Similarly, Bitko and Barik [115] have shown that infection of a lung epithelial cell line by respiratory syncytial virus (RSV) leads to the induction of ER stress, as measured by Bip and calnexin upregulation, the infection-dependent cleavage and activation of caspase-12 and ultimately apoptosis. Remarkably, Bitko and Barik found that treating cells with an antisense construct against caspase-12 prior to RSV infection reduced the percentage of apoptotic cells by almost four-fold, suggesting that the activation of caspase-12 by ER stress may be a critical component of the cell death associated with RSV infection [115].

Although numerous viruses are known to engage ER lumenal chaperones and induce ER stress, it is not yet clear what role the resulting ESR may play in viral pathogenesis. Indeed, some components of the ESR, such as chaperone induction, might be expected to aid virus replication by facilitating the assembly of viral glycoproteins, while other pathways of the ER stress response, such as PERK and caspase-12 activation, may serve to hamper viral protein production and kill infected cells from the inside. It will be interesting to see how the homozygous deletion of known components of the ER stress response (such as Ire1, Xbp-1, PERK and caspase-12) affects the progression of viral infection in cell culture and animal models.

Future Directions: The ESR and Apoptosis

Despite the progress in dissecting the apoptotic pathways induced by ER stress, many important questions remain unanswered. For example, it is not currently clear how signals are transduced from the ER lumen to the cell's apoptotic machinery in the presence of ER stress, or how those signals might be modulated to forestall apoptosis until the ESR has failed to correct the disruption of ER homeostasis. At the level of caspase activation, it is not clear how agents which induce ER stress differently (such as through inhibition of glycosylation, perturbation of the Ca^{+2} or redox homeostasis of the ER lumen or blocking ER to Golgi vesicle transport) all lead to activation of caspase-12, either through identical or distinct but convergent mechanisms. Also, because caspase-12 knockout mice are only partially protected from apoptosis induced by ER stress [60], it seems likely that other components of the cell's apoptotic machinery are capable of responding to the ESR, but those components remain unclear. Finally, little is known about the events that follow caspase activation by the ESR, such as possible substrates of caspase-12 or pathways which amplify the pro-apoptotic cascade initiated by caspase-12 activation.

Cellular dysfunction and apoptosis induced by unfolded protein and ER stress have been implicated in several pathological conditions, including diabetes, Alzheimer's Disease, ischemia, cystic fibrosis and viral infection [78]. Therefore, a better understanding of the apoptotic pathways induced by the ESR will not only clarify the basic cell biology of an evolutionarily conserved signaling network but may also provide insight into how these pathways might be manipulated for therapeutic benefit.

References

1. Kaufman, R.J., *Stress signaling from the lumen of the endoplasmic reticulum: coordination of gene transcriptional and translational controls.* Genes Dev, 1999. **13**(10): p. 1211–33.
2. Mori, K., *Tripartite management of unfolded proteins in the endoplasmic reticulum.* Cell, 2000. **101**(5): p. 451–4.
3. Ma, Y. and L.M. Hendershot, *The unfolding tale of the unfolded protein response.* Cell, 2001. **107**(7): p. 827–30.
4. Harding, H.P., et al., *Transcriptional and translational control in the Mammalian unfolded protein response.* Annu Rev Cell Dev Biol, 2002. **18**: p. 575–99.
5. Ferri, K.F. and G. Kroemer, *Organelle-specific initiation of cell death pathways.* Nat Cell Biol, 2001. **3**(11): p. E255–63.

6. Patil, C. and P. Walter, *Intracellular signaling from the endoplasmic reticulum to the nucleus: the unfolded protein response in yeast and mammals.* Curr Opin Cell Biol, 2001. **13**(3): p. 349–55.
7. Shamu, C.E. and P. Walter, *Oligomerization and phosphorylation of the Ire1p kinase during intracellular signaling from the endoplasmic reticulum to the nucleus.* Embo J, 1996. **15**(12): p. 3028–39.
8. Welihinda, A.A. and R.J. Kaufman, *The unfolded protein response pathway in Saccharomyces cerevisiae. Oligomerization and trans-phosphorylation of Ire1p (Ern1p) are required for kinase activation.* J Biol Chem, 1996. **271**(30): p. 18181–7.
9. Bertolotti, A., et al., *Dynamic interaction of BiP and ER stress transducers in the unfolded-protein response.* Nat Cell Biol, 2000. **2**(6): p. 326–32.
10. Chapman, R.E. and P. Walter, *Translational attenuation mediated by an mRNA intron.* Curr Biol, 1997. **7**(11): p. 850–9.
11. Kawahara, T., et al., *Endoplasmic reticulum stress-induced mRNA splicing permits synthesis of transcription factor Hac1p/Ern4p that activates the unfolded protein response.* Mol Biol Cell, 1997. **8**(10): p. 1845–62.
12. Cox, J.S. and P. Walter, *A novel mechanism for regulating activity of a transcription factor that controls the unfolded protein response.* Cell, 1996. **87**(3): p. 391–404.
13. Sidrauski, C., J.S. Cox, and P. Walter, *tRNA ligase is required for regulated mRNA splicing in the unfolded protein response.* Cell, 1996. **87**(3): p. 405–13.
14. Mori, K., et al., *mRNA splicing-mediated C-terminal replacement of transcription factor Hac1p is required for efficient activation of the unfolded protein response.* Proc Natl Acad Sci U S A, 2000. **97**(9): p. 4660–5.
15. Travers, K.J., et al., *Functional and genomic analyses reveal an essential coordination between the unfolded protein response and ER-associated degradation.* Cell, 2000. **101**(3): p. 249–58.
16. Casagrande, R., et al., *Degradation of proteins from the ER of S. cerevisiae requires an intact unfolded protein response pathway.* Mol Cell, 2000. **5**(4): p. 729–35.
17. Plemper, R.K. and D.H. Wolf, *Retrograde protein translocation: ERADication of secretory proteins in health and disease.* Trends Biochem Sci, 1999. **24**(7): p. 266–70.
18. Plemper, R.K., et al., *Genetic interactions of Hrd3p and Der3p/Hrd1p with Sec61p suggest a retro-translocation complex mediating protein transport for ER degradation.* J Cell Sci, 1999. **112**(Pt 22): p. 4123–34.
19. Hampton, R.Y., R.G. Gardner, and J. Rine, *Role of 26S proteasome and HRD genes in the degradation of 3-hydroxy-3-methylglutaryl-CoA reductase, an integral endoplasmic reticulum membrane protein.* Mol Biol Cell, 1996. **7**(12): p. 2029–44.
20. Knop, M., et al., *Der1, a novel protein specifically required for endoplasmic reticulum degradation in yeast.* Embo J, 1996. **15**(4): p. 753–63.
21. Bordallo, J., et al., *Der3p/Hrd1p is required for endoplasmic reticulum-associated degradation of misfolded lumenal and integral membrane proteins.* Mol Biol Cell, 1998. **9**(1): p. 209–22.
22. Tirasophon, W., A.A. Welihinda, and R.J. Kaufman, *A stress response pathway from the endoplasmic reticulum to the nucleus requires a novel bifunctional protein kinase/endoribonuclease (Ire1p) in mammalian cells.* Genes Dev, 1998. **12**(12): p. 1812–24.
23. Bertolotti, A., et al., *Increased sensitivity to dextran sodium sulfate colitis in IRE1beta-deficient mice.* J Clin Invest, 2001. **107**(5): p. 585–93.
24. Urano, F., A. Bertolotti, and D. Ron, *IRE1 and efferent signaling from the endoplasmic reticulum.* J Cell Sci, 2000. **113 Pt 21**: p. 3697–702.
25. Urano, F., et al., *Coupling of stress in the ER to activation of JNK protein kinases by transmembrane protein kinase IRE1.* Science, 2000. **287**(5453): p. 664–6.
26. Nishitoh, H., et al., *ASK1 is essential for JNK/SAPK activation by TRAF2.* Mol Cell, 1998. **2**(3): p. 389–95.
27. Nishitoh, H., et al., *ASK1 is essential for endoplasmic reticulum stress-induced neuronal cell death triggered by expanded polyglutamine repeats.* Genes Dev, 2002. **16**(11): p. 1345–55.
28. Calfon, M., et al., *IRE1 couples endoplasmic reticulum load to secretory capacity by processing the XBP-1 mRNA.* Nature, 2002. **415**(6867): p. 92–6.
29. Shen, X., et al., *Complementary signaling pathways regulate the unfolded protein response and are required for C. elegans development.* Cell, 2001. **107**(7): p. 893–903.
30. Clauss, I.M., et al., *In situ hybridization studies suggest a role for the basic region-leucine zipper protein hXBP-1 in exocrine gland and skeletal development during mouse embryogenesis.* Dev Dyn, 1993. **197**(2): p. 146–56.

31. Yoshida, H., et al., *XBP1 mRNA is induced by ATF6 and spliced by IRE1 in response to ER stress to produce a highly active transcription factor.* Cell, 2001. **107**(7): p. 881–91.
32. Haze, K., et al., *Mammalian transcription factor ATF6 is synthesized as a transmembrane protein and activated by proteolysis in response to endoplasmic reticulum stress.* Mol Biol Cell, 1999. **10**(11): p. 3787–99.
33. Yoshida, H., et al., *Identification of the cis-acting endoplasmic reticulum stress response element responsible for transcriptional induction of mammalian glucose-regulated proteins. Involvement of basic leucine zipper transcription factors.* J Biol Chem, 1998. **273**(50): p. 33741–9.
34. Ye, J., et al., *ER stress induces cleavage of membrane-bound ATF6 by the same proteases that process SREBPs.* Mol Cell, 2000. **6**(6): p. 1355–64.
35. Hinnebusch, A., *Mechanism and Regulation of Initiator Methionyl-tRNA Binding to Ribosomes*, in *Translational Control of Gene Expression*, N. Sonenberg, Hershey, J.W.B., Mathews, M.B., Editor. 2000, Cold Spring Harbor Laboratory Press: Cold Spring Harbor. p. 185–244.
36. Clemens, M.J., *Initiation factor eIF2 alpha phosphorylation in stress responses and apoptosis.* Prog Mol Subcell Biol, 2001. **27**: p. 57–89.
37. Kaufman, R.J., *Double-stranded RNA-activated Protein Kinase, PKR*, in *Translational Control of Gene Expression*, N. Sonenberg, Hershey, J.W.B., Mathews, M.B., Editor. 2000, Cold Spring Harbor Laboratory Press: Cold Spring Harbor. p. 503–528.
38. Harding, H.P., et al., *Regulated translation initiation controls stress-induced gene expression in mammalian cells.* Mol Cell, 2000. **6**(5): p. 1099–108.
39. Sood, R., et al., *A mammalian homologue of GCN2 protein kinase important for translational control by phosphorylation of eukaryotic initiation factor-2alpha.* Genetics, 2000. **154**(2): p. 787–801.
40. Chen, J.J., *Heme-regulated eIF2a kinase*, in *Translational Control of Gene Expression*, N. Sonenberg, Hershey, J.W.B., Mathews, M.B., Editor. 2000, Cold Spring Harbor Laboratory Press: Cold Spring Harbor. p. 529–546.
41. Ron, D., and Harding, H.P., *PERK and Translational Control by Stress in the Endoplasmic Reticulum*, in *Translational Control of Gene Expression*, N. Sonenberg, Hershey, J.W.B., Mathews, M.B., Editor. 2000, Cold Spring Harbor Laboratory Press: Cold Spring Harbor. p. 547–560.
42. Liu, C.Y., M. Schroder, and R.J. Kaufman, *Ligand-independent dimerization activates the stress response kinases IRE1 and PERK in the lumen of the endoplasmic reticulum.* J Biol Chem, 2000. **275**(32): p. 24881–5.
43. Okamura, K., et al., *Dissociation of Kar2p/BiP from an ER sensory molecule, Ire1p, triggers the unfolded protein response in yeast.* Biochem Biophys Res Commun, 2000. **279**(2): p. 445–50.
44. Fawcett, T.W., et al., *Complexes containing activating transcription factor (ATF)/cAMP- responsive-element-binding protein (CREB) interact with the CCAAT/enhancer-binding protein (C/EBP)-ATF composite site to regulate Gadd153 expression during the stress response.* Biochem J, 1999. **339**(Pt 1): p. 135–41.
45. Hinnebusch, A.G., *Translational regulation of yeast GCN4. A window on factors that control initiator-trna binding to the ribosome.* J Biol Chem, 1997. **272**(35): p. 21661–4.
46. Fernandez, J., et al., *Regulation of internal ribosomal entry site-mediated translation by phosphorylation of the translation initiation factor eIF2alpha.* J Biol Chem, 2002. **277**(21): p. 19198–205.
47. Fernandez, J., et al., *Regulation of internal ribosome entry site-mediated translation by eukaryotic initiation factor-2alpha phosphorylation and translation of a small upstream open reading frame.* J Biol Chem, 2002. **277**(3): p. 2050–8.
48. Tanaka, T., et al., *Targeted disruption of ATF4 discloses its essential role in the formation of eye lens fibres.* Genes Cells, 1998. **3**(12): p. 801–10.
49. Wang, X.Z., et al., *Signals from the stressed endoplasmic reticulum induce C/EBP-homologous protein (CHOP/GADD153).* Mol Cell Biol, 1996. **16**(8): p. 4273–80.
50. Zinszner, H., et al., *CHOP is implicated in programmed cell death in response to impaired function of the endoplasmic reticulum.* Genes Dev, 1998. **12**(7): p. 982–95.
51. Ma, Y., et al., *Two distinct stress signaling pathways converge upon the CHOP promoter during the mammalian unfolded protein response.* J Mol Biol, 2002. **318**(5): p. 1351–65.
52. Ubeda, M., et al., *Stress-induced binding of the transcriptional factor CHOP to a novel DNA control element.* Mol Cell Biol, 1996. **16**(4): p. 1479–89.
53. Ubeda, M., M. Vallejo, and J.F. Habener, *CHOP enhancement of gene transcription by interactions with Jun/Fos AP-1 complex proteins.* Mol Cell Biol, 1999. **19**(11): p. 7589–99.
54. Matsumoto, M., et al., *Ectopic expression of CHOP (GADD153) induces apoptosis in M1 myeloblastic leukemia cells.* FEBS Lett, 1996. **395**(2–3): p. 143–7.

55. Novoa, I., et al., *Feedback inhibition of the unfolded protein response by GADD34-mediated dephosphorylation of eIF2alpha.* J Cell Biol, 2001. **153**(5): p. 1011–22.
56. Fornace, A.J., D.W. Neibert, M.C. Hollander, J.D. Luethy, M. Papathanasiou, and N.J.H. J. Fragoli, *Mammalian genes coordinately regulated by growth arrest signals and DNA-damaging agents.* Mol Cell Biol, 1989. **9**: p. 4196–4203.
57. Lord, K.A., B. Hoffman-Liebermann, D.A. Liebermann, *Sequence of MyD116 cDNA: a novel myeloid differentiation primary response gene induced by IL6.* Nucleic Acids Res, 1990. **18**: p. 2823.
58. Pahl, H.L. and P.A. Baeuerle, *A novel signal transduction pathway from the endoplasmic reticulum to the nucleus is mediated by transcription factor NF-kappa B.* Embo J, 1995. **14**(11): p. 2580–8.
59. Song, L., P. De Sarno, and R.S. Jope, *Central role of glycogen synthase kinase-3b inendoplasmic reticulum stress-induced caspase-3 activation.* J Biol Chem, 2002. **12**: p. 12.
60. Nakagawa, T., et al., *Caspase-12 mediates endoplasmic-reticulum-specific apoptosis and cytotoxicity by amyloid-beta.* Nature, 2000. **403**(6765): p. 98–103.
61. Nakagawa, T. and J. Yuan, *Cross-talk between two cysteine protease families. Activation of caspase-12 by calpain in apoptosis.* J Cell Biol, 2000. **150**(4): p. 887–94.
62. Rao, R.V., et al., *Coupling endoplasmic reticulum stress to the cell death program. Mechanism of caspase activation.* J Biol Chem, 2001. **276**(36): p. 33869–74.
63. Yoneda, T., et al., *Activation of caspase-12, an endoplastic reticulum (ER) resident caspase, through tumor necrosis factor receptor-associated factor 2-dependent mechanism in response to the ER stress.* J Biol Chem, 2001. **276**(17): p. 13935–40.
64. Morishima, N., et al., *An endoplasmic reticulum stress-specific caspase cascade in apoptosis. Cytochrome c-independent activation of caspase-9 by caspase-12.* J Biol Chem, 2002. **277**(37): p. 34287–94.
65. Rao, R.V., et al., *Coupling endoplasmic reticulum stress to the cell death program. An Apaf-1-independent intrinsic pathway.* J Biol Chem, 2002. **277**(24): p. 21836–42.
66. Lee, K., et al., *IRE1-mediated unconventional mRNA splicing and S2P-mediated ATF6 cleavage merge to regulate XBP1 in signaling the unfolded protein response.* Genes Dev, 2002. **16**(4): p. 452–66.
67. Reimold, A.M., et al., *An essential role in liver development for transcription factor XBP-1.* Genes Dev, 2000. **14**(2): p. 152–7.
68. Reimold, A.M., et al., *Plasma cell differentiation requires the transcription factor XBP-1.* Nature, 2001. **412**(6844): p. 300–7.
69. Iwawaki, T., et al., *Translational control by the ER transmembrane kinase/ribonuclease IRE1 under ER stress.* Nat Cell Biol, 2001. **3**(2): p. 158–64.
70. Shi, Y., et al., *Identification and characterization of pancreatic eukaryotic initiation factor 2 alpha-subunit kinase, PEK, involved in translational control.* Mol Cell Biol, 1998. **18**(12): p. 7499–509.
71. Harding, H.P., Y. Zhang, and D. Ron, *Protein translation and folding are coupled by an endoplasmic-reticulum-resident kinase.* Nature, 1999. **397**(6716): p. 271–4.
72. Harding, H.P., et al., *Diabetes mellitus and exocrine pancreatic dysfunction in perk$^{-/-}$mice reveals a role for translational control in secretory cell survival.* Mol Cell, 2001. **7**(6): p. 1153–63.
73. Scheuner, D., et al., *Translational control is required for the unfolded protein response and in vivo glucose homeostasis.* Mol Cell, 2001. **7**(6): p. 1165–76.
74. Zhang, P., et al., *The GCN2 eIF2alpha kinase is required for adaptation to amino acid deprivation in mice.* Mol Cell Biol, 2002. **22**(19): p. 6681–8.
75. Yang, Y.-L.R., L.F. Pavlovic, J. Aguzzi, A. Schafer, R. Kumar, A. Williams, B.R.G. Aguet, M. Weissman, C., *Deficient signaling in mice devoid of double-stranded RNA-dependent protein kinase.* Embo J, 1995. **14**: p. 6095–6106.
76. Abraham, N., et al., *Characterization of transgenic mice with targeted disruption of the catalytic domain of the double-stranded RNA-dependent protein kinase, PKR.* J Biol Chem, 1999. **274**(9): p. 5953–62.
77. Lu, L., A.P. Han, and J.J. Chen, *Translation initiation control by heme-regulated eukaryotic initiation factor 2alpha kinase in erythroid cells under cytoplasmic stresses.* Mol Cell Biol, 2001. **21**(23): p. 7971–80.
78. Aridor, M. and W.E. Balch, *Integration of endoplasmic reticulum signaling in health and disease.* Nat Med, 1999. **5**(7): p. 745–51.
79. Stoss, H., et al., *Wolcott-Rallison syndrome: diabetes mellitus and spondyloepiphyseal dysplasia.* Eur J Pediatr, 1982. **138**(2): p. 120–9.
80. al-Gazali, L.I., et al., *Wolcott-Rallison syndrome.* Clin Dysmorphol, 1995. **4**(3): p. 227–33.
81. Thornton, C.M., D.J. Carson, and F.J. Stewart, *Autopsy findings in the Wolcott-Rallison syndrome.* Pediatr Pathol Lab Med, 1997. **17**(3): p. 487–96.

82. Delepine, M., et al., *EIF2AK3, encoding translation initiation factor 2-alpha kinase 3, is mutated in patients with Wolcott-Rallison syndrome.* Nat Genet, 2000. **25**(4): p. 406–9.
83. Hossmann, K.A., *Disturbances of cerebral protein synthesis and ischemic cell death.* Prog Brain Res, 1993. **96**: p. 161–77.
84. Krause, G.S. and B.R. Tiffany, *Suppression of protein synthesis in the reperfused brain.* Stroke, 1993. **24**(5): p. 747–55; discussion 755–6.
85. Burda, J., et al., *Phosphorylation of the alpha subunit of initiation factor 2 correlates with the inhibition of translation following transient cerebral ischaemia in the rat.* Biochem J, 1994. **302**(Pt 2): p. 335–8.
86. DeGracia, D.J., et al., *Global brain ischemia and reperfusion: modifications in eukaryotic initiation factors associated with inhibition of translation initiation.* J Neurochem, 1996. **67**(5): p. 2005–12.
87. DeGracia, D.J., et al., *Molecular pathways of protein synthesis inhibition during brain reperfusion: implications for neuronal survival or death.* J Cereb Blood Flow Metab, 2002. **22**(2): p. 127–41.
88. DeGracia, D.J., et al., *Effect of brain ischemia and reperfusion on the localization of phosphorylated eukaryotic initiation factor 2 alpha.* J Cereb Blood Flow Metab, 1997. **17**(12): p. 1291–302.
89. Althausen, S., et al., *Changes in the phosphorylation of initiation factor eIF-2alpha, elongation factor eEF-2 and p70 S6 kinase after transient focal cerebral ischaemia in mice.* J Neurochem, 2001. **78**(4): p. 779–87.
90. Paschen, W., *Disturbances of calcium homeostasis within the endoplasmic reticulum may contribute to the development of ischemic-cell damage.* Med Hypotheses, 1996. **47**(4): p. 283–8.
91. DeGracia, D.J., et al., *Eukaryotic initiation factor 2alpha kinase and phosphatase activity during postischemic brain reperfusion.* Exp Neurol, 1999. **155**(2): p. 221–7.
92. Martin de la Vega, C., et al., *Possible mechanisms involved in the down-regulation of translation during transient global ischaemia in the rat brain.* Biochem J, 2001. **357**(Pt 3): p. 819–26.
93. Kumar, R., et al., *Brain ischemia and reperfusion activates the eukaryotic initiation factor 2alpha kinase, PERK.* J Neurochem, 2001. **77**(5): p. 1418–21.
94. Zhang, L., et al., *Dantrolene protects against ischemic, delayed neuronal death in gerbil brain.* Neurosci Lett, 1993. **158**(1): p. 105–8.
95. Berg, M., et al., *Kainic acid-induced seizures and brain damage in the rat: role of calcium homeostasis.* J Neurosci Res, 1995. **40**(5): p. 641–6.
96. Malcolm, C.S., et al., *A prototypic intracellular calcium antagonist, TMB-8, protects cultured cerebellar granule cells against the delayed, calcium-dependent component of glutamate neurotoxicity.* J Neurochem, 1996. **66**(6): p. 2350–60.
97. Frandsen, A. and A. Schousboe, *Mobilization of dantrolene-sensitive intracellular calcium pools is involved in the cytotoxicity induced by quisqualate and N-methyl-D- aspartate but not by 2-amino-3-(3-hydroxy-5-methylisoxazol-4-yl)propionate and kainate in cultured cerebral cortical neurons.* Proc Natl Acad Sci USA, 1992. **89**(7): p. 2590–4.
98. Wei, H. and D.C. Perry, *Dantrolene is cytoprotective in two models of neuronal cell death.* J Neurochem, 1996. **67**(6): p. 2390–8.
99. Katayama, T., et al., *Presenilin-1 mutations downregulate the signalling pathway of the unfolded-protein response.* Nat Cell Biol, 1999. **1**(8): p. 479–85.
100. Katayama, T., et al., *Disturbed activation of endoplasmic reticulum stress transducers by familial Alzheimer's disease-linked presenilin-1 mutations.* J Biol Chem, 2001. **276**(46): p. 43446–54.
101. Sato, N., et al., *Upregulation of BiP and CHOP by the unfolded-protein response is independent of presenilin expression.* Nat Cell Biol, 2000. **2**(12): p. 863–70.
102. Siman, R., et al., *Endoplasmic reticulum stress-induced cysteine protease activation in cortical neurons: effect of an Alzheimer's disease-linked presenilin-1 knock-in mutation.* J Biol Chem, 2001. **276**(48): p. 44736–43.
103. Hammond, C. and A. Helenius, *Folding of VSV G protein: sequential interaction with BiP and calnexin.* Science, 1994. **266**(5184): p. 456–8.
104. Mulvey, M. and D.T. Brown, *Involvement of the molecular chaperone BiP in maturation of Sindbis virus envelope glycoproteins.* J Virol, 1995. **69**(3): p. 1621–7.
105. Gaudin, Y., *Folding of rabies virus glycoprotein: epitope acquisition and interaction with endoplasmic reticulum chaperones.* J Virol, 1997. **71**(5): p. 3742–50.
106. Choukhi, A., et al., *Involvement of endoplasmic reticulum chaperones in the folding of hepatitis C virus glycoproteins.* J Virol, 1998. **72**(5): p. 3851–8.
107. Bolt, G., *The measles virus (MV) glycoproteins interact with cellular chaperones in the endoplasmic reticulum and MV infection upregulates chaperone expression.* Arch Virol, 2001. **146**(11): p. 2055–68.

108. Su, H.L., C.L. Liao, and Y.L. Lin, *Japanese encephalitis virus infection initiates endoplasmic reticulum stress and an unfolded protein response.* J Virol, 2002. **76**(9): p. 4162–71.
109. Tardif, K.D., K. Mori, and A. Siddiqui, *Hepatitis C virus subgenomic replicons induce endoplasmic reticulum stress activating an intracellular signaling pathway.* J Virol, 2002. **76**(15): p. 7453–9.
110. Bolt, G., K. Berg, and M. Blixenkrone-Moller, *Measles virus-induced modulation of host-cell gene expression.* J Gen Virol, 2002. **83**(Pt 5): p. 1157–65.
111. McChesney, M.B., et al., *Measles virus infection of B lymphocytes permits cellular activation but blocks progression through the cell cycle.* J Virol, 1987. **61**(11): p. 3441–7.
112. Naniche, D., S.I. Reed, and M.B. Oldstone, *Cell cycle arrest during measles virus infection: a G0-like block leads to suppression of retinoblastoma protein expression.* J Virol, 1999. **73**(3): p. 1894–901.
113. Brewer, J.W. and J.A. Diehl, *PERK mediates cell-cycle exit during the mammalian unfolded protein response.* Proc Natl Acad Sci U S A, 2000. **97**(23): p. 12625–30.
114. Jordan, R., et al., *Replication of a cytopathic strain of bovine viral diarrhea virus activates PERK and induces endoplasmic reticulum stress-mediated apoptosis of MDBK cells.* J Virol, 2002. **76**(19): p. 9588–99.
115. Bitko, V. and S. Barik, *An endoplasmic reticulum-specific stress-activated caspase (caspase-12) is implicated in the apoptosis of A549 epithelial cells by respiratory syncytial virus.* J Cell Biochem, 2001. **80**(3): p. 441–54.
116. Harding, H., et al., *An Integrated Stress Response Regulates Amino Acid Metabolism and Resistance to Oxidative Stress.* Mol Cell, 2003. **11(3)**: p. 619–33.
117. Scorrano, L., et al., *BAX and BAK Regulation of Endoplasmic Reticulum Ca^{+2}: A Control Point of Apoptosis.* Science, 2003. **300(5616)**: p. 135–9.
118. Kakiuchi, C., et al., *Impaired Feedback Regulation of XBP1 as a Genetic Risk Factor for Bipolar Disorder.* Nat Gen, 2003. Published online 31 August; doi: 10.1038/ng1235.

Chapter 4

The Role of the PI3K Pathway in Anti-IgM (Anti-μ) -Sensitive and -Resistant B-cell Lymphomas

Failure to Disengage PI3K Pathway Signaling Confers Anti-μ Resistance on the CH12 B Cell Lymphoma

GREGORY B. CAREY,* LAURA TONNETTI,*
AND DAVID W. SCOTT*†

ABSTRACT: We and others have firmly established that surface IgM receptor (sIgM-R) crosslinking leads to growth arrest and apoptosis in a series of well characterized B lymphomas. This requires ablation of c-Myc protein expression and the concomitant induction of the CKI, $p27^{Kip1}$. The signaling mechanisms regulating c-Myc and $p27^{Kip1}$ protein expression are poorly understood. However, we recently established that sIgM-R mediated down-modulation of the PI3K pathway directly affected c-Myc and $p27^{Kip1}$ expression and accurately predicted growth arrest or apoptosis in the ECH408 B lymphoma line. Moreover, p27-TAT fusion protein (which is rapidly taken up by these lymphoma lines) disappeared rapidly in growing B-lymphoma lines, but it persisted in the presence of anti-m. We have now extended these studies to the well-characterized CH12 B-lymphoma cell line, which is resistant to anti-m induced growth arrest and apoptosis, despite its sIgM-R^{high} phenotype. The results show that failure to inactivate $p70^{S6K}$ and to activate GSK3-b, results in anti-m resistance. Blockade of PI3K, or an upstream regulator of $p70^{S6K}$, mTOR, reconstitutes or mimics all negative sIgM receptor signaling in these cells. Finally, anti-m or PI3K blockade induces PTEN protein expression especially in the WEHI-231 B-cell line. Together, these results reveal apparently fundamental mechanisms for inducing immune tolerance (PI3K suppression leading to anergy or deletion), or activation (stimulation of or maintenance of PI3K activity), or lymphomagenesis (failure to disengage PI3K signaling).

Abbreviations: Anti-m, anti-mu (IgM) heavy chain; -m, anti-delta (IgD) heavy chain; BCR, B-cell antigen receptor; CKI, cyclin dependent kinase inhibitor; $p27^{Kip1}$, 27kDa cyclin-dependent kinase inhibitor; $p70^{S6K}$, 70kD isoform of ribosomal protein S6, kinase; MIgM, MIgD, membrane; IgM, membrane; PI3K, Phosphatidylinositol 3 kinase.

Department of Immunology, Holland Laboratory of the American Red Cross, Rockville, MD 20855,* and Department of Immunology, George Washington University School of Medicine, Washington, DC 20037† *Correspondence:* Dr. David W. Scott, Department of Immunology, Holland Laboratory of the American Red Cross, 15601 Crabbs Branch Way, Rockville, MD 20855, USA. Fax: (301) 517-0344; E-mail: ScottD@usa.redcross.org

Introduction

Crosslinking membrane IgM (mIgM) leads to growth arrest and apoptosis in a panel of B lymphoma lines (1–16). Growth arrest and apoptosis is preceded by the loss of c-Myc and a striking induction of $p27^{Kip1}$ proteins, which block cell cycle progress. The mechanism(s) regulating c-Myc and $p27^{Kip1}$ protein expression are poorly understood; nonetheless, we recently established a critical role of the PI3K signaling pathway in their control (17). Brennan *et al.* have shown that the PI3K effector, PKB, regulates CDK, which can negatively regulate $p27^{Kip1}$ protein expression in IL-2 stimulated T-cells (18). CDK phosphorylates $p27^{Kip1}$, which results in its ubiquitination and proteosomal degradation (19–21). Hence, our results are consistent with those of others showing that the loss of PI3K and PKB, leads to loss of negative regulation of $p27^{Kip1}$ and therefore, promotes its accumulation. Moreover, we also showed that loss of PI3K activity, *via* anti-m or *via* a PI3K specific inhibitor, leads to $p27^{Kip1}$ accumulation (17 and vide infra).

The mRNA for the c-Myc oncogene belongs to a specialized class, whose translation is highly regulated through the $p70^{S6K}$/S6 pathway (22–25). Indeed, our results from anti-m stimulated ECH408 cells established that the loss of PI3K signaling leads to the loss of $p70^{S6K}$ activation, which is immediately followed by a decline in c-Myc protein expression (17). Therefore, the results showed in our system for the first time, that uncoupling of PI3K signaling by anti-m results in the both the loss of c-Myc protein expression and the induction of $p27^{Kip1}$ protein. We also show that p27 protein expression, effected via uptake of p27-TAT fusion protein, is tightly regulated by this signaling pathway.

Our working hypothesis is that PI3K disengagement is a critical requirement for the anti-m induced changes on Myc and $p27^{Kip1}$ protein expression and eventual growth arrest and apoptosis. To further test this hypothesis, we examined whether failure to disengage PI3K signaling components gave rise to anti-m resistance in the CH12 lymphoma line. The results presented herein show that, indeed, $p70^{S6K}$ is not regulated by anti-m and PI3K in those cells. Therefore, c-Myc protein expression is stable and, consequently, these cells do not undergo apoptosis. Furthermore, Rapamycin, which induces growth arrest and apoptosis, unlike anti-m, activates the pro-apoptotic enzyme, GSK3-b in CH12 cells. All together, the results suggest that a signaling disconnect between mIgM and $p70^{S6K}$ (and possibly mTOR) might lead to the anti-m resistant phenotype observed in CH12 cells.

Materials and Methods

Cell Lines and Culture Conditions

The WEHI-231, CH31, and ECH408 murine B-lymphoma cell lines are defined as immature based in part on its high surface IgM expression phenotype and their sensitivity to anti-m induced growth arrest and apoptosis (3, 13, 15). ECH408 also expresses sIgD, which allows us to study differential signaling by this isotype. CH12 cells which are also $sIgM^{high}$, are resistant to anti-m mediated growth arrest and apoptosis and hence are considered to be of the 'mature' phenotype (26). Therefore, parallel studies utilizing the WEHI-231 (for example) and CH12 cells allows the study of signaling differences in the two cell lines in order to explain the different outcomes following anti-m stimulation, while ECH408 can be utilized for analysis of IgM *versus* IgD signaling in the same cell. Furthermore, use of these cell lines also permits us to explore whether or not a loss of BCR to PI3K signaling could play a role in the resistance of the $sIgM^{high}$ CH12 cell to anti-m. It is important to note all of these lymphoma cells have been extensively studied and characterized over the past 20 years (3, 15, 26). They are maintained at 37°C, in a humidified, 7% CO2 atmosphere as previously described (3, 7, 15, 17, 27), in supplemented RPMI-1640 with 5% FCS (Bio-Whittaker, Walkersville MD), and new vials are thawed after 20 passages *in vitro*.

Reagents and Inhibitors

The PI3K specific inhibitor LY294002 (2-4-morpholinyl-8-phenyl-4H-1-benzopyran-4-one (28), was obtained from LC Laboratories, Woburn, MA and used at 0 to 20 mM. Rapamycin, (RAP), a specific inhibitor of $p70^{S6K}$ activation (22, 23, 29–31), was obtained from Sigma-Aldrich (St. Louis, MO), and was used at final concentrations of 0.1 to 100 nM. Stock solutions of inhibitors were prepared to ensure that final DMSO carrier concentrations were at or below 0.05%. Additionally, appropriately diluted vehicle was added to control samples.

TATp27 wild type plasmid was a kind gift from Dr. Steven F. Dowdy (Washington University, St. Louis, MO) and the fusion protein was purified as previously described (31a). TATp27 (T187A) was generated by inserting a KpnI-EcoRI fragment from p27 (T187A), obtained from Dr. Michele Pagano (NYU, New York), into the pTAT vector and the fusion protein was purified using the same conditions described for wild type TATp27.

Antibody Treatment of Lymphoma Lines

Maximal anti-m-mediated changes in c-Myc and $p27^{Kip1}$ protein expression and ultimately growth arrest and apoptosis in all our anti-m sensitive cell lines (e.g. CH31, ECH408 and WEHI-231 cells) is observed at ~1 mg/ml antibody (Ab) and between 18 and 32 hours, depending on the cell line (see references above). Therefore, the cells were stimulated with or without monoclonal rat anti mouse anti-m (clone B7.6) which is routinely purified by protein G affinity chromatography in our laboratory. All antibody batches were titrated for biologically effective doses. Comparable results were observed with commercially obtained monoclonal and polyclonal antibodies, respectively.

Gel Electrophoresis and Western Blotting

For Western blot analysis, cells were harvested and prepared essentially as previously described by our laboratory (7, 17, 32), using a modified Nonidet-P40 (N-P40) lysis buffer to minimize the impact of serine/threonine and tyrosine protein phosphatases (as described by Carey and Scott (17)). Fifty micrograms (50 mg) total cellular proteins were resolved per lane on 8 or 12.5% SDS-PAGE gels as previously described (7, 17, 27). Importantly, for analyzing phosphorylated proteins using phospho-specific antibodies, resulting filters were blocked with 2–3% sonicated, filtered BSA in 1X TBST at a final pH of 7.4.

All incubations included the same blocking solution. Primary antibodies included: rabbit polyclonal anti-rat $p70^{S6K}$ (C-18, catalogue SC-230, 1:200 dilution, Santa Cruz Biotechnology, Santa Cruz, CA), anti-phospho Thr389-$p70^{S6K}$ (activation specific) antibody (Cell Signaling/New England Biolabs, Beverly, MA; catalogue #9205S, used at 1:500 dilution), polyclonal rabbit anti-mouse p27Kip1 (C-19, cat. no. SC-528; Santa Cruz, at a 1:200 dilution), and rabbit anti-mouse c-Myc (rabbit polyclonal, catalogue SC-764, Santa Cruz, at 1:200 dilution). Anti phospho GSK3-a /b (Ser21/9) was also obtained from Cell

Signaling, Beverly, MA. Following probing with the primary antibodies, the filters were washed and developed exactly as previously described (7, 17). Densitometric analysis of specific bands on resulting exposed films was performed using Un-Scanit™ software (Silk Scientific, Orem, UT).

Cell Cycle Analysis

FACS analysis employed 0.5 to 1×10^6 cells and was performed using ethanol fixation essentially as previously described (7, 17, 32). The cell cycle was also analyzed using the saponin permeabilization technique described by Jacob *et al.* (33). Saponin/PI solution contained: 0.1% saponin, 1 mg/ml RNAase and 50 mg/ml PI diluted in phosphate buffered saline. After saponin permeabilization, the samples were incubated for at least 15′ at 37°C and DNA content assessed within 2h. Although forward and side light scatter properties change dramatically with this method, no differences in cell cycle profiles (DNA content) are seen in comparing ethanol fixation or saponin permeabilization methods. The cell cycle was assessed with FACScalibur™ and FAScan flow cytometers (Becton Dickinson, San Diego, CA) after the method of Jacob *et al.* (33). The resulting data was analyzed using Cell Quest™ software (Becton Dickinson).

Results and Discussion

Anti-μ Mediated Growth-Arrest and Apoptosis Follows the Loss of c-Myc Protein Expression and Induction of p27^{Kip1}

To illustrate the fundamental differences in anti-meffects on c-Myc and p27^{Kip1} proteins and their relationship with eventual outcomes, CH12, WEHI-231, CH31 and ECH408 B lymphoma cells were stimulated with anti-m for 24 h and c-Myc and p27^{Kip1} protein expression assessed by Western Blot analysis. The c-Myc protein is depressed and p27^{Kip1} is elevated in all lines that undergo growth-arrest and apoptosis (Figure 1). This pattern is not observed in the CH12 cell line, which does not undergo growth-arrest and apoptosis following mIgM crosslinking. These results have been reproducibly demonstrated during the last two decades in a number of labs (13, 15, 26, 34–40).

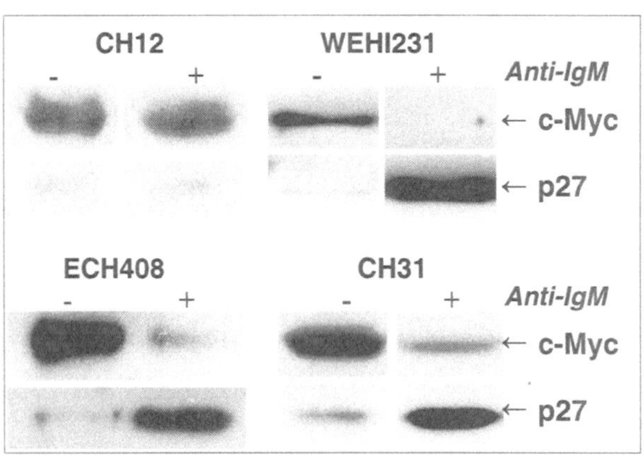

Figure 1. Effect of BCR signaling on c-Myc and p27 protein levels in anti-m sensitive (WEHI231, ECH408 and CH31) versus resistant CH12 B-lymphoma cells.

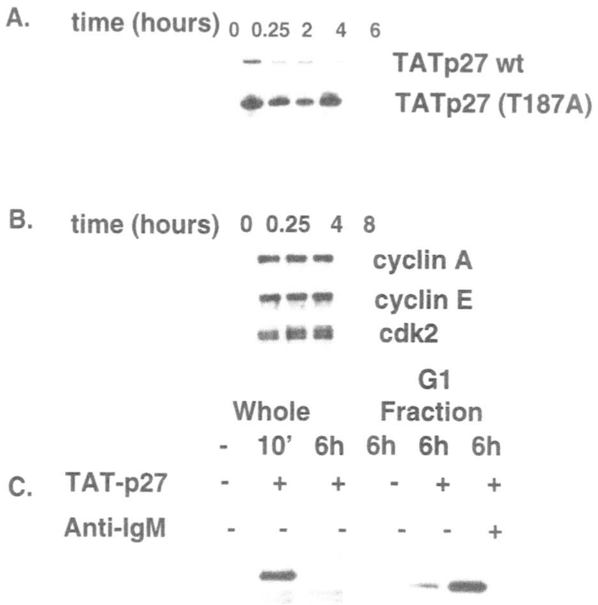

Figure 2. Regulation of p27 expression in WEHI-231 B-lymphoma cells. (A) Expression of TATp27 fusion protein, WT and TATp27(T187) (B) WEHI-231 cells were incubated with 200 nM of TATp27 (T187) for the indicates times and immune complexes were then loaded on a SDS-PAGE and hybridized with anti-cyclin A, anti-cyclin E and anti-Cdk2. (C) Anti-m prolongs the survival of TATp27 in elutriated G1 cells.

Expression and Degradation of p27-TAT Fusion Protein in B-lymphoma Cells: Influence of Anti-IgM and T187 Phosphorylation

To establish the role of p27 in growth arrest and apoptosis, we used a bacterial expression vector, pTAT-HA, to produce TAT fusion wild type and mutant p27 proteins in-frame. As shown in Figure 2A, TATp27 wild type fusion protein was rapidly taken up by cycling the cells but its amount decreased by 60% or more within hours. A mutant form of p27, in which threonine 187 was substituted with alanine (T187A), and which is not degraded by the ubiquitinin-dependent pathway, was also expressed in the pTAT-HA vector and the resultant fusion protein transduced into the cells. Unlike the wild type, p27 (T187A) mutant did not disappear from the cells and, at 6 hours, was still highly expressed. We further verified that the mutation on T187 only affects p27 phosphorylation and not its affinity for the cyclin/CDK complex, by testing the ability of p27 (T187A) to form aggregates with cyclin A, cyclin E and Cdk2. As shown in Figure 2B, TATp27 (T187A) fusion protein efficiently binds as well to the cyclin-CDK complex in WEHI-231 cells as wild type p27. We next demonstrated that the maintenance of the TATp27 fusion protein is prolonged by treatment of these cells with anti-m (Figure 2C) which we know modulates cyclin-kinase activity (12, 39).

Inactivation of PI3K and p70^{S6K} Leads to Loss of c-Myc and Accumulation of p27 in Sensitive B-Lymphoma Cells but Not in Resistant Lines

We have found that anti-m mediated changes in c-Myc and p27^{Kip1} proteins are required the down-regulation of PI3K and p70^{S6K} activities and that disengagement of PI3K signaling

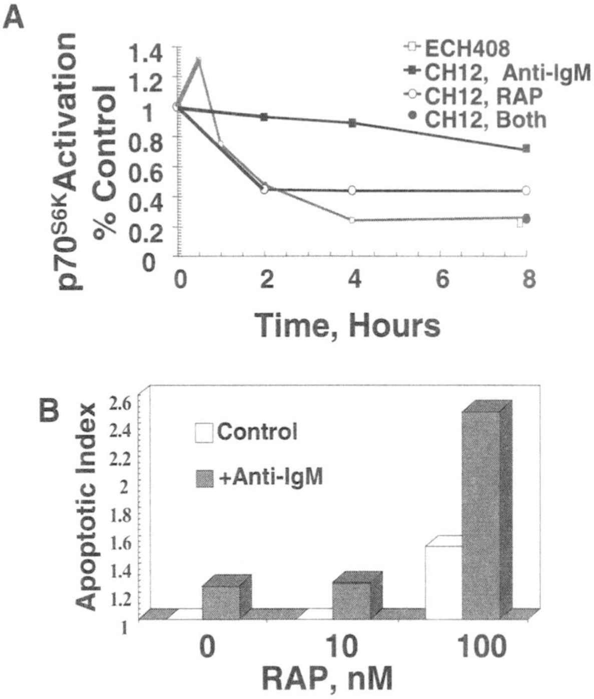

Figure 3. Effect of anti-m and Rapamycin on p70S6K activity in B lymphoma cells lines. Bottom: Rapamycin and anti-m syngerzie to induce apoptosis in resistant CH12 cells.

is required for growth arrest and apoptosis (17). To further test the role of the PI3K pathway in anti-m sensitive and resistant B-lymphomas, ECH408 cells were treated with anti-m or anti-d and CH12 cells treated with anti-m, and $p70^{S6K}$ phosphorylation (activation) assessed. The results presented in figure 3A clearly show that anti-m leads prolonged to inactivation $p70^{S6K}$ in ECH408 cells (anti-d does not; see ref. 17); moreover, anti-m does not affect $p70^{S6K}$ in resistant CH12 cells.

Our previous data from anti-m sensitive ECH408 cells show that c-Myc protein expression exactly parallels the activation state of $p70^{S6K}$ and that suppression of PI3K or PI3K effector signaling is required for the accumulation of $p27^{Kip1}$ and the ablation of c-Myc protein expression (17). Thus, we next tested if failure to modulate this pathway gives rise to the anti-m resistant phenotype observed in CH12 cells. Therefore, CH12 cells were treated with and without anti-m in the presence or absence of PI3K inhibitor LY294002 (not shown) or the $p70^{S6K}$ activation specific inhibitor, Rapamycin (RAP). The results presented in Figure 3B clearly show that Rapamycin, which causes induction of p27 in CH12 cells and a modest loss of c-Myc (not shown; see Carey and Scott, in preparation), in concert with anti-m, rapamycin leads to apoptosis in CH12. Interestingly, Rapamycin and LY, also induce $p70^{S6K}$ and PKB dephosphorylation, respectively, in WEHI-231 and ECH408 cells to mimic and enhance the effect of anti-m on sensitive B-lymphoma cells, including the loss of c-Myc and accumulation of p27 (see ref. 17 for ECH408 data, WEHI-231 data not shown). Because RAP blocks mTOR, the upstream regulator of $p70^{S6K}$, these data suggest that there is a signaling disconnect between mIgM and PI3K and $p70^{S6K}$ or most likely between PI3K and other intermediates and $p70^{S6K}$ in CH12 cells.

To determine whether CH12 cells had dysfunctional PI3K/PKB or a dysfunctional PI3K/p70^{S6K} pathway, these cells were treated with or without PI3K inhibitor and PKB and p70^{S6K} phosphorylation (Thr389 for p70^{S6K} and anti-phospho Ser473 for PKB/Akt) assessed following immunoprecipitation and Western blotting. We found in preliminary experiments that a suboptimal LY concentration causes the slow inactivation of PKB ($t_{1/2} \approx$ 6 hours; data not shown). Interestingly, anti-m also transiently inactivates PKB/Akt which parallels its modest, transient inductive effect on p27^{Kip1} expression. LY and serum deprivation inactivate PKB/Akt. LY (PI3K blockade) does not inactivate p70^{S6K}. Importantly, serum deprivation results in complete loss of activation specific phosphorylation of p70^{S6K}, indicating that signaling to p70^{S6K} and not p70^{S6K} itself, is dysfunctional. Thus, anti-m appears to transiently or weakly modulate a pro-apoptotic pathway in CH12 cells but these cells recover from this signal. Thus, failure to disengage pathways governing *both* p27^{Kip1} and c-Myc, results in failure to invoke growth arrest and apoptosis.

Effect of Anti-μ on Activation of GSK-3 (α and β in Anti-μ Sensitive and Resistant B-Lymphoma Cell Lines

Anti-m inactivates PKB in WEHI-231 and ECH408 cells (data not shown). Activated PKB is required to phosphorylate glycogen synthase 3 (GSK3), which results in GSK3 inactivation. In growth-factor deprived cells, loss of PKB activation results in GSK3 dephosphorylation and activation for pro-apoptotic signaling. Activated GSK3-b is known to phosphorylate b-catenin and cyclin D1 which results in their ubiquitination and proteosomal targeting. High levels of b-catenin/LEF transcriptional activity stimulate c-Myc promoters (44–46). Hence, GSK-3 activation could hypothetically lead to the loss of c-Myc and cyclin D1 protein expression. Therefore, to examine if anti-IgM treatment led to the activation (by de-phosphorylation) of GSK-a/b, exponentially proliferating WEHI231, ECH408 and CH12 cells were treated with or without anti-m, GSK3-a/b immunoprecipitated and their phosphorylation examined using anti phospho GSK3-b or GSK3-a&b (Ser21 of GSK3-a and Ser9 of GSK3-b) antibodies essentially as described for phospho-PKB and phospho-p70^{S6K} (17). In anti-m sensitive WEHI231 (not shown) and ECH408 (Figure 4A) cells, anti-m treatment results in the dephosphorylation of both GSK3 alpha and beta by 4 h. These results correspond to the loss of b-catenin protein expression and loss of c-Myc protein expression at 8h. Anti-d leads to only a transient activation of GSK3, consistent with its transient effect on PI3K/PKB signaling (fig. 4A and ref. 17). However, no activation occurs with anti-m in CH12 (data not shown). Nonetheless, RAP treatment, which reduces c-Myc and induces p27^{Kip1}, also leads to the activation of both GSK3a/b and GSK3-b in CH12 cells (Figure 4B). Note that RAP only reduces c-Myc in ECH408 cells, but does not regulate p27^{Kip1} except in the presence of anti-m (data not shown). Therefore, these results suggest that RAP, which inactivates p70^{S6K}, induces p27^{Kip1} and suppresses c-Myc and induces growth arrest and apoptosis in CH12 cells, does so *via* mimicking the effect of anti-m in the anti-IgM sensitive cell lines. Indeed, it can even cause apoptosis in anti-m resistant lines and this seems to be dependent on activation of GSK3-b.

It is well established that anti-m induces growth arrest and apoptosis B-cell lymphomas, which is always associated with down-regulation of c-Myc and increased p27^{Kip1} protein expression. In contrast, crosslinking of the sIgD receptor on immature B-cell lymphomas transiently modulates c-Myc protein expression, but does not result in a dramatic loss of c-Myc protein expression, nor in increased p27^{Kip1} protein expression; thus, anti-d does not

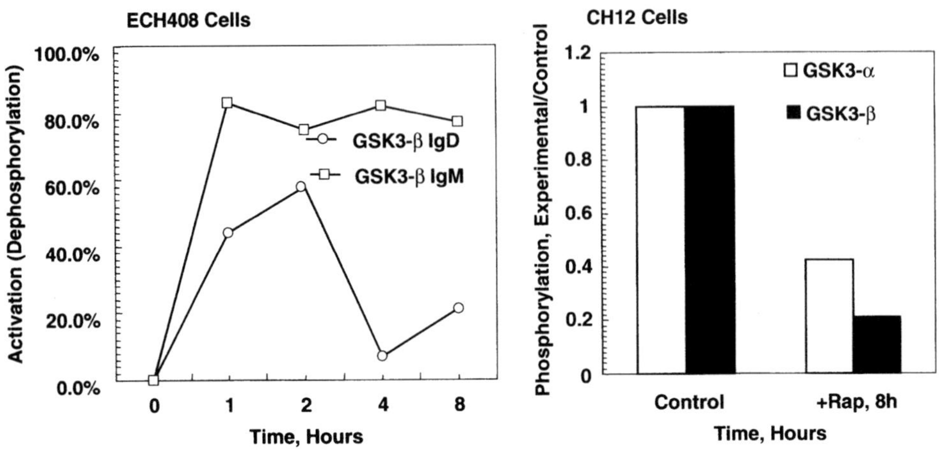

Figure 4. BCR signaling effects on GSK3 in different B cell lines. (A) Effect of anti-m (IgM) or anti-d (IgD) on GSK3-b activity in ECH408. (B) Effect of Rapamycin on GSK activation in CH12.

lead to growth arrest and apoptosis. Indeed, a similar pattern of functional results occurs with anti-IgM signaling in the resistant B-cell lymphoma, CH12. Our goal herein was to discern whether similar programs are invoked in the resistant to BCR signaling by CH12 with anti-m and ECH408 with anti-d.

We have found that pharmacological interference with PI3K/ $p70^{S6K}$ in all tested anti-m sensitive B-cell lymphomas leads to an increase in $p27^{Kip1}$ and synergy with anti-m (ECH408 and WEHI-231 cell lines). In fact, treatment with LY also leads to growth arrest and apoptosis in an anti-m resistant CH12 B-cell lymphoma. We propose that there is a disconnect or that a block exists between the sIgM receptor and the signals that lead to a downregulation of the PI3K pathway in CH12 (see Figure 5). On the other hand, anti-m mimicry by LY in both anti-m sensitive and resistant B-cell lines, supports the critical role of downregulation of PI3K (and its sequelae) being sufficient for B-cell growth arrest and apoptosis. A critical feature appears to be the activation of GSK3b, which can modulate

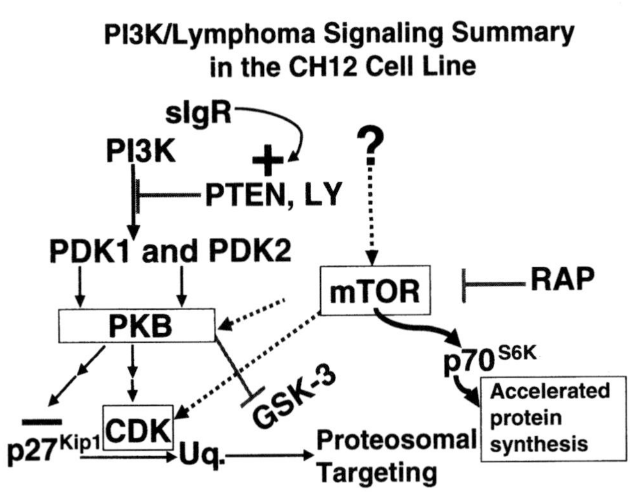

Figure 5. Scheme of PI3K signaling targets in anti-m resistant CH12 B-lymphoma cells.

the expression of p27 and possibly c-Myc (Figure 5). The nature of this pathway and of the block in CH12 is under investigation.

ACKNOWLEDGMENTS: The work reported herein has been supported by USPHS grants CA55644, CA55664S1, and CA94027, as well as by funds from the American Red Cross.

References

1. Green, D. R. 1997. A Myc-induced apoptosis pathway surfaces [comment]. *Science 278:1246.*
2. Green, D. R., and D. W. Scott. 1994. Activation-induced apoptosis in lymphocytes. *Curr Opin Immunol 6:476.*
3. Ales-Martinez, J. E., G. L. Warner, and D. W. Scott. 1990. Lymphoma models for B cell activation and tolerance. VIII. Cross-desensitization by sIgM and sIgD and its effects on growth regulation by anti-isotype antibodies. *Cell Immunol 127:527.*
4. DeFranco, A. L., J. T. Kung, and W. E. Paul. 1982. Regulation of growth and proliferation in B cell subpopulations. *Immunol Rev 64:161.*
5. DeFranco, A. L. 1997. The complexity of signaling pathways activated by the BCR. *Curr Opin Immunol 9:296.*
6. Donjerkovic, D., and D. W. Scott. 2000. Activation induced death in B lymphocytes. *Cell Research 10:179.*
7. Donjerkovic, D., L. Zhang, and D. W. Scott. 1999. Regulation of p27Kip1 accumulation in murine B-lymphoma cells: role of c-Myc and calcium. *Cell Growth Differ 10:695.*
8. Gottschalk, A. R., and J. Quintans. 1995. Apoptosis in B lymphocytes: the WEHI-231 perspective. *Immunol Cell Biol 73:8.*
9. Kim, K. M., T. Yoshimura, H. Watanabe, T. Ishigami, M. Nambu, D. Hata, Y. Higaki, M. Sasaki, T. Tsutsui, M. Mayumi, and et al. 1991. Growth regulation of a human mature B cell line, B104, by anti-IgM and anti-IgD antibodies. *J Immunol 146:819.*
10. Maddox, B. R., and D. W. Scott. 1996. Differential susceptibility to anti-receptor induced apoptosis in adult murine B-cells: Role of B1 cells. *Front Biosci 1:a39.*
11. Scott, D. W. 1993. B-Lymphoma models for tolerance: The good, the bad and the apoptotic. *ImmunoMethods 2:105.*
12. Scott, D. W., T. Brunner, D. Donjerkovich, S. Ezhevsky, T. Grdina, D. R. Green, Y. B. Shi, and X. R. Yao. 1997. Murder and suicide: A Tale of T and B cell apoptosis. In *Programmed Cell Death.* Y. B. Shi, D. W. Scott, and X. Yu, eds. Plenum Press, New York.
13. Hasbold, J., and G. G. Klaus. 1990. Anti-immunoglobulin antibodies induce apoptosis in immature B cell lymphomas. *Eur J Immunol 20:1685.*
14. Scott, D. W., D. Donjerkovic, B. Maddox, S. Ezhevsky, and T. Grdina. 1997. Role of c-myc and p27 in anti-IgM induced B-lymphoma apoptosis. *Curr Top Microbiol Immunol 224:103.*
15. Scott, D. W., J. H. Chace, G. L. Warner, A. O'Garra, G. G. Klaus, and H. Quill. 1987. Role of T cell-derived lymphokines in two models of B-cell tolerance. *Immunol Rev 99:153.*
16. Tisch, R., C. M. Roifman, and N. Hozumi. 1988. Functional differences between immunoglobulins M and D expressed on the surface of an immature B-cell line. *Proc Natl Acad Sci U S A 85:6914.*
17. Carey, G., and D. W. Scott. 2001. Role of PI3K in anti-IgM and anti-IgD induced apoptosis in B-cell lymphomas. *J Immunol 166:1618.*
18. Brennan, P., J. W. Babbage, B. M. Burgering, B. Groner, K. Reif, and D. A. Cantrell. 1997. Phosphatidylinositol 3-kinase couples the interleukin-2 receptor to the cell cycle regulator E2F. *Immunity 7:679.*
19. Alessandrini, A., D. S. Chiaur, and M. Pagano. 1997. Regulation of the cyclin-dependent kinase inhibitor p27 by degradation and phosphorylation. *Leukemia 11:342.*
20. Shirane, M., Y. Harumiya, N. Ishida, A. Hirai, C. Miyamoto, S. Hatakeyama, K. Nakayama, and M. Kitagawa. 1999. Down-regulation of p27(Kip1) by two mechanisms, ubiquitin-mediated degradation and proteolytic processing. *J Biol Chem 274:13886.*
21. Casagrande, F., D. Bacqueville, M. J. Pillaire, F. Malecaze, S. Manenti, M. Breton-Douillon, and J. M. Darbon. 1998. G1 phase arrest by the phosphatidylinositol 3-kinase inhibitor LY 294002 is correlated to up-regulation of p27Kip1 and inhibition of G1 CDKs in choroidal melanoma cells. *FEBS Lett 422:385.*
22. Dufner, A., and G. Thomas. 1999. Ribosomal S6 kinase signaling and the control of translation. *Exp Cell Res 253:100.*

23. Jefferies, H. B., S. Fumagalli, P. B. Dennis, C. Reinhard, R. B. Pearson, and G. Thomas. 1997. Rapamycin suppresses 5'TOP mRNA translation through inhibition of p70s6k. *Embo J 16:3693.*
24. West, M. J., M. Stoneley, and A. E. Willis. 1998. Translational induction of the c-myc oncogene via activation of the FRAP/TOR signalling pathway. *Oncogene 17:769.*
25. Brown, E. J., and S. L. Schreiber. 1996. A signaling pathway to translational control. *Cell 86:517.*
26. Arnold, L. W., N. J. LoCascio, P. M. Lutz, C. A. Pennell, D. Klapper, and G. Haughton. 1983. Antigen-induced lymphomagenesis: identification of a murine B cell lymphoma with known antigen specificity. *J Immunol 131:2064.*
27. Ezhevsky, S. A., H. Toyoshima, T. Hunter, and D. W. Scott. 1996. Role of cyclin A and p27 in anti-IgM induced G1 growth arrest of murine B-cell lymphomas. *Mol Biol Cell 7:553.*
28. Vlahos, C. J., W. F. Matter, K. Y. Hui, and R. F. Brown. 1994. A specific inhibitor of phosphatidylinositol 3-kinase, 2-(4-morpholinyl)-8-phenyl-4H-1-benzopyran-4-one (LY294002). *J Biol Chem 269:5241.*
29. Abraham, R. T., and G. J. Wiederrecht. 1996. Immunopharmacology of Rapamycin. *Annu Rev Immunol 14:483.*
30. Gottschalk, A. R., L. H. Boise, C. B. Thompson, and J. Quintans. 1994. Identification of immunosuppressant-induced apoptosis in a murine B-cell line and its prevention by bcl-x but not bcl-2. *Proc Natl Acad Sci U S A 91:7350.*
31. Carey, G. B., and J. P. Liberti. 1995. Stimulation of receptor-associated kinase, tyrosine kinase, and MAP kinase is required for prolactin-mediated macromolecular biosynthesis and mitogenesis in Nb2 lymphoma. *Arch Biochem Biophys 316:179.*
31a. Nagahara, H., A. M. Vocero-Akabani, E. L. Snyder, A. Ho, D. G. Latham, N. A. Lissy, M. Becker-Hapak, S. A. Ezhevsky, and F. Dowdy. 1998. Transduction of full-length TAT fusion protein into mammalian cells: TAT-p27^{Kip1} induces cell migration. *Nat. Med.* 12:1449–1452.
32. Donjerkovic, D., C. M. Mueller, and D. W. Scott. 2000. Steroid- and retinoid-mediated growth arrest and apoptosis in WEHI-231 cells: role of NF-kappaB, c-Myc and CKI p27(Kip1). *Eur J Immunol 30:1154.*
33. Jacob, M. C., M. Favre, and J. C. Bensa. 1991. Membrane cell permeabilization with saponin and multiparametric analysis by flow cytometry. *Cytometry 12:550.*
34. McCormack, J. E., V. H. Pepe, R. B. Kent, M. Dean, A. Marshak-Rothstein, and G. E. Sonenshein. 1984. Specific regulation of c-myc oncogene expression in a murine B-cell lymphoma. *Proc Natl Acad Sci U S A 81:5546.*
35. Wu, M., M. Arsura, R. E. Bellas, M. J. FitzGerald, H. Lee, S. L. Schauer, D. H. Sherr, and G. E. Sonenshein. 1996. Inhibition of c-myc expression induces apoptosis of WEHI 231 murine B cells. *Mol Cell Biol 16:5015.*
36. Wu, M., W. Yang, R. E. Bellas, S. L. Schauer, M. J. FitzGerald, H. Lee, and G. E. Sonenshein. 1997. c-myc promotes survival of WEHI 231 B lymphoma cells from apoptosis. *Curr Top Microbiol Immunol 224:91.*
37. Pennell, C. A., and D. W. Scott. 1986. Lymphoma models for B cell activation and tolerance. IV. Growth inhibition by anti-Ig of CH31 and CH33 B lymphoma cells. *Eur J Immunol 16:1577.*
38. Warner, G. L., J. W. Ludlow, D. A. Nelson, A. Gaur, and D. W. Scott. 1992. Anti-immunoglobulin treatment of murine B-cell lymphomas induces active transforming growth factor beta but pRB hypophosphorylation is transforming growth factor beta independent. *Cell Growth Differ 3:175.*
39. Joseph, L. F., S. Ezhevsky, and D. W. Scott. 1995. Lymphoma models for B-cell activation and tolerance: anti-immunoglobulin M treatment induces growth arrest by preventing the formation of an active kinase complex which phosphorylates retinoblastoma gene product in G1. *Cell Growth Differ 6:51.*
40. LoCascio, N. J., L. W. Arnold, R. B. Corley, and G. Haughton. 1984. Induced differentiation of a B cell lymphoma with known antigen specificity. *J Mol Cell Immunol 1:177.*
41. Lloyd, R. V., L. A. Erickson, L. Jin, E. Kulig, X. Qian, J. C. Cheville, and B. W. Scheithauer. 1999. p27kip1: a multifunctional cyclin-dependent kinase inhibitor with prognostic significance in human cancers. *Am J Pathol 154:313.*
42. Stewart, M. J., and G. Thomas. 1994. Mitogenesis and protein synthesis: a role for ribosomal protein S6 phosphorylation? *Bioessays 16:809.*
43. Polakiewicz, R. D., S. M. Schieferl, A. C. Gingras, N. Sonenberg, and M. J. Comb. 1998. mu-Opioid receptor activates signaling pathways implicated in cell survival and translational control. *J Biol Chem 273:23534.*
44. Yoganathan, T. N., P. Costello, X. Chen, M. Jabali, J. Yan, D. Leung, Z. Zhang, A. Yee, S. Dedhar, and J. Sanghera. 2000. Integrin-linked kinase (ILK): a "hot" therapeutic target [In Process Citation]. *Biochem Pharmacol 60:1115.*
45. Delcommenne, M., C. Tan, V. Gray, L. Rue, J. Woodgett, and S. Dedhar. 1998. Phosphoinositide-3-OH kinase-dependent regulation of glycogen synthase kinase 3 and protein kinase B/AKT by the integrin-linked kinase. *Proc Natl Acad Sci U S A 95:11211.*

46. Aoki, M., A. Hecht, U. Kruse, R. Kemler, and P. K. Vogt. 1999. Nuclear endpoint of Wnt signaling: neoplastic transformation induced by transactivating lymphoid-enhancing factor 1. *Proc Natl Acad Sci U S A 96:139.*
47. Wishart, M. J., G. S. Taylor, J. T. Slama, and J. E. Dixon. 2001. PTEN and myotubularin phosphoinositide phosphatases: bringing bioinformatics to the lab bench. *Curr Opin Cell Biol 13:172.*
48. Maehama, T., G. S. Taylor, and J. E. Dixon. 2001. PTEN AND MYOTUBULARIN: Novel Phosphoinositide Phosphatases. *Annu Rev Biochem 70:247.*
49. Ales-Martinez, J. E., G. L. Warner, and D. W. Scott. 1988. Immunoglobulins D and M mediate signals that are qualitatively different in B cells with an immature phenotype. *Proc Natl Acad Sci U S A 85:6919.*
50. Wu, M., R. E. Bellas, J. Shen, W. Yang, and G. E. Sonenshein. 1999. Increased p27Kip1 cyclin-dependent kinase inhibitor gene expression following anti-IgM treatment promotes apoptosis of WEHI 231 B cells. *J Immunol 163:6530.*
51. Fischer, G., S. C. Kent, L. Joseph, D. R. Green, and D. W. Scott. 1994. Lymphoma models for B cell activation and tolerance. X. Anti-mu-mediated growth arrest and apoptosis of murine B cell lymphomas is prevented by the stabilization of myc. *J Exp Med 179:221.*
52. Carpenter, C. L., and L. C. Cantley. 1996. Phosphoinositide kinases. *Curr Opin Cell Biol 8:153.*
53. Chan, T. O., S. E. Rittenhouse, and P. N. Tsichlis. 1999. AKT/PKB and other D3 phosphoinositide-regulated kinases: kinase activation by phosphoinositide-dependent phosphorylation. *Annu Rev Biochem 68:965.*
54. Vanhaesebroeck, B., and D. R. Alessi. 2000. The PI3K-PDK1 connection: more than just a road to PKB. *Biochem J 346 Pt 3:561.*
55. Downward, J. 1998. Mechanisms and consequences of activation of protein kinase B/Akt. *Curr Opin Cell Biol 10:262.*
56. Amati, B., K. Alevizopoulos, and J. Vlach. 1998. Myc and the cell cycle [In Process Citation]. *Front Biosci 3:D250.*
57. D'Amours, D., S. Desnoyers, I. D'Silva, and G. G. Poirier. 1999. Poly(ADP-ribosyl)ation reactions in the regulation of nuclear functions. *Biochem J 342:249.*
58. Pawson, T. 1994. SH2 and SH3 domains in signal transduction. *Adv Cancer Res 64:87.*
59. Campbell, K. S. 1999. Signal transduction from the B cell antigen-receptor. *Curr Opin Immunol 11:256.*
60. Fruman, D. A., R. E. Meyers, and L. C. Cantley. 1998. Phosphoinositide kinases. *Annu Rev Biochem 67:481.*
61. Heikkila, R., J. G. Iversen, and T. Godal. 1985. No correlation between membrane potential and increased cytosolic free Ca2+ concentration, 86Rb+ influx or subsequent [3H]-thymidine incorporation in neoplastic human B cells stimulated with antibodies to surface immunoglobulin. *Acta Physiol Scand 124:107.*
62. Carey, G. B., D. Donjerkovic, C. M. Mueller, S. Liu, J. A. Hinshaw, L. Tonnetti, W. Davidson, and D. W. Scott. 2000. B-cell receptor and Fas-mediated signals for life and death. *Immunol Rev 176:105.*
63. Beckwith, M., R. G. Fenton, I. M. Katona, and D. L. Longo. 1996. Phosphatidylinositol-3-kinase activity is required for the anti-ig-mediated growth inhibition of a human B-lymphoma cell line. *Blood 87:202.*
64. Prendergast, G. C. 1999. Mechanisms of apoptosis by c-Myc. *Oncogene 18:2967.*
65. Eder, A. M., L. Dominguez, T. F. Franke, and J. D. Ashwell. 1998. Phosphoinositide 3-kinase regulation of T cell receptor-mediated interleukin-2 gene expression in normal T cells. *J Biol Chem 273:28025.*
66. Klippel, A., M. A. Escobedo, M. S. Wachowicz, G. Apell, T. W. Brown, M. A. Giedlin, W. M. Kavanaugh, and L. T. Williams. 1998. Activation of phosphatidylinositol 3-kinase is sufficient for cell cycle entry and promotes cellular changes characteristic of oncogenic transformation. *Mol Cell Biol 18:5699.*
67. Cantley, L. C., and B. G. Neel. 1999. New insights into tumor suppression: PTEN suppresses tumor formation by restraining the phosphoinositide 3-kinase/AKT pathway. *Proc Natl Acad Sci U S A 96:4240.*
68. Wang, A., A. Pierce, K. Judson-Kremer, S. Gaddis, C. M. Aldaz, D. G. Johnson, and M. C. MacLeod. 1999. Rapid analysis of gene expression (RAGE) facilitates universal expression profiling. *Nucleic Acids Res 27:4609.*

Chapter 5

Signaling for Inducible Fas-Resistance in Primary B Lymphocytes

THOMAS L. ROTHSTEIN*

ABSTRACT: The sensitivity of B cells to Fas (APO-1, CD95)-mediated apoptosis is modulated by signals derived from other cell surface receptors. Engagement of CD40 upregulates Fas expression and markedly increases susceptibility to cell death induced by Fas triggering, whereas engagement of the B cell antigen receptor (or the IL-4 receptor) diminishes susceptibility to Fas killing, even in otherwise Fas-sensitive, CD40-stimulated targets, and so produces a state of Fas-resistance. Fas-resistance likely functions to protect B cells during interaction with activated (Fas ligand-bearing) T cells. B cell receptor signaling for Fas-resistance bypasses Btk but requires protein kinase C and NF-kB, eventuating in the upregulation of anti-apoptotic molecules including FLIP, Bcl-xL, and possibly others. The Btk-independence of BCR-mediated NF-kB induction is unexpected but explained by prior CD40 engagement. After CD40L treatment, IkBa is degraded, and NF-kB is induced, as a result of BCR engagement, even in *xid* B cells in which Btk is mutated. Thus, CD40 signaling appears to create new conditions that provide BCR access to IkB/NF-kB in a Btk-independent fashion, which does not occur in the absence of CD40 stimulation. These results suggest that B cells play a key role in determining their own susceptibility to Fas-mediated apoptosis, and that the process of inducing Fas-sensitivity alters the signaling requirements for Fas-resistance.

Key Words: Apoptosis, Fas, CD95, B lymphocytes, surface immunoglobulin, CD40, Btk, NF-kB

Introduction

Lymphocytes express several surface molecules that act as death receptors. These TNFR family members contain extracellular cysteine-rich pseudorepeats and intracellular C-terminal death domains, and, upon engagement, transmit signals that produce target cell apoptosis (Chinnaiyan et al., 1996; Chaudhary et al., 1997; Nagata, 1997; Schneider et al., 1997a; Yeh et al., 1998). First amongst the death receptors in terms of regulating immune function is CD95 (or Fas, APO-1) (Trauth et al., 1989; Yonehara et al., 1989). Mutation of this molecule results in a clinical syndrome that includes autoantibody production and

*Departments of Medicine and Microbiology, Boston University School of Medicine, and the Immunobiology Unit, Evans Memorial Department of Clinical Research, Boston University Medical Center, Boston, MA 02118, U.S.A. *Correspondence:* Thomas L. Rothstein, Immunobiology Unit, Evans Biomedical Research Center, Room 437, Boston Medical Center, 650 Albany Street, Boston, MA 02118, U.S.A. Tel: 617-638-7028; FAX: 617-638-7140. e-mail: trothstein@medicine.bu.edu autoimmunity, as observed in murine *lpr* or human ALPS disease (Cohen and Eisenberg, 1991; Watanabe-Fukunaga et al., 1992; Fisher et al., 1995; Nagata and Suda, 1995; Rieux-Laucat et al., 1995; Drappa et al., 1996; Le Deist et al., 1996). A similar autoimmune constellation occurs with mutation of the ligand for CD95 in murine *gld* disease and some cases of ALPS (Cohen and Eisenberg, 1991; Takahashi et al., 1994; Nagata and Suda, 1995; Wu et al., 1996). Various mixing and expression studies indicate that loss of Fas function produces an intrinsic B cell defect in which autoreactive B cells are poorly regulated and escape normal controls (Perkins et al., 1987; Nemazee et al., 1991; Fukuyama et al., 1998; Sobel et al., 1998). These results alone suggest that immune homeostasis depends on the integrity of Fas-mediated apoptosis. This notion is supported by adoptive transfer experiments demonstrating that autoreactive B cells are deleted in a Fas-dependent manner (Rathmell et al., 1995).

The severe consequences of Fas derangement raise the question of whether Fas function operates constitutively or is regulated during the course of B cell activity. We hypothesized that cell surface receptors beyond Fas might influence the outcome of Fas signaling for cell death, and beginning several years ago we sought to test this idea.

Early Results

We initially evaluated two B cell mitogenic receptors, the multicomponent B cell antigen receptor (BCR) and CD40, the receptor for direct T cell help. In previous work we had established that engagement of these receptors on primary B cells produces many downstream effects in common, such as induction and activation of NF-kB, AP-1, and NF-AT (Chiles et al., 1991; Liu et al., 1991; Rooney et al., 1991; Lalmanach-Girard et al., 1993; Berberich et al., 1994; Venkataraman et al., 1994; Francis et al., 1995); however, we and others had found that the fine specificity of the protein constituents of transcription factor complexes varies somewhat according to the receptor triggered (Huo and Rothstein, 1995; Neumann et al., 1996; Francis et al., 1998), suggesting the potential for different transcriptional outcomes. To elucidate differential effects of CD40 and BCR stimulation on Fas-mediated apoptosis, we stimulated primary B cells with the CD40L/CD8a fusion protein described by Lane et al. (Lane et al., 1993), crosslinked with anti-CD8 antibody (CD40L), or with $F(ab')_2$ fragments of goat anti-mouse IgM antibody (anti-Ig). After various periods of time B cells were tested for susceptibility to Fas-mediated apoptosis. To do so B cells were labeled with ^{51}Cr and incubated with a T cell line that expresses FasL, along with lectin to maintain contact between the effector T cell and the target B cell. The level of ^{51}Cr (now intracellular and protein-bound) released at the end of the culture period was determined and used to calculate percent specific cell lysis.

Within 24 hours, CD40 triggering produced marked upregulation of surface Fas expression, to levels that greatly exceeded the small amount present on naïve B cells, as detected by immunofluorescent staining and flow cytometric analysis. In conjunction with this B cells acquired exquisite sensitivity to Fas-mediated apoptosis. These changes were evident after 2 and 3 days of CD40L treatment as well (Rothstein et al., 1995; Foote et al., 1996b; and unpublished observations), and generally paralleled results that had been reported before that time for LPS-stimulated B cells, although the levels of Fas expression and Fas-sensitivity were greater after CD40L than after LPS stimulation (Daniel and Krammer, 1994; Ju et al., 1994; Stalder et al., 1994; Onel et al., 1995; Watanabe et al., 1995). In direct contrast, BCR stimulation for 1–3 days failed to enhance B cell susceptibility to Fas signaling for cell death (Rothstein et al., 1995). Although Fas expression was upregulated by anti-Ig treatment, the levels attained remained below those produced by CD40L, which may have contributed to the lower degree of Fas-sensitivity observed after anti-Ig as opposed to CD40L treatment (Rothstein et al., 1995).

Alternatively, the BCR might have acted to generate signals that protect B cells from Fas-mediated apoptosis in a dominant fashion. To examine this possibility, we stimulated B cells with the combination of CD40L and anti-Ig concurrently, and compared the resultant level of Fas-sensitivity to that observed with B cells stimulated by CD40L and by anti-Ig individually. B cells stimulated with both anti-Ig and CD40L together were not susceptible to Fas killing (Rothstein et al., 1995; Foote et al., 1996b). In fact, they were as little susceptible to Fas-induced cytotoxicity as B cells stimulated with anti-Ig alone, despite expressing levels of surface Fas that were upregulated to the same extent as in B cells stimulated with CD40L alone (Rothstein et al., 1995). Typically, the level of specific cell lysis at the highest effector:target cell ratio (9:1) utilized, over a 4 hour period, amounted to 40–60% for B cells stimulated by CD40L alone, but only 0–15% for B cells dual stimulated by anti-Ig plus CD40L. A large number of effector cell titrations demonstrated that anti-Ig produced an approximately 20-fold level of protection against Fas-mediated apoptosis, meaning that a 20-fold greater T effector:B target cell ratio was required to produce the same degree of specific cell lysis in B cells stimulated through both BCR and CD40 as in B cells stimulated through CD40 alone. Further, experiments with B cells expressing transgenic antigen receptors indicated that physiologic BCR stimulation with specific antigen also produced protection against Fas signaling for cell death (Rothstein et al., 1995; Foote et al., 1998). Thus, BCR engagement initiates signals that oppose upregulated susceptibility to Fas-mediated apoptosis induced by CD40L, and in so doing produces a state of Fas-resistance. The dichotomy between CD40 and BCR signaling in relation to the regulation of Fas-sensitivity has been confirmed in subsequent in vitro and in vivo experiments carried out here and elsewhere (Lagresle et al., 1996; Rathmell et al., 1996; Wang et al., 2000). A similar dichotomy involves IL-4, which, like anti-Ig, produces resistance against Fas killing in CD40L-treated B cells, although the level of protection (about 10-fold) is not as complete as with BCR triggering (Foote et al., 1996a). These results are depicted schematically in Figure 1.

Role of Fas-Resistance

Protection against Fas-mediated apoptosis may represent a key element in the generation of normal immune responses. Activated T cells that express CD40L likely also express FasL (Armitage et al., 1992; Lane et al., 1992; Noelle et al., 1992; Alderson et al., 1995; Brunner et al., 1995; Dhein et al., 1995; Ju et al., 1995; Suda et al., 1995), the latter constituting a threat to the viability of CD40-triggered B cells that are highly sensitive to Fas killing. We hypothesize that Fas-resistance acts to insure that B cells stimulated through antigen receptors survive contact with activated T cells, and in so doing fosters antigen-specificity in serological immunity. The observation that Fas-resistant anti-Ig-stimulated B cells are more effective in presenting antigen to T cells than Fas-sensitive CD40L-stimulated B cells (Ozdemirli et al., 1996) is consistent with the hypothesis outlined above and likely reflects the ability of Fas-resistant B cells to efficiently survive an initial interaction with T cells and re-cycle to present antigen anew.

However, aberrant or exaggerated acquisition of Fas-resistance may interfere with Fas-dependent deletion of autoreactive B cells, potentially leading to autoimmunity, as occurs in mice in which Fas or FasL expression is diminished (notably, however, transgenic studies suggest that in the normal situation autoreactive B cells do not efficiently acquire

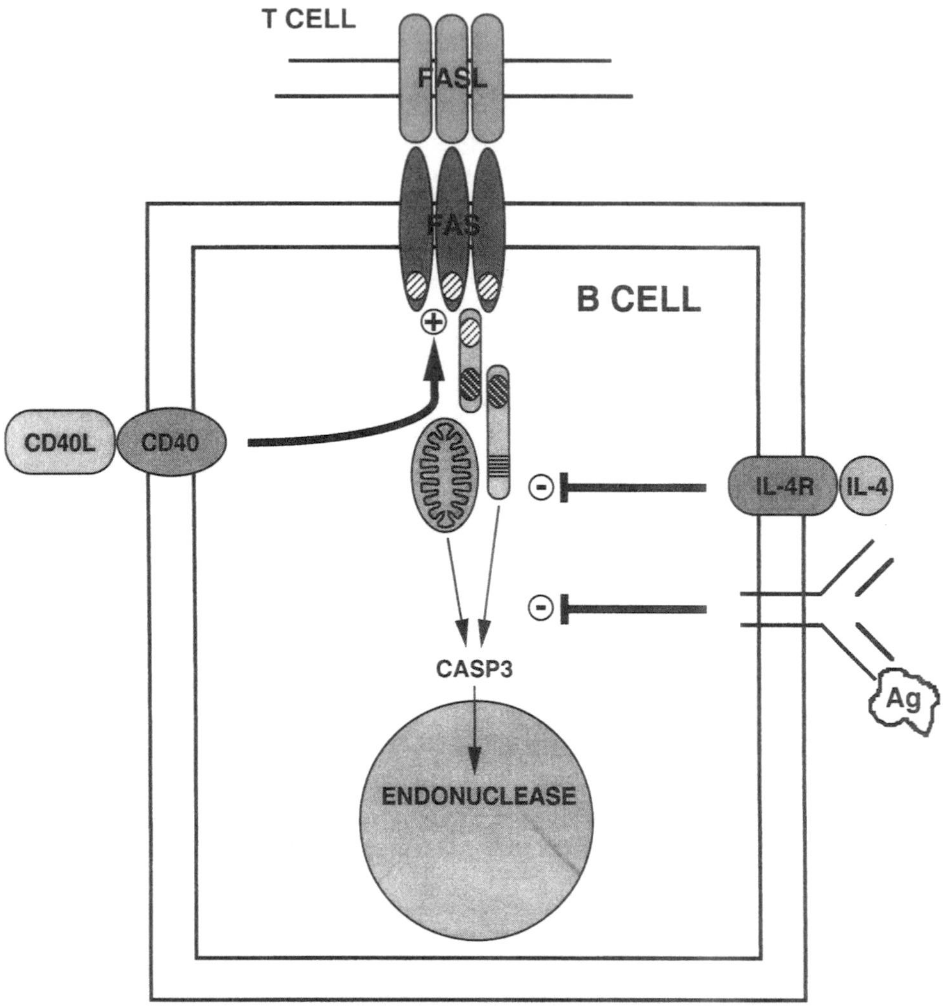

Figure 1. B cell sensitivity to Fas-mediated apoptosis is regulated in a receptor-specific fashion. Initial results obtained with murine splenic B cells stimulated and tested for susceptibility to Fas-mediated apoptosis in vitro are summarized in diagrammatic form. CD40 stimulation by soluble CD40L/CD8a fusion protein cross-linked with anti-CD8 antibody (CD40L) results (within 24 hours) in greatly increased surface Fas expression (FAS), and markedly increased sensitivity to Fas-mediated apoptosis produced by FasL (FASL)-bearing Th1 (T CELL) effector cells (+). Fas signaling for cell death is represented by unlabeled symbols denoting FADD (Fas-associating protein with death domain), caspase 8, and a mitochondrion, noting that the relative contributions of FADD-caspase 8 activation, and cytochrome c-caspase 9 activation, in directing primary B cell apoptosis remains uncertain. Terminal events represented here include activation of caspase 3 (CASP3) and caspase-activated DNase (ENDONUCLEASE). Homologous death domains in Fas and FADD are indicated by lightly hatched circles; homologous death effector domains in FADD and caspase 8 are indicated by darkly hatched ovals; and a caspase domain in caspase 8 is indicated by a series of horizontal lines. Stimulation of BCR by $F(ab')_2$ fragments of anti-IgM antibody (not shown) or by soluble antigen (Ag), and stimulation of IL-4R by IL-4, either concurrently with, or during the last 12–24 hours of, CD40L stimulation (for 48 hours), produces marked reduction in the level of apoptosis brought about by subsequent Fas engagement (−), without any change in the upregulated level of Fas expression attributable to CD40 signaling. The points at which negative signals from BCR and IL-4R intersect the Fas pathway are arbitrarily drawn because the level at which Fas-resistance occurs is unknown at present.

Fas-resistance as a result of antigen binding (Foote et al., 1998)). Further, malignant cells that express Fas but do not succumb to Fas-mediated apoptosis may have activated in a constitutive manner the normal mechanisms utilized by BCR signaling to produce Fas-resistance (Owen-Schaub et al., 1994; Natoli et al., 1995; Panayiotidis et al., 1995; Keane et al., 1996; Xerri et al., 1997; Hedlund et al., 1998; Plumas et al., 1998; Ungefroren et al., 1998; Baldwin et al., 1999; Xerri et al., 1999). Thus, a deeper understanding of the means by which inducible Fas-resistance is established may elucidate important facets of normal, autoimmune, and tumor immune responses.

Initial Signaling Studies

The intracellular signaling pathways responsible for induction of Fas-resistance differ for anti-Ig and IL-4. Experiments in which B cell protein kinase C (PKC) activity was reduced by long term depletion with the phorbol ester, phorbol myristate acetate (PMA) (Mond et al., 1987; Kawakami and Parker, 1993), indicate that Fas-resistance produced by BCR engagement is PKC-dependent, whereas that produced by IL-4 is not (Foote et al., 1996a; Foote et al., 1996b). Experiments with B cells obtained from Signal Transducers and Activators of Transcription (STAT)6-deficient animals (Kaplan et al., 1996) indicate that Fas-resistance produced by IL-4R engagement is STAT6-dependent, whereas that produced by anti-Ig is not (Wurster et al., 2002; and unpublished observations). The role of PKC in anti-Ig-induced Fas-resistance suggests that the well-described pathway for BCR signaling that relies on the production of second messenger molecules by phospholipase C (PLC) is involved; in keeping with this, Fas-resistance is produced by the mitogenic combination of PMA plus ionomycin (P/I) (Foote et al., 1996b), pharmacologic agents that mimic the effects of diacylglycerol (DAG) and Ca++ resulting from PLC activation and metabolism of phosphatidylinositol-bisphosphate (PIP2).

Because naïve B cells are little affected by Fas engagement (at least as measured by chromium release assay), induction of Fas-resistance is best seen against the background of Fas-sensitivity produced by CD40L, and, as noted above, Fas-resistance is established when anti-Ig, P/I, or IL-4 are added to B cells concurrently with CD40L. In addition, Fas-resistance is produced when anti-Ig, P/I, or IL-4 are added sequentially as much as 24–36 hours after initiation of CD40L treatment (Foote et al., 1996a; Foote et al., 1996b). Notably, B cell stimulation with CD40L for 24 hours is sufficient to upregulate Fas expression and sensitivity to Fas-mediated apoptosis. Thus, induction of Fas-resistance when reagents are added sequentially demonstrates that anti-Ig, P/I, and IL-4 do not simply prevent the acquisition of Fas-sensitivity, but rather, reverse Fas-sensitivity that has already been established.

Experiments in which the time of reagent addition was varied showed that the duration of exposure required for induction of Fas-resistance depends on the nature of the stimulus. Fas-resistance produced by anti-Ig and P/I developed rapidly and reached a maximum within 12 hours, whereas Fas-resistance produced by IL-4 developed more slowly and only reached a maximum after 24 hours (Foote et al., 1996a; Foote et al., 1996b). The rapid production of Fas-resistance by anti-Ig has made it possible to dissect the signaling pathways involved through the use of metabolic inhibitors that do not affect B cell viability over short periods of time, but are toxic after longer term exposure. Experiments of this sort in which H-7 was added 30 minutes prior to anti-Ig confirmed the PKC-dependence of Fas-resistance produced by BCR engagement (Foote et al., 1996b). In a similar fashion, experiments

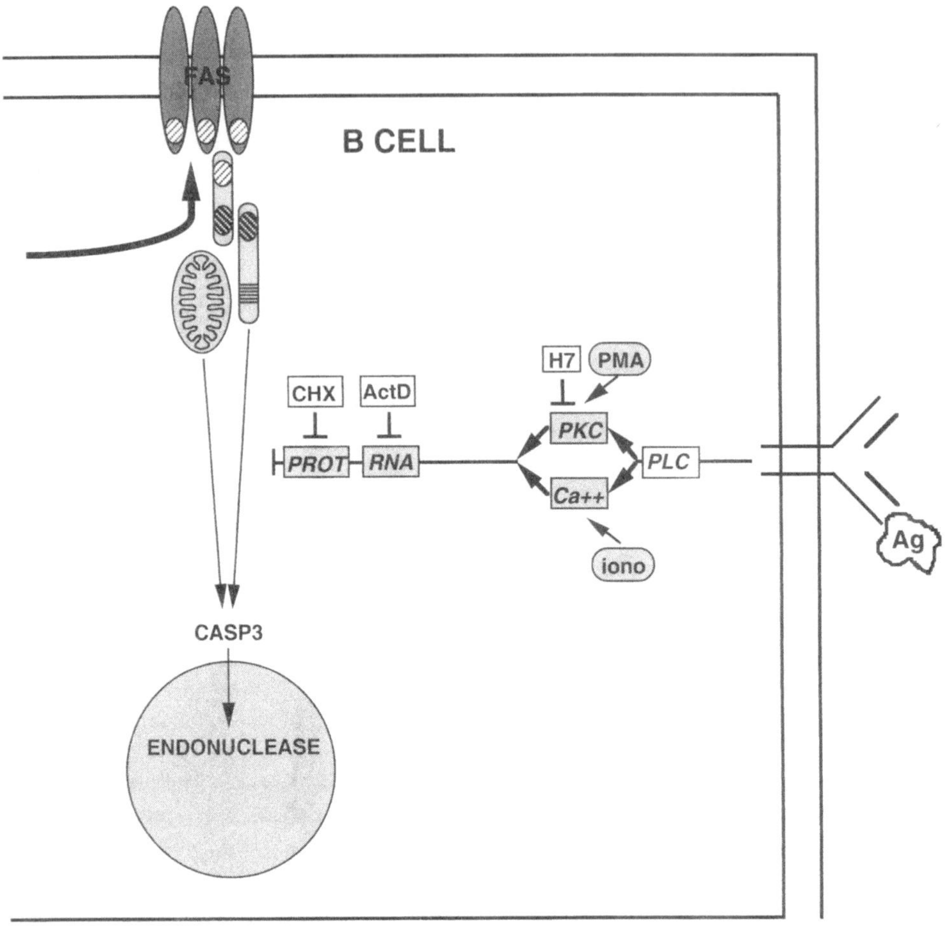

Figure 2. BCR signaling for Fas-resistance is mimicked by the combination of PMA plus ionomycin and requires new macromolecular synthesis. The results of a series of experiments designed to elucidate the signaling pathway leading from BCR to Fas-resistance are summarized in the diagram, which enlarges on the model presented in Figure 1. BCR signaling for Fas-resistance is blocked by inhibition (H7) or depletion (not shown) of protein kinase C (PKC) and is replicated by the combination of a PKC-activating phorbol ester such as PMA and a calcium ionophore such as ionomycin (iono), implying that Ca++ signals as well as PKC signals derived from phospholipase C (PLC) activation mediate BCR-induced Fas-resistance. Induction of Fas-resistance is blocked by inhibition of RNA synthesis with Actinomycin D (ActD) and by inhibition of protein (PROT) synthesis with cycloheximide (CHX).

with cycloheximide and actinomycin D demonstrated that Fas-resistance produced by BCR engagement requires new macromolecular synthesis (Foote et al., 1996b; and unpublished observations). See Figure 2.

Btk and NF-k B

To provide a more complete understanding of anti-Ig-induced Fas-resistance, we examined signaling molecules both proximal and distal to PKC. Bruton's tyrosine kinase (Btk) is a key intermediary for BCR signaling that contributes greatly to PLC activation and the generation of second messenger molecules that activate PKC (Sieckmann et al., 1978;

Rigley et al., 1989; Takata and Kurosaki, 1996; Fluckiger et al., 1998). Btk is mutated (in the Pleckstrin Homology domain) and non-functional in *xid* mice that, as a result, contain a somewhat reduced number of B cells (Rawlings et al., 1993; Satterthwaite and Witte, 2000). Further, the B cells that are present in *xid* mice respond poorly to anti-Ig stimulation (Sieckmann et al., 1978; Rigley et al., 1989). Similar defects are observed in genetically engineered Btk-deficient (knock-out) mice (Kerner et al., 1995; Khan et al., 1995). We used *xid* and Btk-deficient mice to test the role of Btk in BCR-induced Fas-resistance. Mutation of Btk in *xid* B cells, and loss of Btk in knock-out animals, had no effect on upregulation of Fas expression nor on acquisition of Fas-sensitivity following B cell stimulation with CD40L (Tumang et al., 2002). Surprisingly, we found that Btk mutation and Btk deficiency also had no effect on the efficiency with which anti-Ig produced Fas-resistance in CD40L-stimulated B cells, despite the important role of Btk in BCR signaling (Tumang et al., 2002). This was true whether the Fas trigger was supplied by FasL-bearing Th1 cells or by recombinant, soluble FasL. Thus, Fas-resistance produced by anti-Ig is Btk-independent.

NF-kB is a key transcription factor activated by BCR signaling downstream of PKC that opposes apoptosis in certain situations (Liu et al., 1991; Rooney et al., 1991; Beg et al., 1995; Beg and Baltimore, 1996; Liu et al., 1996; Schauer et al., 1996; Van Antwerp et al., 1996; Wang et al., 1996). NF-kB is normally located in the cytosol in a quiescent form bound to an inhibitor protein, IkB (eg, IkBa); following stimulation, IkB becomes phosphorylated and ubiquitinated, and then degraded in the proteasome, releasing active NF-kB which translocates to the nucleus and binds DNA (Baldwin, 1996; Karin and Ben-Neriah, 2000). We used several inhibitors of NF-kB induction to probe the need for NF-kB in BCR-induced Fas-resistance, including lactacystin (LC), which blocks NF-kB induction through inhibition of proteasomal function, and pyrrolidinedithiocarbmate (PDTC), an anti-oxidant that interferes with activation of NF-kB through an undefined mechanism (Schreck et al., 1992; Verma et al., 1995; Baldwin, 1996; Phillips and Ghosh, 1997). These inhibitors display toxicity to primary B cells after prolonged treatment; for this reason the period of anti-Ig stimulation used to induce Fas-resistance in these experiments was limited to 12.5 hours. BCR engagement by anti-Ig for the last 12 hours of 48 hour cultures with CD40L produced marked protection against Fas killing which was reversed by concurrent addition of LC or PDTC (Schram and Rothstein, 2003). Neither treatment altered the elevated level of Fas expression produced by CD40L, and neither treatment, added 12.5 hours before the end of 48 hour cultures, substantially increased the Fas-sensitivity of B cells treated with CD40L alone (B.R. Schram and T.L. Rothstein, 2003). These results, obtained with two different inhibitors of NF-kB activation, strongly suggest that NF-kB is required for BCR-induced Fas-resistance.

Because chemical inhibitors may not be perfectly selective, we constructed a TAT fusion protein version of mutant, dominant negative IkBa using the pTAT-HA bacterial expression vector, kindly provided by Dr. S. Dowdy (Washington University, St. Louis, MO) (Nagahara et al., 1998). IkBa is normally targeted for destruction by IKK-mediated phosphorylation of serines 32 and 36; mutation of these serines to alanines yields a molecule that acts in a dominant negative fashion to block IkB degradation and hence NF-kB activation (Van Antwerp et al., 1996). Fusion proteins of a certain size that incorporate a peptide sequence from HIV TAT are taken up by virtually all cells, including primary B cells (Vocero-Akbani et al., 2000). As with LC and PDTC, addition of TAT-IkBa DN in conjunction with anti-Ig during the last 12 hours of 48 hour B cell cultures with C40L eliminated Fas-resistance attributable to anti-Ig (B.R. Schram and T.L. Rothstein, submitted). At the same time TAT-IkBa DN did not alter the elevated level of Fas expression produced by CD40L treatment, and did not increase the Fas-sensitivity of B cells treated with CD40L alone. Further, TAT-IkBa DN eliminated Fas-resistance induced by anti-Ig in B cells obtained from C3H/HeJ mice that are defective in Toll-like receptor (TLR)-4 signaling (Poltorak et al., 1998; Hoshino et al., 1999; Qureshi et al., 1999), suggesting that contamination by bacterial products played no role in the observed effects (Schram and Rothstein, 2003). A TAT fusion protein incorporating green fluorescence protein (GFP) rather than IkBa DN had no effect on anti-Ig-induced Fas-resistance, showing that the TAT moiety is not responsible for reversal of Fas-resistance (Schram and Rothstein, 2003). Moreover, separate EMSA results showed that TAT-IkBa DN specifically blocked the increase in nuclear NF-kB produced by adding anti-Ig to CD40L-treated B cells without affecting transcrption factor binding to a consensus AP-1 site (Schram and Rothstein, 2003). The TAT-IkBa DN data further emphasize the need for NF-kB in signaling for BCR-induced Fas-resistance. See Figure 3.

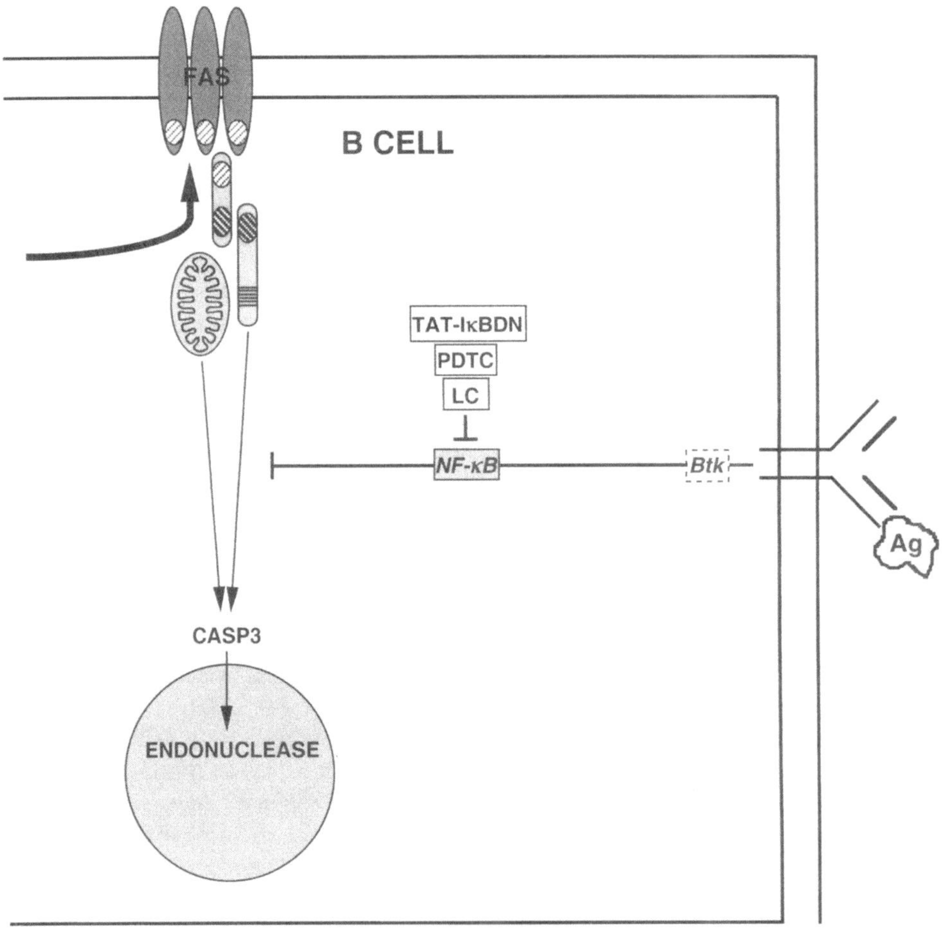

Figure 3. BCR signaling for Fas-resistance bypasses Btk but requires NF-kB. The results of experiments testing the roles of Btk and NF-kB in BCR-induced Fas-resistance are summarized in the diagram, which enlarges on the model presented in Figure 1. BCR signaling for Fas-resistance is not affected by (and thus bypasses) mutation and deletion of Btk. BCR signaling for Fas-resistance is completely interrupted by inhibition of NF-kB induction with lactacystin (LC) and pyrrolidinedithiocarbmate (PDTC) as well as by a TAT fusion protein that incorporates a dominant negative form of IkBa (TAT-IkBDN).

Two anti-apoptotic molecules whose expression has been reported to be NF-kB dependent have been implicated as terminal mediators of BCR-induced Fas-resistance: Bcl-xL and FLIP (Schneider et al., 1997b; Chen et al., 2000; Wang et al., 2000; Micheau et al., 2001). Thus, it might be speculated that expression of anti-apoptotic Bcl-xL and FLIP would be blocked by interruption of NF-kB induction. We tested this notion by western blotting extracts following B cell stimulation by CD40L and anti-Ig, the latter in the presence or absence of TAT-IkBa DN fusion protein. Expression of both Bcl-xL and FLIP was upregulated following anti-Ig stimulation of CD40L-treated B cells, as expected, and elevated expression of both was blocked by addition of TAT-IkBa DN one-half hour before anti-Ig (Schram and Rothstein, 2003). These results are consistent with the general concept that BCR signaling for induction of Fas-resistance requires NF-kB.

It is not entirely clear, however, that both Bcl-xL and FLIP are necessary for inducible Fas-resistance. In a different approach to testing the role of NF-kB, Liou and colleagues constructed c-Rel–deficient (knock-out) animals (Tumang et al., 1998). The absence of c-Rel in B cells from these mice had little effect on CD40 signaling for Fas-sensitivity or on the degree to which anti-Ig induced protection against Fas-mediated apoptosis after anti-CD40 treatment, but interrupted induction of Bcl-xL (although upregulated FLIP expression was preserved) (Owyang et al., 2001). Thus, in this model, BCR engagement successfully signaled Fas-resistance even in the absence of upregulated Bcl-xL, suggesting that Bcl-xL does not play a substantial role in mediating BCR-induced Fas-resistance, or that loss of Bcl-xL is fully compensated by FLIP and/or other BCR-induced anti-apoptotic molecules. Apparently the loss of c-Rel differs from the loss of all DNA-binding NF-kB in terms of inducible Fas-resistance and expression of the molecules that bring it about.

Confusing Facts about BCR Signaling for Fas-Resistance

These results indicate that BCR signaling for Fas-resistance bypasses Btk but requires PKC (and presumably, PLC), and eventuates in the activation of NF-kB and the transactivation, in turn, of anti-apoptotic genes encoding FLIP and possibly other molecules that may or may not include Bcl-xL. See Figure 4. However, this brief summary constitutes a confusing set of facts, because BCR signaling for NF-kB has been reported to absolutely require Btk (Bajpai et al., 2000; Petro et al., 2000). Thus, our results would appear to present a conundrum, in that we found BCR signaling for Fas-resistance requires NF-kB but not Btk, whereas two other laboratories found that BCR cannot trigger NF-kB without Btk. As a result we re-evaluated the stimulatory regimen used here to induce Fas-resistance. As noted above, in most of our work anti-Ig has been added to B cells already treated with CD40L, typically at the 24 hour mark of 48 hour cultures, but in some cases, as when metabolic inhibitors were tested, at the 36 hour mark. Thus a potentially important difference between the experiments alluded to above is that in one case anti-Ig was added to B cells after CD40L treatment whereas in the other anti-Ig was added to naïve B cells. This led us to hypothesize that CD40L treatment alters the signaling requirements for BCR activation of NF-kB.

Btk-Independent BCR Signaling for NF-kB

We tested the effect of prior CD40 engagement on BCR signaling through a series of experiments using *xid* B cells. Nuclear extracts were prepared before and after B cell

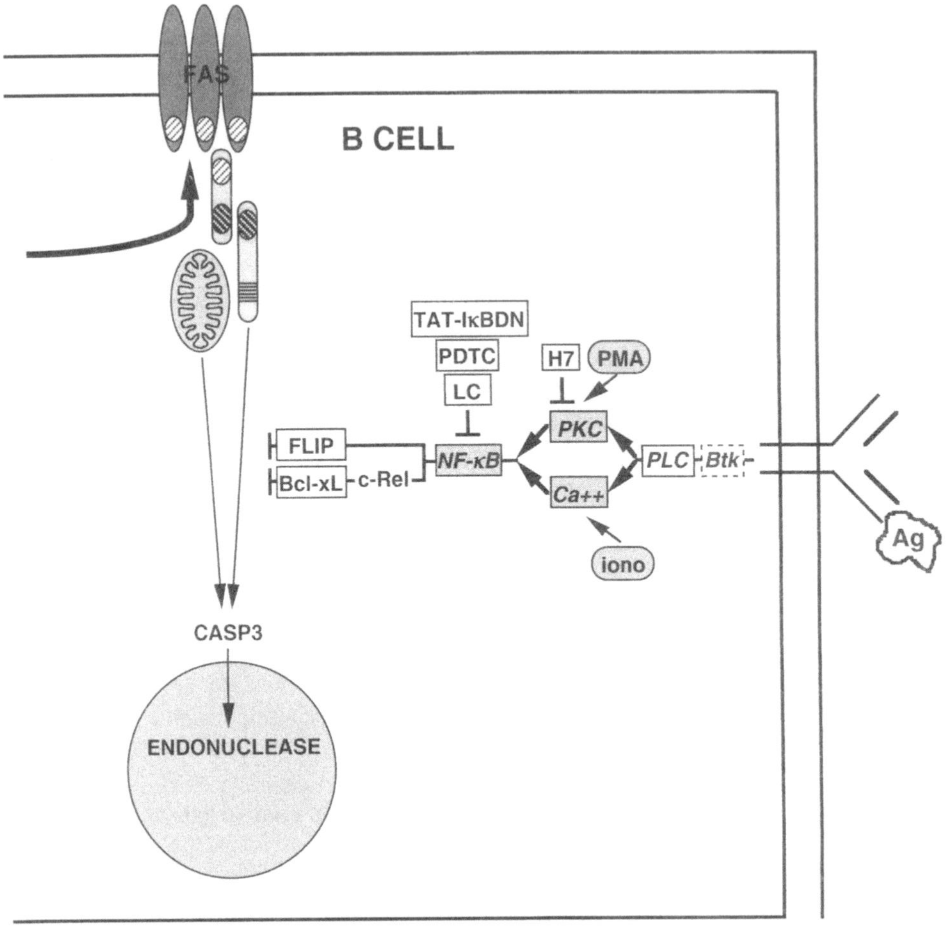

Figure 4. BCR signaling for Fas-resistance eventuates in expression of NF-kB-dependent anti-apoptotic molecules. The results of experiments examining the terminal mediators of BCR-induced Fas-resistance are summarized in the diagram, which incorporates elements of Figures 2 and 3. BCR signaling for Fas-resistance is accompanied by upregulated expression of NF-kB-dependent anti-apoptotic molecules FLIP and Bcl-xL, each of which is implicated by previous studies of primary B cell Fas-resistance. Inhibition of BCR-induced NF-kB is associated with inhibition of BCR-induced FLIP and Bcl-xL expression, and loss of Fas-resistance. However, deletion of c-Rel does not affect BCR signaling for Fas-resistance although upregulated expression of Bcl-xL is interrupted.

stimulation with CD40L and/or anti-Ig and tested for NF-kB by binding to a radiolabeled kB-site-containing oliogonucleotide, assessed by electrophoretic mobility shift assay (EMSA). Mature, primary B cells expressed nuclear NF-kB constitutively, in the absence of any stimulation, as previously reported (Liu et al., 1991; Rooney et al., 1991). In wild-type control B cells this "resting" level of nuclear NF-kB was substantially increased by anti-Ig stimulation, as expected (Liu et al., 1991; Rooney et al., 1991). In contrast, nuclear NF-kB was not increased at all when naïve *xid* B cells were treated with anti-Ig, as expected from previous reports (Bajpai et al., 2000; Petro et al., 2000) (notably, NF-kB was induced in *xid* B cells by the combination of PMA plus ionomycin, which together stimulate B cells distal to Btk, demonstrating that *xid* B cells are not inherently impaired in NF-kB responses).

However, we found that nuclear NF-kB was substantially induced in *xid* B cells when anti-Ig was applied after treatment with CD40L (Mizuno and Rothstein, 2003). The increase in NF-kB produced by anti-Ig stimulation of CD40L-treated B cells was as great in *xid* B cells as in control B cells. Although CD40L stimulated NF-kB activation in naïve B cells, both *xid* and control, the level of nuclear NF-kB had returned to baseline by the time anti-Ig was added at 45 hours (for an additional 3 hours). Moreover, CD40L-treated and washed *xid* B cells responded to anti-Ig with an increase in nuclear NF-kB (T. Mizuno and T.L. Rothstein, unpublished observations), indicating that BCR-mediated NF-kB activation in *xid* B cells does not depend on concurrent CD40 receptor signaling but can operate sequentially. Thus, it appears that prior treatment with CD40L alters the accessibility of NF-kB to BCR signaling, such that BCR signaling now triggers NF-kB activation in the absence of Btk.

The specificity of the kB-binding activity detected by EMSA was confirmed by "cold" competition analysis in which unlabeled oligonucleotide containing a mutant kB-binding site failed to interfere with the formation of nucleoprotein complexes (consisting of labeled wild-type kB-binding site-containing oligonucleotide and nuclear extract protein) whereas unlabeled wild-type kB oligonucleotide did so, for both *xid* and control B cell nuclear extract protein (Mizuno and Rothstein, 2003). The nature of the BCR-induced kB-binding activity was further evaluated by supershift analyses which showed that the principal Rel-related proteins constituting NF-kB were p52 and c-Rel, along with lesser amounts of p50, RelB and p65, for both *xid* and control B cell nuclear extract protein (Mizuno and Rothstein, 2003). Thus, nuclear NF-kB obtained from *xid* B cells after sequential stimulation by CD40L and anti-Ig is similar to that obtained from control B cells in competition and composition analyses. Further, there was no difference between *xid* and control B cells in relative amounts of various Rel-related proteins after CD40L (and before anti-Ig) treatment, ruling out the possibility that induction of NF-kB by BCR stimulation of *xid* B cells depends on any unusual increase in NF-kB content (Mizuno and Rothstein, 2003).

Inasmuch as the accepted mechanism for NF-kB activation involves IkB degradation, we determined levels of IkBa before and after addition of anti-Ig to *xid* and control B cells with or without CD40L pretreatment. Addition of anti-Ig produced a progressive, time-dependent decline in cellular content of IkBa in naïve control B cells whereas the same treatment produced little change in IkBa in naïve *xid* B cells over a 2 hour period, as expected on the basis of previous reports (Bajpai et al., 2000; Petro et al., 2000). However, prior treatment of *xid* B cells with CD40L altered the behavior of IkBa, so that subsequent anti-Ig stimulation produced a loss of IkBa that was as great as that observed with control B cells pretreated with CD40L and stimulated with anti-Ig (Mizuno and Rothstein, 2003). This result further supports the notion that the elements directing NF-kB induction become accessible to BCR signaling in the absence of Btk after CD40L stimulation.

To determine the functionality of NF-kB induced by anti-Ig in *xid* B cells pretreated with CD40L, we monitored the expression of the NF-kB-dependent gene, Bcl-xL, by RNase protection assay applied to RNA extracted from *xid* and control B cells treated in various ways. Anti-Ig induced upregulated expression of Bcl-xL in naïve, wild-type control B cells (about 3-fold), whereas anti-Ig failed completely to upregulate Bcl-xL expression in naïve xid B cells. However, BCR signaling upregulated expression of Bcl-xL in *xid* B cells (about 2-fold) when anti-Ig was added after CD40L treatment (Tumang et al., 2002). Thus, NF-kB induced by BCR in a Btk-independent fashion appears to be fully functional. See Figure 5.

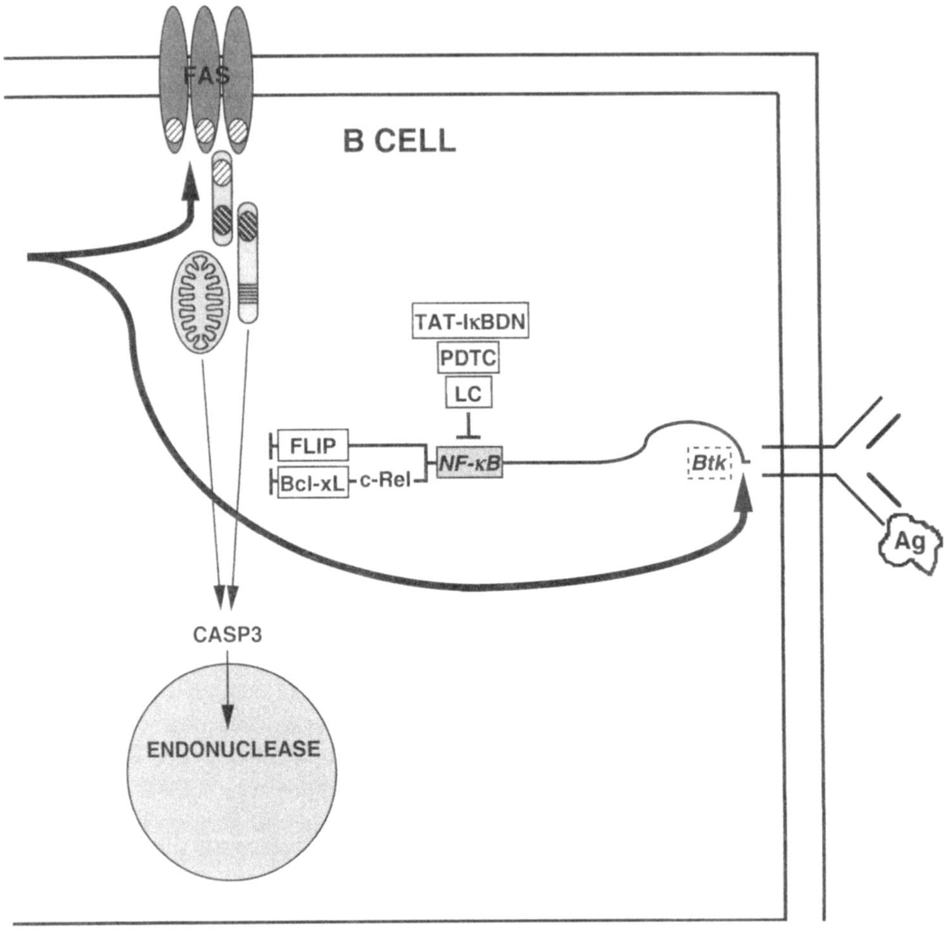

Figure 5. CD40 engagement eliminates the need for Btk in BCR signaling for NF-kB. The results of experiments examining the effect of CD40L treatment on BCR signaling are summarized in the diagram, which incorporates elements of Figure 4. BCR signaling for NF-kB induction fails in naïve B cells obtained from *xid* mice in which Btk is mutated; however, after treatment of *xid* B cells with CD40L, BCR engagement produces IkB degradation (not depicted) and NF-kB induction in a Btk-independent fashion. Intracellular signals derived from CD40 engagement are represented by thick lines with arrows. BCR-mediated induction of NF-kB in CD40L-treated *xid* B cells eventuates in upregulated Bcl-xL expression; upregulated expression of NF-kB-dependent FLIP is presumed to occur as well but has not been directly tested.

To determine whether Btk-independent BCR signaling extends to other aspects of B cell behavior, we tested proliferative responses by measuring tritiated thymidine uptake after B cell stimulation. Anti-Ig and CD40L were tested alone and in combination. *Xid* and control B cells responded similarly to stimulation by LPS and by PMA in conjunction with ionomycin (Mond et al., 1982; Klaus et al., 1986; Rothstein et al., 1986). However, *xid* B cells failed to incorporate thymidine in response to anti-Ig, as expected, whereas control B cells did so (Sieckmann et al., 1978). Remarkably, the combination of CD40L and anti-Ig acted in synergy on *xid* B cells to produce a level of thymidine incorporation that was much greater than the sum of the *xid* B cell responses to CD40L and anti-Ig alone (the latter of which was negligible) (Mizuno and Rothstein, 2003). Although the absolute magnitude of the response to combined treatment was still lower in *xid* B cells than in control B cells, the degree of synergy in *xid* B cells was much higher than that in control B cells (because, at least in part, control B cells respond vigorously to both anti-Ig and CD40L applied individually, whereas *xid* B cells respond only to CD40L and not to anti-Ig, applied individually). These results indicate that, in *xid* B cells, prior CD40L treatment reconstitutes not only defective NF-kB activation, but also poor S phase entry, in response to BCR crosslinking.

Summary and Discussion

Perhaps the single most important conclusion to be drawn from the work described herein is that B cells modulate their intrinsic level of susceptibility to Fas-mediated apoptosis in a receptor-specific fashion. As a result B cells should not be viewed as mere passive targets for FasL-bearing T cells, but rather as active participants in directing and regulating the cell death process. In keeping with this, B cells respond to environmental cues, such as the presence of antigen, with dramatic changes in sensitivity to Fas signaling for cell death. Downregulation of Fas-sensitivity produced by BCR engagement, or by pharmacologic treatment with PMA and ionomycin, is of such magnitude as to constitute protection against Fas killing or Fas-resistance, and appears to depend at least in part on NF-kB-mediated upregulation of anti-apoptotic molecules such as FLIP and possibly Bcl-xL. Particularly with respect to BCR engagement, Fas-resistance most likely functions to foster antigen-specific B cell responses.

CD40 engagement triggers NF-kB activation as does BCR engagement, which raises the question of why CD40L-stimulated B cells are Fas-sensitive if BCR-induced NF-kB is responsible for Fas-resistance. Most simply, the anti-apoptotic effect of NF-kB may be overwhelmed by CD40-mediated upregulation of pro-apoptotic molecules, which may not occur after anti-Ig stimulation. Alternatively, NF-kB may be necessary but not sufficient for anti-apoptosis, with critical contributions derived from transcription factors other than NF-kB that are induced following BCR, but not CD40, engagement. Finally, CD40-induced NF-kB may differ in important ways from anti-Ig-induced NF-kB. Supporting this latter suggestion is previous work indicating that subtle differences exist in the Rel-related protein composition of kB-binding activity after B cell stimulation via CD40 in comparison with BCR (Neumann et al., 1996; Francis et al., 1998). Notably, NF-kB-dependent FLIP is upregulated after BCR, but not CD40, engagement, which may reflect receptor-specificity in NF-kB induction (Wang et al., 2000; and unpublished data). Although this issue remains unresolved at the present time, the requirement for NF-kB rests securely on the loss of BCR-induced Fas-resistance in the face of 3 different inhibitors of NF-kB induction.

Consideration of the signaling pathway responsible for BCR-induced Fas-resistance produced the unexpected result that Btk is not needed, whereas previous reports indicate that BCR signaling for NF-kB is Btk-dependent (Bajpai et al., 2000; Petro et al., 2000). This potential conundrum is now resolved by the demonstration that CD40 engagement alters the signals required for coupling BCR and NF-kB so that the former activates the latter in a Btk-independent fashion. By circumventing the need for Btk in BCR signaling for NF-kB, these results recast the recognized dependence of BCR signaling on Btk as a malleable, initial condition that is subject to change as a result of CD40 triggering. This work further suggests the more general principle, yet to be tested, that once a B cell is treated to induce

Fas-sensitivity (as with CD40L), the capacity of other receptors (such as BCR) to signal for Fas-resistance is altered.

The nature of the mechanism by which CD40 stimulation provides access to NF-kB for BCR in the absence of Btk remains uncertain. BCR-induced IkBa degradation occurred similarly in CD40L-treated *xid* and wild-type B cells, suggesting that the new Btk-independent BCR pathway merges with the classical Btk-dependent BCR pathway upstream of IkB, although the level at which this happens is unknown. Importantly, the Btk-independent BCR pathway would appear to represent more than just enhancement of CD40 signaling for NF-kB induction because nuclear NF-kB and cytosolic IkBa are back to baseline values by the time anti-Ig is added 2 days after initiation of CD40L treatment (Mizuno and Rothstein, 2003; and unpublished observations). Moreover, the notion of enhancement still means that BCR signaling affects NF-kB, which does not take place in the absence of Btk unless B cells have been treated with CD40L. Thus, CD40 engagement creates a new pathway, or markedly facilitates a previously underutilized or inaccessible pathway, for BCR signaling. In so doing the genetic block in BCR-triggered signaling produced by Btk mutation is alleviated, at least in terms of downstream NF-kB induction. These results raise the possibility that other loss-of-function signaling mutations might be normalized through stimulation of specific receptors.

ACKNOWLEDGMENTS: This work was supported was supported by United States Public Health Service grants AI40181 and AI45112 awarded by the National Institutes of Health.

References

Alderson, M. R., Tough, T. W., Davis-Smith, T., Braddy, S., Falk, B., Schooley, K. A., Goodwin, R. G., Smith, C. A., Ramsdell, F., and Lynch, D. H. (1995). Fas ligand mediates activation-induced cell death in human T lymphocytes. J Exp Med *181*, 71–77.

Armitage, R. J., Fanslow, W. C., Strockbine, L., Sato, T. A., Clifford, K. N., Macduff, B. M., Anderson, D. M., Gimpel, S. D., Davis-Smith, T., Maliszewski, C. R., and et al. (1992). Molecular and biological characterization of a murine ligand for CD40. Nature *357*, 80–82.

Bajpai, U. D., Zhang, K., Teutsch, M., Sen, R., and Wortis, H. H. (2000). Bruton's Tyrosine Kinase Links the B Cell Receptor to Nuclear Factor kappaB Activation. J Exp Med *191*, 1735–1744.

Baldwin, A. S., Jr. (1996). The NF-kappa B and I kappa B proteins: new discoveries and insights. Annu Rev Immunol *14*, 649–683.

Baldwin, R. L., Tran, H., and Karlan, B. Y. (1999). Primary ovarian cancer cultures are resistant to Fas-mediated apoptosis. Gynecol Oncol *74*, 265–271.

Beg, A. A., and Baltimore, D. (1996). An essential role for NF-kappa B in preventing TNF-alpha-induced cell death. Science *274*, 782–784.

Beg, A. A., Sha, W. C., Bronson, R. T., Ghosh, S., and Baltimore, D. (1995). Embryonic lethality and liver degeneration in mice lacking the RelA component of NF-kappa B. Nature *376*, 167–170.

Berberich, I., Shu, G. L., and Clark, E. A. (1994). Cross-linking CD40 on B cells rapidly activates nuclear factor-kappa B. J Immunol *153*, 4357–4366.

Brunner, T., Mogil, R. J., LaFace, D., Yoo, N. J., Mahboubi, A., Echeverri, F., Martin, S. J., Force, W. R., Lynch, D. H., Ware, C. F., and et al. (1995). Cell-autonomous Fas (CD95)/Fas-ligand interaction mediates activation-induced apoptosis in T-cell hybridomas. Nature *373*, 441–444.

Chaudhary, P. M., Eby, M., Jasmin, A., Bookwalter, A., Murray, J., and Hood, L. (1997). Death receptor 5, a new member of the TNFR family, and DR4 induce FADD-dependent apoptosis and activate the NF-kappa B pathway. Immunity *7*, 821–830.

Chen, C., Edelstein, L. C., and Gelinas, C. (2000). The Rel/NF-kappa B family directly activates expression of the apoptosis inhibitor Bcl-x(L). Mol Cell Biol *20*, 2687–2695.

Chiles, T. C., Liu, J. L., and Rothstein, T. L. (1991). Cross-linking of surface Ig receptors on murine B lymphocytes stimulates the expression of nuclear tetradecanoyl phorbol acetate-response element-binding proteins. J Immunol *146*, 1730–1735.

Chinnaiyan, A. M., O'Rourke, K., Yu, G. L., Lyons, R. H., Garg, M., Duan, D. R., Xing, L., Gentz, R., Ni, J., and Dixit, V. M. (1996). Signal transduction by DR3, a death domain-containing receptor related to TNFR-1 and CD95. Science *274*, 990–992.

Cohen, P. L., and Eisenberg, R. A. (1991). Lpr and gld: single gene models of systemic autoimmunity and lymphoproliferative disease. Annu Rev Immunol *9*, 243–269.

Daniel, P. T., and Krammer, P. H. (1994). Activation induces sensitivity toward APO-1 (CD95)-mediated apoptosis in human B cells. J Immunol *152*, 5624–5632.

Dhein, J., Walczak, H., Baumler, C., Debatin, K. M., and Krammer, P. H. (1995). Autocrine T-cell suicide mediated by APO-1/(Fas/CD95). Nature *373*, 438–441.

Drappa, J., Vaishnaw, A. K., Sullivan, K. E., Chu, J. L., and Elkon, K. B. (1996). Fas gene mutations in the Canale-Smith syndrome, an inherited lymphoproliferative disorder associated with autoimmunity. N Engl J Med *335*, 1643–1649.

Fisher, G. H., Rosenberg, F. J., Straus, S. E., Dale, J. K., Middleton, L. A., Lin, A. Y., Strober, W., Lenardo, M. J., and Puck, J. M. (1995). Dominant interfering Fas gene mutations impair apoptosis in a human autoimmune lymphoproliferative syndrome. Cell *81*, 935–946.

Fluckiger, A. C., Li, Z., Kato, R. M., Wahl, M. I., Ochs, H. D., Longnecker, R., Kinet, J. P., Witte, O. N., Scharenberg, A. M., and Rawlings, D. J. (1998). Btk/Tec kinases regulate sustained increases in intracellular Ca2+ following B-cell receptor activation. Embo J *17*, 1973–1985.

Foote, L. C., Howard, R. G., Marshak-Rothstein, A., and Rothstein, T. L. (1996a). IL-4 induces Fas resistance in B cells. J Immunol *157*, 2749–2753.

Foote, L. C., Marshak-Rothstein, A., and Rothstein, T. L. (1998). Tolerant B lymphocytes acquire resistance to Fas-mediated apoptosis after treatment with interleukin 4 but not after treatment with specific antigen unless a surface immunoglobulin threshold is exceeded. J Exp Med *187*, 847–853.

Foote, L. C., Schneider, T. J., Fischer, G. M., Wang, J. K., Rasmussen, B., Campbell, K. A., Lynch, D. H., Ju, S. T., Marshak-Rothstein, A., and Rothstein, T. L. (1996b). Intracellular signaling for inducible antigen receptor-mediated Fas resistance in B cells. J Immunol *157*, 1878–1885.

Francis, D. A., Karras, J. G., Ke, X. Y., Sen, R., and Rothstein, T. L. (1995). Induction of the transcription factors NF-kappa B, AP-1 and NF-AT during B cell stimulation through the CD40 receptor. Int Immunol *7*, 151–161.

Francis, D. A., Sen, R., Rice, N., and Rothstein, T. L. (1998). Receptor-specific induction of NF-kappa B components in primary B cells. Int Immunol *10*, 285–293.

Fukuyama, H., Adachi, M., Suematsu, S., Miwa, K., Suda, T., Yoshida, N., and Nagata, S. (1998). Transgenic expression of Fas in T cells blocks lymphoproliferation but not autoimmune disease in MRL-lpr mice. J Immunol *160*, 3805–3811.

Hedlund, T. E., Duke, R. C., Schleicher, M. S., and Miller, G. J. (1998). Fas-mediated apoptosis in seven human prostate cancer cell lines: correlation with tumor stage. Prostate *36*, 92–101.

Hoshino, K., Takeuchi, O., Kawai, T., Sanjo, H., Ogawa, T., Takeda, Y., Takeda, K., and Akira, S. (1999). Cutting edge: Toll-like receptor 4 (TLR4)-deficient mice are hyporesponsive to lipopolysaccharide: evidence for TLR4 as the Lps gene product. J Immunol *162*, 3749–3752.

Huo, L., and Rothstein, T. L. (1995). Receptor-specific induction of individual AP-1 components in B lymphocytes. J Immunol *154*, 3300–3309.

Ju, S. T., Cui, H., Panka, D. J., Ettinger, R., and Marshak-Rothstein, A. (1994). Participation of target Fas protein in apoptosis pathway induced by CD4+ Th1 and CD8+ cytotoxic T cells. Proc Natl Acad Sci U S A *91*, 4185–4189.

Ju, S. T., Panka, D. J., Cui, H., Ettinger, R., el-Khatib, M., Sherr, D. H., Stanger, B. Z., and Marshak-Rothstein, A. (1995). Fas(CD95)/FasL interactions required for programmed cell death after T-cell activation. Nature *373*, 444–448.

Kaplan, M. H., Schindler, U., Smiley, S. T., and Grusby, M. J. (1996). Stat6 is required for mediating responses to IL-4 and for development of Th2 cells. Immunity *4*, 313–319.

Karin, M., and Ben-Neriah, Y. (2000). Phosphorylation meets ubiquitination: the control of NF-kB activity. Annu Rev Immunol *18*, 621–663.

Kawakami, K., and Parker, D. C. (1993). Antigen and helper T lymphocytes activate B lymphocytes by distinct signaling pathways. Eur J Immunol *23*, 77–84.

Keane, M. M., Ettenberg, S. A., Lowrey, G. A., Russell, E. K., and Lipkowitz, S. (1996). Fas expression and function in normal and malignant breast cell lines. Cancer Res *56*, 4791–4798.

Kerner, J. D., Appleby, M. W., Mohr, R. N., Chien, S., Rawlings, D. J., Maliszewski, C. R., Witte, O. N., and Perlmutter, R. M. (1995). Impaired expansion of mouse B cell progenitors lacking Btk. Immunity *3*, 301–312.

Khan, W. N., Alt, F. W., Gerstein, R. M., Malynn, B. A., Larsson, I., Rathbun, G., Davidson, L., Muller, S., Kantor, A. B., Herzenberg, L. A., and et al. (1995). Defective B cell development and function in Btk-deficient mice. Immunity *3*, 283–299.

Klaus, G. G., O'Garra, A., Bijsterbosch, M. K., and Holman, M. (1986). Activation and proliferation signals in mouse B cells. VIII. Induction of DNA synthesis in B cells by a combination of calcium ionophores and phorbol myristate acetate. Eur J Immunol *16*, 92–97.

Lagresle, C., Mondiere, P., Bella, C., Krammer, P. H., and Defrance, T. (1996). Concurrent engagement of CD40 and the antigen receptor protects naive and memory human B cells from APO-1/Fas-mediated apoptosis. J Exp Med *183*, 1377–1388.

Lalmanach-Girard, A. C., Chiles, T. C., Parker, D. C., and Rothstein, T. L. (1993). T cell-dependent induction of NF-kappa B in B cells. J Exp Med *177*, 1215–1219.

Lane, P., Brocker, T., Hubele, S., Padovan, E., Lanzavecchia, A., and McConnell, F. (1993). Soluble CD40 ligand can replace the normal T cell-derived CD40 ligand signal to B cells in T cell-dependent activation. J Exp Med *177*, 1209–1213.

Lane, P., Traunecker, A., Hubele, S., Inui, S., Lanzavecchia, A., and Gray, D. (1992). Activated human T cells express a ligand for the human B cell-associated antigen CD40 which participates in T cell-dependent activation of B lymphocytes. Eur J Immunol *22*, 2573–2578.

Le Deist, F., Emile, J. F., Rieux-Laucat, F., Benkerrou, M., Roberts, I., Brousse, N., and Fischer, A. (1996). Clinical, immunological, and pathological consequences of Fas-deficient conditions. Lancet *348*, 719–723.

Liu, J. L., Chiles, T. C., Sen, R. J., and Rothstein, T. L. (1991). Inducible nuclear expression of NF-kappa B in primary B cells stimulated through the surface Ig receptor. J Immunol *146*, 1685–1691.

Liu, Z. G., Hsu, H., Goeddel, D. V., and Karin, M. (1996). Dissection of TNF receptor 1 effector functions: JNK activation is not linked to apoptosis while NF-kappa B activation prevents cell death. Cell *87*, 565–576.

Micheau, O., Lens, S., Gaide, O., Alevizopoulos, K., and Tschopp, J. (2001). NF-kappa B signals induce the expression of c-FLIP. Mol Cell Biol *21*, 5299–5305.

Mizuno, T., and Rothstein, T. L. (2003). Cutting edge: CD40 engagement eliminates the need for Bruton's tyrosine kinase in B cell receptor signaling for NF-kappa B. J Immunol *170*, 2806–2810.

Mond, J. J., Feuerstein, N., Finkelman, F. D., Huang, F., Huang, K. P., and Dennis, G. (1987). B-lymphocyte activation mediated by anti-immunoglobulin antibody in the absence of protein kinase C. Proc Natl Acad Sci U S A *84*, 8588–8592.

Mond, J. J., Scher, I., Cossman, J., Kessler, S., Mongini, P. K., Hansen, C., Finkelman, F. D., and Paul, W. E. (1982). Role of the thymus in directing the development of a subset of B lymphocytes. J Exp Med *155*, 924–936.

Nagahara, H., Vocero-Akbani, A. M., Snyder, E. L., Ho, A., Latham, D. G., Lissy, N. A., Becker-Hapak, M., Ezhevsky, S. A., and Dowdy, S. F. (1998). Transduction of full-length TAT fusion proteins into mammalian cells: TAT-p27Kip1 induces cell migration. Nat Med *4*, 1449–1452.

Nagata, S. (1997). Apoptosis by death factor. Cell *88*, 355–365.

Nagata, S., and Suda, T. (1995). Fas and Fas ligand: lpr and gld mutations. Immunol Today *16*, 39–43.

Natoli, G., Ianni, A., Costanzo, A., De Petrillo, G., Ilari, I., Chirillo, P., Balsano, C., and Levrero, M. (1995). Resistance to Fas-mediated apoptosis in human hepatoma cells. Oncogene *11*, 1157–1164.

Nemazee, D., Guiet, C., Buerki, K., and Marshak-Rothstein, A. (1991). B lymphocytes from the autoimmune-prone mouse strain MLR/lpr manifest an intrinsic defect in tetraparental MRL/lpr in equilibrium DBA/2 chimeras. J Immunol *147*, 2536–2539.

Neumann, M., Wohlleben, G., Chuvpilo, S., Kistler, B., Wirth, T., Serfling, E., and Schimpl, A. (1996). CD40, but not lipopolysaccharide and anti-IgM stimulation of primary B lymphocytes, leads to a persistent nuclear accumulation of RelB. J Immunol *157*, 4862–4869.

Noelle, R. J., Roy, M., Shepherd, D. M., Stamenkovic, I., Ledbetter, J. A., and Aruffo, A. (1992). A 39-kDa protein on activated helper T cells binds CD40 and transduces the signal for cognate activation of B cells. Proc Natl Acad Sci U S A *89*, 6550–6554.

Onel, K. B., Tucek-Szabo, C. L., Ashany, D., Lacy, E., Nikolic-Zugic, J., and Elkon, K. B. (1995). Expression and function of the murine CD95/FasR/APO-1 receptor in relation to B cell ontogeny. Eur J Immunol *25*, 2940–2947.

Owen-Schaub, L. B., Radinsky, R., Kruzel, E., Berry, K., and Yonehara, S. (1994). Anti-Fas on nonhematopoietic tumors: levels of Fas/APO-1 and bcl-2 are not predictive of biological responsiveness. Cancer Res *54*, 1580–1586.

Owyang, A. M., Tumang, J. R., Schram, B. R., Hsia, C. Y., Behrens, T. W., Rothstein, T. L., and Liou, H. C. (2001). c-Rel is required for the protection of B cells from antigen receptor-mediated, but not Fas-mediated, apoptosis. J Immunol *167*, 4948–4956.

Ozdemirli, M., El-Khatib, M., Foote, L. C., Wang, J. K., Marshak-Rothstein, A., Rothstein, T. L., and Ju, S. T. (1996). Fas (CD95)/Fas ligand interactions regulate antigen-specific, major histocompatibility complex-restricted T/B cell proliferative responses. Eur J Immunol *26*, 415–419.

Panayiotidis, P., Ganeshaguru, K., Foroni, L., and Hoffbrand, A. V. (1995). Expression and function of the FAS antigen in B chronic lymphocytic leukemia and hairy cell leukemia. Leukemia *9*, 1227–1232.

Perkins, D. L., Michaelson, J., Glaser, R. M., and Marshak-Rothstein, A. (1987). Selective elimination of non-lpr lymphoid cells in mice undergoing lpr-mediated graft-vs-host disease. J Immunol *139*, 1406–1413.

Petro, J. B., Rahman, S. M., Ballard, D. W., and Khan, W. N. (2000). Bruton's tyrosine kinase is required for activation of Ikappa B kinase and nuclear factor kappa B in response to B cell receptor engagement. J Exp Med *191*, 1745–1754.

Phillips, R. J., and Ghosh, S. (1997). Regulation of Ikappa B beta in WEHI 231 mature B cells. Mol Cell Biol *17*, 4390–4396.

Plumas, J., Jacob, M. C., Chaperot, L., Molens, J. P., Sotto, J. J., and Bensa, J. C. (1998). Tumor B cells from non-Hodgkin's lymphoma are resistant to CD95 (Fas/Apo-1)-mediated apoptosis. Blood *91*, 2875–2885.

Poltorak, A., He, X., Smirnova, I., Liu, M. Y., Huffel, C. V., Du, X., Birdwell, D., Alejos, E., Silva, M., Galanos, C., *et al.* (1998). Defective LPS signaling in C3H/HeJ and C57BL/10ScCr mice: mutations in Tlr4 gene. Science *282*, 2085–2088.

Qureshi, S. T., Lariviere, L., Leveque, G., Clermont, S., Moore, K. J., Gros, P., and Malo, D. (1999). Endotoxin-tolerant mice have mutations in Toll-like receptor 4 (Tlr4). J Exp Med *189*, 615–625.

Rathmell, J. C., Cooke, M. P., Ho, W. Y., Grein, J., Townsend, S. E., Davis, M. M., and Goodnow, C. C. (1995). CD95 (Fas)-dependent elimination of self-reactive B cells upon interaction with CD4+ T cells. Nature *376*, 181–184.

Rathmell, J. C., Townsend, S. E., Xu, J. C., Flavell, R. A., and Goodnow, C. C. (1996). Expansion or elimination of B cells in vivo: dual roles for CD40- and Fas (CD95)-ligands modulated by the B cell antigen receptor. Cell *87*, 319–329.

Rawlings, D. J., Saffran, D. C., Tsukada, S., Largaespada, D. A., Grimaldi, J. C., Cohen, L., Mohr, R. N., Bazan, J. F., Howard, M., Copeland, N. G., and et al. (1993). Mutation of unique region of Bruton's tyrosine kinase in immunodeficient XID mice. Science *261*, 358–361.

Rieux-Laucat, F., Le Deist, F., Hivroz, C., Roberts, I. A., Debatin, K. M., Fischer, A., and de Villartay, J. P. (1995). Mutations in Fas associated with human lymphoproliferative syndrome and autoimmunity. Science *268*, 1347–1349.

Rigley, K. P., Harnett, M. M., Phillips, R. J., and Klaus, G. G. (1989). Analysis of signaling via surface immunoglobulin receptors on B cells from CBA/N mice. Eur J Immunol *19*, 2081–2086.

Rooney, J. W., Dubois, P. M., and Sibley, C. H. (1991). Cross-linking of surface IgM activates NF-kappa B in B lymphocyte. Eur J Immunol *21*, 2993–2998.

Rothstein, T. L., Baeker, T. R., Miller, R. A., and Kolber, D. L. (1986). Stimulation of murine B cells by the combination of calcium ionophore plus phorbol ester. Cell Immunol *102*, 364–373.

Rothstein, T. L., Wang, J. K., Panka, D. J., Foote, L. C., Wang, Z., Stanger, B., Cui, H., Ju, S. T., and Marshak-Rothstein, A. (1995). Protection against Fas-dependent Th1-mediated apoptosis by antigen receptor engagement in B cells. Nature *374*, 163–165.

Satterthwaite, A. B., and Witte, O. N. (2000). The role of Bruton's tyrosine kinase in B-cell development and function: a genetic perspective. Immunol Rev *175*, 120–127.

Schauer, S. L., Wang, Z., Sonenshein, G. E., and Rothstein, T. L. (1996). Maintenance of nuclear factor-kappa B/Rel and c-myc expression during CD40 ligand rescue of WEHI 231 early B cells from receptor-mediated apoptosis through modulation of I kappa B proteins. J Immunol *157*, 81–86.

Schneider, P., Thome, M., Burns, K., Bodmer, J. L., Hofmann, K., Kataoka, T., Holler, N., and Tschopp, J. (1997a). TRAIL receptors 1 (DR4) and 2 (DR5) signal FADD-dependent apoptosis and activate NF-kappa B. Immunity *7*, 831–836.

Schneider, T. J., Grillot, D., Foote, L. C., Nunez, G. E., and Rothstein, T. L. (1997b). Bcl-x protects primary B cells against Fas-mediated apoptosis. J Immunol *159*, 4834–4839.

Schram, B. R., and Rothstein, T. L. (2003). NF-kappa B is required for surface immunoglobulin-induced Fas-resistance in B cells. J Immunol *170*, 3118–3124.

Schreck, R., Meier, B., Mannel, D. N., Droge, W., and Baeuerle, P. A. (1992). Dithiocarbamates as potent inhibitors of nuclear factor kappa B activation in intact cells. J Exp Med *175*, 1181–1194.

Sieckmann, D. G., Asofsky, R., Mosier, D. E., Zitron, I. M., and Paul, W. E. (1978). Activation of mouse lymphocytes by anti-immunoglobulin. I. Parameters of the proliferative response. J Exp Med *147*, 814–829.

Sobel, E. S., Kakkanaiah, V. N., Schiffenbauer, J., Reap, E. A., Cohen, P. L., and Eisenberg, R. A. (1998). Novel immunoregulatory B cell pathways revealed by lpr-+ mixed chimeras. J Immunol *160*, 1497–1503.

Stalder, T., Hahn, S., and Erb, P. (1994). Fas antigen is the major target molecule for CD4+ T cell-mediated cytotoxicity. J Immunol *152*, 1127–1133.

Suda, T., Okazaki, T., Naito, Y., Yokota, T., Arai, N., Ozaki, S., Nakao, K., and Nagata, S. (1995). Expression of the Fas ligand in cells of T cell lineage. J Immunol *154*, 3806–3813.

Takahashi, T., Tanaka, M., Brannan, C. I., Jenkins, N. A., Copeland, N. G., Suda, T., and Nagata, S. (1994). Generalized lymphoproliferative disease in mice, caused by a point mutation in the Fas ligand. Cell *76*, 969–976.

Takata, M., and Kurosaki, T. (1996). A role for Bruton's tyrosine kinase in B cell antigen receptor-mediated activation of phospholipase C-gamma 2. J Exp Med *184*, 31–40.

Trauth, B. C., Klas, C., Peters, A. M., Matzku, S., Moller, P., Falk, W., Debatin, K. M., and Krammer, P. H. (1989). Monoclonal antibody-mediated tumor regression by induction of apoptosis. Science *245*, 301–305.

Tumang, J. R., Negm, R. S., Solt, L. A., Schneider, T. J., Colarusso, T. P., Hastings, W. D., Woodland, R. T., and Rothstein, T. L. (2002). BCR engagement induces Fas resistance in primary B cells in the absence of functional Bruton's tyrosine kinase. J Immunol *168*, 2712–2719.

Tumang, J. R., Owyang, A., Andjelic, S., Jin, Z., Hardy, R. R., Liou, M. L., and Liou, H. C. (1998). c-Rel is essential for B lymphocyte survival and cell cycle progression. Eur J Immunol *28*, 4299–4312.

Ungefroren, H., Voss, M., Jansen, M., Roeder, C., Henne-Bruns, D., Kremer, B., and Kalthoff, H. (1998). Human pancreatic adenocarcinomas express Fas and Fas ligand yet are resistant to Fas-mediated apoptosis. Cancer Res *58*, 1741–1749.

Van Antwerp, D. J., Martin, S. J., Kafri, T., Green, D. R., and Verma, I. M. (1996). Suppression of TNF-alpha-induced apoptosis by NF-kappa B. Science *274*, 787–789.

Venkataraman, L., Francis, D. A., Wang, Z., Liu, J., Rothstein, T. L., and Sen, R. (1994). Cyclosporin-A sensitive induction of NF-AT in murine B cells. Immunity *1*, 189–196.

Verma, I. M., Stevenson, J. K., Schwarz, E. M., Van Antwerp, D., and Miyamoto, S. (1995). Rel/NF-kappa B/I kappa B family: intimate tales of association and dissociation. Genes Dev *9*, 2723–2735.

Vocero-Akbani, A., Lissy, N. A., and Dowdy, S. F. (2000). Transduction of full-length Tat fusion proteins directly into mammalian cells: analysis of T cell receptor activation-induced cell death. Methods Enzymol *322*, 508–521.

Wang, C. Y., Mayo, M. W., and Baldwin, A. S., Jr. (1996). TNF- and cancer therapy-induced apoptosis: potentiation by inhibition of NF-kappa B. Science *274*, 784–787.

Wang, J., Lobito, A. A., Shen, F., Hornung, F., Winoto, A., and Lenardo, M. J. (2000). Inhibition of Fas-mediated apoptosis by the B cell antigen receptor through c-FLIP. Eur J Immunol *30*, 155–163.

Watanabe, D., Suda, T., and Nagata, S. (1995). Expression of Fas in B cells of the mouse germinal center and Fas-dependent killing of activated B cells. Int Immunol *7*, 1949–1956.

Watanabe-Fukunaga, R., Brannan, C. I., Copeland, N. G., Jenkins, N. A., and Nagata, S. (1992). Lymphoproliferation disorder in mice explained by defects in Fas antigen that mediates apoptosis. Nature *356*, 314–317.

Wu, J., Wilson, J., He, J., Xiang, L., Schur, P. H., and Mountz, J. D. (1996). Fas ligand mutation in a patient with systemic lupus erythematosus and lymphoproliferative disease. J Clin Invest *98*, 1107–1113.

Wurster, A. L., Rodgers, V. L., White, M. F., Rothstein, T. L., and Grusby, M. J. (2002). IL-4 mediated protection of parimary B cells from apoptosis through STAT6-dependent upregulation of Bcl-xL. J Biol Chem *277*, 27169–27175.

Xerri, L., Devilard, E., Bouabdallah, R., Stoppa, A. M., Hassoun, J., and Birg, F. (1999). FADD expression and caspase activation in B-cell lymphomas resistant to Fas-mediated apoptosis. Br J Haematol *106*, 652–661.

Xerri, L., Devilard, E., Hassoun, J., Haddad, P., and Birg, F. (1997). Malignant and reactive cells from human lymphomas frequently express Fas ligand but display a different sensitivity to Fas-mediated apoptosis. Leukemia *11*, 1868–1877.

Yeh, W. C., Pompa, J. L., McCurrach, M. E., Shu, H. B., Elia, A. J., Shahinian, A., Ng, M., Wakeham, A., Khoo, W., Mitchell, K., *et al.* (1998). FADD: essential for embryo development and signaling from some, but not all, inducers of apoptosis. Science *279*, 1954–1958.

Yonehara, S., Ishii, A., and Yonehara, M. (1989). A cell-killing monoclonal antibody (anti-Fas) to a cell surface antigen co-downregulated with the receptor of tumor necrosis factor. J Exp Med *169*, 1747–1756.

Chapter 6

Apoptosis and Autoimmune Diseases

YOUHAI H. CHEN*

ABSTRACT: Apoptosis plays critical roles in the initiation, progression and remission of autoimmune diseases. On the one hand, apotosis of resident tissue cells in diseased organs contributes to the pathology of autoimmune diseases. On the other hand, apoptosis of inflammatory cells is essential for disease recovery and is the major goal of therapeutic interventions for autoimmune diseases. In this review, I will discuss the roles of apoptosis in two common autoimmune diseases: multiple sclerosis and rheumatoid arthritis. I will also examine the potential roles of transcription factor p53 and the tumor necrosis factor family of proteins in autoimmune diseases.

Key Words: RA; MS; Fas; TRAIL; Transcription Factors, TNF

Multiple Sclerosis (MS) and Experimental Autoimmune (or Allergic) Encephalomyelitis (EAE)

Multiple sclerosis is a chronic inflammatory disease of the central nervous system. The active inflammatory process in MS is confined to the white matter in the central nervous system, not affecting the peripheral nervous tissues. One of the hallmarks of MS pathology is demyelination and death of oligodendrocytes that produce the myelin sheath (Dowling et al., 1997; Ffrench-Constant, 1994; Ozawa et al., 1994). Using in situ TUNEL technique, a method that sensitively detects apoptosis at the single cell level, Dowling et al. (Dowling et al., 1997) showed that the acute MS plaques contained massive numbers of inflammatory and glial cells undergoing apoptosis. The presence of apoptotic cells in the MS plaques was also confirmed using other techniques such as confocal microscopy and electrophoresis of the DNA isolated from MS brains (Dowling et al., 1997). These results corroborate with earlier reports by Ozawa et al. that death of glial cells was an important pathological feature of MS (Ozawa et al., 1994). The mechanisms by which inflammation and apoptosis are initiated and regulated in MS are not well understood. In acute MS plaques, both infiltrating leukocytes and resident microglial cells are activated, and produce various inflammatory and cytotoxic molecules such as interleukin (IL)-1, IL-6, IL-12, interferon (IFN)-g , nitric oxide, tumor necrosis factor (TNF)-a , and Fas-ligand (FasL) (D'Souza et al., 1996; Dowling et al., 1996; Tanaka et al., 1995). Delineation of the molecular pathways involved in the

*Associate Professor, Department of Pathology and Laboratory Medicine, University of Pennsylvania School of Medicine, Philadelphia, PA 19104. *Correspondence:* Youhai H. Chen, M.D., Ph.D., 653 BRB-II/III, Department of Pathology and Laboratory Medicine, University of Pennsylvania School of Medicine, 421 Curie Blvd., Philadelphia, PA 19104. Phone: 215-898-4671. Fax: 215-573-8606. E-Mail: yhc@mail.med.upenn.edu pathogenesis of MS will not only help our understanding of the pathogenesis of the disease but also aid in developing specific treatment for the disease.

EAE is a putative animal model for multiple sclerosis. The disease can be induced in susceptible strains of animals by immunization with myelin antigens such as myelin basic protein (MBP), proteolipid protein (PLP) or myelin oligodendrocyte glycoprotein (MOG) (Miller and Karpus, 1994; Miller et al., 1995; Zamvil and Steinman, 1990). EAE is mediated by encephalitogenic TH1 cells secreting IL-2, IFN-g and TNF-a. Myelin-specific TH2 cells are less encephalitogenic and may suppress TH1 cell-mediated EAE (Chen et al., 1994; Kuchroo et al., 1995). As in MS, apoptosis may directly contribute to the pathology of EAE. By combined immunohistochemistry and in situ nick labeling, Schmied et al. and Pender et al. detected large numbers of apoptotic cells in the brain and spinal cord of Lewis rats during the acute phase of EAE (Pender et al., 1992; Schmied et al., 1993); surprisingly, up to 50% of T lymphocytes in EAE lesions showed signs of apoptosis (Pender et al., 1992). Recently, using several mouse models of EAE, we and others have also detected large numbers of apoptotic cells in the central nervous system, both at peak of the disease and during disease recovery (Chen et al., 1998; Hilliard et al., 2001; Sabelko et al., 1997; Waldner et al., 1997).

It must be emphasized that apoptosis in MS and EAE may be a "double-edged" sword, which can be either beneficial or detrimental to the CNS tissue. Apoptosis of inflammatory cells promotes resolution of inflammation and prevents neural tissue injury. By contrast, apoptosis of oligodendrocytes leads to demyelination and exacerbates the disease. Interestingly, both immune cells and neural cells are equipped with similar cell death machinery (e.g., caspases and death receptors) that can be activated by TNF, FasL, reactive oxygen species and cytotoxic enzymes; and recent studies suggest that many of these apoptosis inducing agents have dual roles in EAE (Das et al., 1998; Dittel et al., 1999; Kassiotis et al., 2001; Korner et al., 1997; Sabelko-Downes et al., 1999; Suvannavejh et al., 2000). Understanding the molecular mechanisms whereby apoptosis is regulated in the CNS is therefore crucial for the development of effective strategies to manipulate the disease.

Rheumatoid Arthritis (RA) and Collagen-Induced Arthritis (CIA)

Rheumatoid arthritis is a chronic inflammatory disease of the joints that afflicts approximately 1% of the population. Although it can affect children and adolescents, the disease usually strikes adults and the incidence of clinical illness is greatest among those aged 40 to 60 years. Rheumatoid arthritis is an extremely disabling disease that carries a high mortality. Up to 7% of patients are disabled to some extent 5 years after disease onset, and 50% are disabled to work 10 years after disease onset (Feldmann et al., 1996; Miller-Blair and Robbins, 1993; Panayi, 1993). Histopathologically, rheumatoid arthritis is characterized by hyperplasia of the synovial membrane and hyper-activation of synovial cells which lead to progressive cartilage and bone destruction through their release of degradative enzymes such as matrix metalloproteinases. Although the etiology of rheumatoid arthritis is not clear, activation of autoreactive lymphocytes may play an important role in the initiation of the disease (Brennan et al., 1995; Salmon and Gaston, 1995). In rheumatoid arthritis patients, activation of T cells specific for cartilage antigens such as collagen type II or activation of B cells specific for Fc region of the immunoglobulin (Ig) G or filaggrin is common (Brennan et al., 1995; Chou et al., 2001; Cutolo et al., 1993; Firestein et al., 1995; Maini et al., 1995; Salmon and Gaston, 1995; Schumacher et al., 1994; Sewell and Trentham, 1993). Moreover, the disease is associated with HLA DR1 and DR4 which are crucial for antigen-presentation to CD4+ T cells (Feldmann et al., 1996; Miller-Blair and Robbins, 1993; Panayi, 1993).

Collagen-induced arthritis (CIA) is an animal model for human rheumatoid arthritis (Myers, 1993; Sartor et al., 1996). The disease can be induced in susceptible strains of animals (e.g., DBA1 and B10.Q mice) by immunization with type II collagen (Myers, 1993; Sartor et al., 1996). Similar to rheumatoid arthritis, collagen-induced arthritis is characterized by massive infiltration of synovial joints by inflammatory cells and hyperplasia of the synovial membrane. Both collagen-specific T and B lymphocytes are involved in the induction of the arthritis. The effector mechanisms that lead to joint tissue destruction appear to be similar to rheumatoid arthritis and a number of inflammatory cell types have been implicated (Myers, 1993; Sartor et al., 1996). These include fibroblast-like synoviocytes, bone marrow-derived macrophages, granulocytes, and dendritic cells as well as lymphocytes.

The Transcription Factor p53 and Autoimmune Diseases

Development of autoimmune diseases requires coordinated expression of a myriad of genes. These include genes that encode antigen receptors, costimulatory molecules, cytokines, chemokines, adhesion molecules, cytotoxic enzymes, cell cycle regulators and apoptotic proteins. These molecules in turn play crucial roles in the initiation, progression and resolution of autoimmune inflammation. The outcomes of autoimmune diseases are largely dictated by the relative contributions of these molecules. To date, little is known about the molecular events leading to the expression of these genes in autoimmune diseases. Elucidation of the molecular mechanisms whereby these genes are regulated is essential for our understanding of the pathogenesis of autoimmune diseases.

p53 (transformation-related protein 53) is a member of the p53 transcription factor family which also includes p63 and p73 (Arrowsmith, 1999; Burns and El-Deiry, 1999; De Laurenzi and Melino, 2000; Irwin and Kaelin, 2001; Kaelin, 1999; Levrero et al., 2000; Mills et al., 1999; Morrison and Kinoshita, 2000; Yang et al., 2000). It is ubiquitously expressed at low levels in a variety of tissues and is significantly upregulated in tumors, inflamed or damaged tissues (Arrowsmith, 1999; De Laurenzi and Melino, 2000; Eizenberg et al., 1995; Irwin and Kaelin, 2001; Kaelin, 1999; Levrero et al., 2000; Mills et al., 1999; Moon et al., 2000; Sakhi et al., 1994; Yamanishi et al., 2002; Yang et al., 2000). Each p53 polypeptide contains a transactivation domain at its N-terminus, a tetramerization domain at the C-terminus, and a DNA-binding domain at the center. The C-terminus also holds a regulatory domain, which can negatively regulate the central DNA-binding domain. p53 has a short half-life and is normally present as a tetramer in association with the inhibitor protein called MDM2. MDM2 binds to the N-terminus of p53 and blocks its transcriptional activity. It also facilitates the export of p53 from nucleus to cytoplasm where it is degraded through the ubiquitin pathway (Arrowsmith, 1999; De Laurenzi and Melino, 2000; Irwin and Kaelin, 2001; Kaelin, 1999; Levrero et al., 2000; Mills et al., 1999; Yang et al., 2000).

A wide variety of stimuli including cytokines, growth factors, oncogenic stimuli, hypoxia, and irradiation can stabilize and activate p53. Although the exact mechanisms of p53 protein activation are not well characterized, phosphorylation and acetylation of p53 are believed to be important steps. Once activated, p53 can bind to specific DNA sequences located in the promoter regions of target genes. To date, a large number ($\sim$110) of p53 target genes have been identified in various cell types. These include genes that regulate 1) apoptosis (such as Fas, death receptor 5, Bax, PIG3, p85, PAG608, IGF-Bp3) (Bennett et al.,

1998; Buckbinder et al., 1995; Guan et al., 2001; Miyashita and Reed, 1995; Muller et al., 1998; Polyak et al., 1997; Wu et al., 1997), and 2) cell cycle (such as p21, GADD45, B99, 14-3-3s) (el-Deiry et al., 1994; el-Deiry et al., 1993; Hermeking et al., 1997; Utrera et al., 1998; Zhan et al., 1994). The spectrum of genes activated by p53 may be context and cell type dependent. It should be noted that in addition to transactivating genes, p53 may also repress expression of certain genes such as c-fos, c-myc, IL-4 and IL-6 (Pesch et al., 1996; Santhanam et al., 1991). The mechanism of p53-mediated gene repression is not clear and may require the presence of the C-terminal domain. Additionally, it has been reported that p53 may be capable of inhibiting nuclear DNA replication by directly binding to DNA (Arrowsmith, 1999; De Laurenzi and Melino, 2000; Irwin and Kaelin, 2001; Kaelin, 1999; Levrero et al., 2000; Mills et al., 1999; Yang et al., 2000).

Despite intense investigations of p53 during the past few years, the physiological and pathological roles of p53 in vivo are still not well understood. Somatic mutations of p53 gene have been detected in many tumor cells and synovial cells of rheumatoid arthritis (RA) patients (Arrowsmith, 1999; De Laurenzi and Melino, 2000; Firestein et al., 1997; Irwin and Kaelin, 2001; Kaelin, 1999; Levrero et al., 2000; Mills et al., 1999; Yang et al., 2000). Germline mutations of p53 gene in humans and mice significantly increase the incidence of tumors of various cell lineages (De Laurenzi and Melino, 2000; Jacks et al., 1994; Kaelin, 1999). These and other observations have led to the conclusion that p53 is responsible for preventing oncogenesis, presumably by inducing cell cycle arrest or apoptosis of proliferating cells. However, p53 gene expression is upregulated in a number of conditions not directly related to oncogenesis. These include inflammation, trauma, hypoxia and infections (Arrowsmith, 1999; De Laurenzi and Melino, 2000; Eizenberg et al., 1995; Irwin and Kaelin, 2001; Kaelin, 1999; Levrero et al., 2000; Mills et al., 1999; Moon et al., 2000; Sakhi et al., 1994; Yamanishi et al., 2002; Yang et al., 2000). Additionally, as pointed out above, p53 gene mutation has also been detected in synovial cells of rheumatoid arthritis patients (Firestein et al., 1997), and p53-deficiency in mice accelerates, whereas p53 gene transfer ameliorates, autoimmune arthritis (Yamanishi et al., 2002; Yao et al., 2001). Using microdissected RA synovial tissue sections, Firestein and colleagues recently reported that p53 mutations were located mainly in the synovial intimal lining rather than the sublining. Regions with high rates of p53 mutations contained significantly greater amounts of IL-6 mRNA compared with the low mutation samples (Yamanishi et al., 2002), suggesting that p53 gene mutation may exacerbate arthritis by enhancing the production of inflammatory cytokines. Thus, in addition to oncogenesis, p53 may also play important roles in inflammatory disorders (Eizenberg et al., 1995; Eizenberg et al., 1996; Ladiwala et al., 1999). In the case of CNS inflammation, a role for p53 is strongly suggested by the following observations: 1) p53 gene expression is significantly upregulated during CNS inflammation, which correlates to the degree of apoptosis in the CNS (Moon et al., 2000); and 2) oligodendrocytes upregulate p53 expression upon treatment with TNF and undergo apoptosis upon enforced p53 expression (Eizenberg et al., 1995; Eizenberg et al., 1996; Ladiwala et al., 1999).

TRAIL, a New Member of the TNF Superfamily That Is Capable of Inducing Apoptosis

The TNF/TNF receptor families consist of approximately 20 ligands and 30 receptors. These proteins play crucial roles in regulating lymphocyte activation and apoptosis, and therefore, are important for immune homeostasis and self-tolerance (Argiles et al., 1997; Baker and Reddy, 1996; Beutler and Bazzoni, 1998; Lynch et al., 1996; Magnusson and Vaux, 1999; Mountz et al., 1996; Ruddle, 1992; Wallach et al., 1996; Ware et al., 1996). TRAIL, the **T**NF-**r**elated **a**poptosis-**i**nducing **l**igand, is a newly identified member of the TNF family (Meng et al., 2000; Pan et al., 1997; Pan et al., 1998; Pan et al., 1997; Schneider et al., 1997; Screaton et al., 1997; Sheikh et al., 1998; Sheikh et al., 1999; Sheridan et al., 1997; Walczak et al., 1997; Wiley et al., 1995). In humans, TRAIL can interact with two death receptors [*death receptor 4 (DR4, TRAIL-R1) and death receptor 5 (DR5, TRAIL-R2)*] and two decoy receptors [*decoy receptor 1 (DcR1, TRAIL-R3, TRID) and decoy receptor 2 (DcR2, TRAIL-R4, TRUNDD)*] (Pan et al., 1997; Pan et al., 1998; Schneider et al., 1997; Screaton et al., 1997; Sheikh et al., 1998; Sheridan et al., 1997; Walczak et al., 1997). In mice, only one TRAIL death receptor (DR5) has been characterized (Wu et al., 1999), and no TRAIL decoy receptor has been cloned.

In vitro, TRAIL can induce apoptosis of many, but not all, tumor cell lines (Pan et al., 1997; Sheridan et al., 1997). This appears to be mediated by the death receptor DR4 and DR5, which possess intracellular death domains similar to those of TNF receptor I and CD95 (Fas/Apo-1). The DR4 and DR5 death domains can activate both mitochondria-dependent and mitochondria-independent pathways of apoptosis through FADD/caspase 8, leading to the activation of the caspase cascade (Bodmer et al., 2000; Deng et al., 2002; Sprick et al., 2000; Zhang et al., 2001). The decoy receptors DcR1 and DcR2, which do not contain functional death domains, can block TRAIL-induced apoptosis (Pan et al., 1997; Sheridan et al., 1997). Although both TRAIL and TRAIL receptors are constitutively expressed in various tissues (Abulencia et al., 2002; Pan et al., 1997; Schneider et al., 1997; Wiley et al., 1995; Wu et al., 1999) and are upregulated upon cell activation (Jeremias et al., 1998; Mariani and Krammer, 1998; Sheikh et al., 1998), TRAIL may not induce apoptosis of most non-transformed cells (Pan et al., 1997; Sheridan et al., 1997). In vivo administration of recombinant TRAIL selectively kills tumor cells, but not normal cells, leaving most host tissues unharmed (Ashkenazi et al., 1999; Walczak et al., 1999). However, recent studies suggest that unlike most normal cells, thymocytes, neurons and human hepatocytes are extremely sensitive to TRAIL-induced apoptosis (Jo et al., 2000; Martin-Villalba et al., 1999; Simon et al., 2001).

In addition to inducing apoptosis, TRAIL may also promote cell survival through activating NF-kB or c-Jun pathway of signal transduction (Chaudhary et al., 1997; Schneider et al., 1997). This appears to be mediated through TRAF-2 and/or RIP as recently demonstrated by several laboratories (Chaudhary et al., 1997; Hu et al., 1999; Lin et al., 2000). Using gene microarray technology, Kumar-Sinha et al. recently showed that similar to FasL and TNF, TRAIL was able to activate a large number of genes in tumor cells (Kumar-Sinha et al., 2002). These include NF-kB-dependent genes such as cIAP2, A20, and E-selectin (Kumar-Sinha et al., 2002). The biological relevance of these findings to TRAIL function in vivo is not clear.

The possibility of both promoting and preventing apoptosis by a single ligand/receptor pair is a recurrent theme in TNF/TNF receptor family research and provides an extremely exciting opportunity to study the regulation of cell death and survival in the context of immunity and tolerance. We started our TRAIL project four years ago when essentially nothing was known about the roles of TRAIL in vivo. Our initial approach involved blocking or overexpressing TRAIL in animal models of autoimmune diseases. We found that chronic blockade of TRAIL in mice exacerbated autoimmune arthritis, and that intra-articular TRAIL gene transfer ameliorated the disease (Song et al., 2000). In vivo, TRAIL-blockade led to profound hyper-proliferation of synovial cells and arthritogenic lymphocytes, and heightened the production of cytokines and autoantibodies. In vitro, TRAIL inhibited DNA synthesis and prevented cell cycle progression of lymphocytes. Interestingly, TRAIL had no effect on apoptosis of inflammatory cells either in vivo or in vitro (Song et al., 2000). Thus, unlike other members of the tumor necrosis factor superfamily, TRAIL is a prototype inhibitor protein that inhibits autoimmune inflammation by blocking cell cycle progression (Goke et al., 2000; Goke et al., 2001; Hilliard et al., 2001; Song et al., 2000).

Fas Ligand, the Closest Homologue of TRAIL

FasL shares the highest sequence homology with TRAIL and is one of the most extensively studied members of the TNF family (Nagata and Suda, 1995). FasL is normally expressed by a small number of cell types including activated lymphocytes and cells of immune privileged organs (such as eye, testis, brain and spinal cord) (French and Tschopp, 1996; Nagata and Suda, 1995; Saas et al., 1997; Stuart et al., 1997). Its receptor Fas (CD95) is a type I membrane protein of the TNF-receptor family. Unlike FasL, Fas is expressed constitutively in most tissues and is dramatically up-regulated at sites of inflammation. Fas/FasL interaction activates FADD, which in turn triggers the activation of the IL-1 converting enzyme (ICE) family of caspases, leading to DNA fragmentation and cell death. However, Fas/FasL interaction does not always lead to apoptosis. Under certain conditions, Fas/FasL interaction can also activate target cells, presumably through the nuclear factor (NF)-kB pathway (Abreu-Martin et al., 1995; Malinin et al., 1997). In this case, Fas may transmit activating signals similar to those of TNF-receptors, leading to secretion of pro-inflammatory cytokines such as IL-1 and IL-8 (Abreu-Martin et al., 1995; Malinin et al., 1997).

Fas/FasL have been reported to both inhibit and promote autoimmune inflammation. Mutations of genes encoding Fas or FasL lead to lymphocytic proliferation and autoimmune inflammatory diseases in both humans and mice (Cohen and Eisenberg, 1992; Nagata and Golstein, 1995; Sneller et al., 1997). Under these conditions, T cells of presumably autoimmune origin accumulate in extremely large numbers and exhibit a peculiar phenotype, i.e., CD4-CD8-B220+ or CD4+CD8-B220+. In the late stages of the disease, these aberrant cells become functionally inactive, or anergic. While these observations have led to the recognition that Fas and FasL are essential for maintaining self-tolerance, presumably by deleting autoreactive cells through activation-induced cell death (AICD), recent studies suggest that Fas/FasL interaction can also contribute to autoimmune inflammation. Kang et al., Chervonsky et al. and Giordano et al. reported that Fas/FasL interaction may contribute to the pathogenesis of autoimmune thyroiditis and diabetes (Chervonsky et al., 1997; Giordano et al., 1997; Kang et al., 1997). Similarly, Waldner et al. and Sabelko et al. first reported that EAE was diminished and apoptosis inhibited in mice carrying the lpr (Fas) or gld (FasL) mutation, suggesting that Fas/FasL interaction may play an active role in the pathogenesis of EAE (Sabelko et al., 1997; Waldner et al., 1997). In EAE, FasL is expressed by microglial cells and neurons as well as activated T cells infiltrating the central nervous system (D'Souza et al., 1996; Saas et al., 1997; Xerri et al., 1997). By adoptive cell transfer, Sabelko-Downes et al. recently showed that Fas expressed by recipient mice, but not T cells, is involved in the pathogenesis of EAE (Sabelko-Downes et al., 1999).

While FasL may promote EAE by directly killing neural cells, it can also inhibit EAE by deleting inflammatory cells. Miller and colleagues reported that EAE was significantly enhanced in Fas-deficient SJL lpr/lpr mice, which displayed significantly increased mean peak clinical scores, reduced remission rates, and increased mortality when compared with their SJL +/lpr littermates (Suvannavejh et al., 2000). Similarly, we found that spontaneous EAE was dramatically exacerbated in MBP-specific TCR transgenic mice carrying Fas or FasL gene mutation (Liu et al., 2000). The exacerbation of EAE was evidenced primarily by an increase in disease incidence and a decrease in spontaneous disease recovery (Liu et al., 2000).

ACKNOWLEDGMENT: This work was supported by grants from the National Institutes of Health (AI50059, NS40188, and NS40447).

References

Abreu-Martin, M. T., Vidrich, A., Lynch, D. H., and Targan, S. R. (1995). Divergent induction of apoptosis and IL-8 secretion in HT-29 cells in response to TNF-alpha and ligation of Fas antigen. Journal of Immunology *155*, 4147–54.

Abulencia, J. P., Gaspard, R., Quackenbush, J., and konstantopoulos, K. (2002). Discovery and characterization of differentially regulated genes in the chondrocytic cell line T/C-28a2 under dynamic fluid shear. FASEB Journal *16*, A656.7.

Argiles, J. M., Lopez-Soriano, J., Busquets, S., and Lopez-Soriano, F. J. (1997). Journey from cachexia to obesity by TNF. Faseb J *11*, 743–51.

Arrowsmith, C. H. (1999). Structure and function in the p53 family. Cell Death & Differentiation *6*, 1169–73.

Ashkenazi, A., Pai, R. C., Fong, S., Leung, S., Lawrence, D. A., Marsters, S. A., Blackie, C., Chang, L., McMurtrey, A. E., Hebert, A., DeForge, L., Koumenis, I. L., Lewis, D., Harris, L., Bussiere, J., Koeppen, H., Shahrokh, Z., and Schwall, R. H. (1999). Safety and antitumor activity of recombinant soluble Apo2 ligand. Journal of Clinical Investigation *104*, 155–62.

Baker, S. J., and Reddy, E. P. (1996). Transducers of life and death: TNF receptor superfamily and associated proteins. Oncogene *12*, 1–9.

Bennett, M., Macdonald, K., Chan, S. W., Luzio, J. P., Simari, R., and Weissberg, P. (1998). Cell surface trafficking of Fas: a rapid mechanism of p53-mediated apoptosis. Science *282*, 290–3.

Beutler, B., and Bazzoni, F. (1998). TNF, apoptosis and autoimmunity: a common thread? Blood Cells Mol Dis *24*, 216–30.

Bodmer, J. L., Holler, N., Reynard, S., Vinciguerra, P., Schneider, P., Juo, P., Blenis, J., and Tschopp, J. (2000). TRAIL receptor-2 signals apoptosis through FADD and caspase-8. Nat Cell Biol *2*, 241–3.

Brennan, F. M., Maini, R. N., and Feldmann, M. (1995). Cytokine expression in chronic inflammatory disease. British Medical Bulletin *51*, 368–84.

Buckbinder, L., Talbott, R., Velasco-Miguel, S., Takenaka, I., Faha, B., Seizinger, B. R., and Kley, N. (1995). Induction of the growth inhibitor IGF-binding protein 3 by p53. Nature *377*, 646–9.

Burns, T. F., and El-Deiry, W. S. (1999). The p53 pathway and apoptosis. Journal of Cellular Physiology *181*, 231–9.

Chaudhary, P. M., Eby, M., Jasmin, A., Bookwalter, A., Murray, J., and Hood, L. (1997). Death receptor 5, a new member of the TNFR family, and DR4 induce FADD-dependent apoptosis and activate the NF-kappaB pathway. Immunity *7*, 821–30.

Chen, Y., Hancock, W. W., Marks, R., Gonnella, P., and Weiner, H. L. (1998). Mechanisms of recovery from experimental autoimmune encephalomyelitis: T cell deletion and immune deviation in myelin basic protein T cell receptor transgenic mice. Journal of Neuroimmunology *82*, 149–59.

Chen, Y., Kuchroo, V. K., Inobe, J.-I., Hafler, D. A., and Weiner, H. L. (1994). Regulatory T cell clones induced by oral tolerance: suppression of autoimmune encephalomyelitis. Science *265*, 1237–1240.

Chervonsky, A. V., Wang, Y., Wong, F. S., Visintin, I., Flavell, R. A., Janeway, C. A., Jr., and Matis, L. A. (1997). The role of Fas in autoimmune diabetes. Cell *89*, 17–24.

Chou, C. T., Yang, J. S., and Lee, M. R. (2001). Apoptosis in rheumatoid arthritis—expression of Fas, Fas-L, p53, and Bcl-2 in rheumatoid synovial tissues. Journal of Pathology *193*, 110–6.

Cohen, P. L., and Eisenberg, R. A. (1992). The lpr and gld genes in systemic autoimmunity: life and death in the Fas lane. Immunology Today *13*, 427–8.

Cutolo, M., Sulli, A., Barone, A., Seriolo, B., and Accardo, S. (1993). Macrophages, synovial tissue and rheumatoid arthritis. Clinical & Experimental Rheumatology *11*, 331–9.

D'Souza, S. D., Bonetti, B., Balasingam, V., Cashman, N. R., Barker, P. A., Troutt, A. B., Raine, C. S., and Antel, J. P. (1996). Multiple sclerosis: Fas signaling in oligodendrocyte cell death. Journal of Experimental Medicine *184*, 2361–70.

Das, M. P., Howard, E. D., Weiner, H. L., Sobel, R. A., Kuchroo, V. K., and Sean Riminton, D. (1998). Challenging cytokine redundancy: inflammatory cell movement and clinical course of experimental autoimmune encephalomyelitis are normal in lymphotoxin-deficient, but not tumor necrosis factor-deficient, mice. Journal of Immunology *161*, 3299–306.

De Laurenzi, V., and Melino, G. (2000). Evolution of functions within the p53/p63/p73 family. Annals of the New York Academy of Sciences *926*, 90–100.

Deng, Y., Lin, Y., and Wu, X. (2002). TRAIL-induced apoptosis requires Bax-dependent mitochondrial release of Smac/DIABLO. Genes Dev *16*, 33–45.

Dittel, B. N., Merchant, R. M., and Janeway, C. A., Jr. (1999). Evidence for Fas-dependent and Fas-independent mechanisms in the pathogenesis of experimental autoimmune encephalomyelitis. Journal of Immunology *162*, 6392–400.

Dowling, P., Husar, W., Menonna, J., Donnenfeld, H., Cook, S., and Sidhu, M. (1997). Cell death and birth in multiple sclerosis brain. Journal of the Neurological Sciences *149*, 1–11.

Dowling, P., Shang, G., Raval, S., Menonna, J., Cook, S., and Husar, W. (1996). Involvement of the CD95 (APO-1/Fas) receptor/ligand system in multiple sclerosis brain. Journal of Experimental Medicine *184*, 1513–8.

Eizenberg, O., Faber-Elman, A., Gottlieb, E., Oren, M., Rotter, V., and Schwartz, M. (1995). Direct involvement of p53 in programmed cell death of oligodendrocytes. EMBO Journal *14*, 1136–44.

Eizenberg, O., Faber-Elman, A., Gottlieb, E., Oren, M., Rotter, V., and Schwartz, M. (1996). p53 plays a regulatory role in differentiation and apoptosis of central nervous system-associated cells. Molecular & Cellular Biology *16*, 5178–85.

el-Deiry, W. S., Harper, J. W., O'Connor, P. M., Velculescu, V. E., Canman, C. E., Jackman, J., Pietenpol, J. A., Burrell, M., Hill, D. E., Wang, Y., and et al. (1994). WAF1/CIP1 is induced in p53-mediated G1 arrest and apoptosis. Cancer Res *54*, 1169–74.

el-Deiry, W. S., Tokino, T., Velculescu, V. E., Levy, D. B., Parsons, R., Trent, J. M., Lin, D., Mercer, W. E., Kinzler, K. W., and Vogelstein, B. (1993). WAF1, a potential mediator of p53 tumor suppression. Cell *75*, 817–25.

Feldmann, M., Brennan, F. M., and Maini, R. N. (1996). Rheumatoid arthritis. Cell *85*, 307–10.

Ffrench-Constant, C. (1994). Pathogenesis of multiple sclerosis. Lancet *343*, 271–5.

Firestein, G. S., Echeverri, F., Yeo, M., Zvaifler, N. J., and Green, D. R. (1997). Somatic mutations in the p53 tumor suppressor gene in rheumatoid arthritis synovium. Proceedings of the National Academy of Sciences of the United States of America *94*, 10895–900.

Firestein, G. S., Yeo, M., and Zvaifler, N. J. (1995). Apoptosis in rheumatoid arthritis synovium. Journal of Clinical Investigation *96*, 1631–8.

French, L. E., and Tschopp, J. (1996). Constitutive Fas ligand expression in several non-lymphoid mouse tissues: implications for immune-protection and cell turnover. Behring Institute Mitteilungen *97*, 156–60.

Giordano, C., Stassi, G., De Maria, R., Todaro, M., Richiusa, P., Papoff, G., Ruberti, G., Bagnasco, M., Testi, R., and Galluzzo, A. (1997). Potential involvement of Fas and its ligand in the pathogenesis of Hashimoto's thyroiditis. Science *275*, 960–3.

Goke, R., Goke, A., Goke, B., and Chen, Y. (2000). Regulation of TRAIL-induced apoptosis by transcription factors. Cellular Immunology *201*, 77–82.

Goke, R., Goke, A., Goke, B., El-Deiry, W. S., and Chen, Y. (2001). Pioglitazone inhibits growth of carcinoid cells and promotes TRAIL-induced apoptosis by induction of p21waf1/cip1. Digestion *64*, 75–80.

Guan, B., Yue, P., Clayman, G. L., and Sun, S. Y. (2001). Evidence that the death receptor DR4 is a DNA damage-inducible, p53-regulated gene. Journal of Cellular Physiology *188*, 98–105.

Hermeking, H., Lengauer, C., Polyak, K., He, T. C., Zhang, L., Thiagalingam, S., Kinzler, K. W., and Vogelstein, B. (1997). 14-3-3 sigma is a p53-regulated inhibitor of G2/M progression. Mol Cell *1*, 3–11.

Hilliard, B., Wilmen, A., Seidel, C., Liu, T. S., Goke, R., and Chen, Y. (2001). Roles of TNF-related apoptosis-inducing ligand in experimental autoimmune encephalomyelitis. Journal of Immunology *166*, 1314–9.

Hu, W. H., Johnson, H., and Shu, H. B. (1999). Tumor necrosis factor-related apoptosis-inducing ligand receptors signal NF-kappaB and JNK activation and apoptosis through distinct pathways. J Biol Chem *274*, 30603–10.

Irwin, M. S., and Kaelin, W. G. (2001). p53 family update: p73 and p63 develop their own identities. Cell Growth & Differentiation *12*, 337–49.

Jacks, T., Remington, L., Williams, B. O., Schmitt, E. M., Halachmi, S., Bronson, R. T., and Weinberg, R. A. (1994). Tumor spectrum analysis in p53-mutant mice. Current Biology *4*, 1–7.

Jeremias, I., Herr, I., Boehler, T., and Debatin, K. M. (1998). TRAIL/Apo-2-ligand-induced apoptosis in human T cells. European Journal of Immunology *28*, 143–52.

Jo, M., Kim, T. H., Seol, D. W., Esplen, J. E., Dorko, K., Billiar, T. R., and Strom, S. C. (2000). Apoptosis induced in normal human hepatocytes by tumor necrosis factor-related apoptosis-inducing ligand. Nat Med *6*, 564–7.

Kaelin, W. G., Jr. (1999). The p53 gene family. Oncogene *18*, 7701–5.

Kang, S. M., Schneider, D. B., Lin, Z., Hanahan, D., Dichek, D. A., Stock, P. G., and Baekkeskov, S. (1997). Fas ligand expression in islets of Langerhans does not confer immune privilege and instead targets them for rapid destruction. Nature Medicine *3*, 738–43.

Kassiotis, G., Kollias, G., and Calida, D. M. (2001). Uncoupling the proinflammatory from the immunosuppressive properties of tumor necrosis factor (TNF) at the p55 TNF receptor level: implications for pathogenesis and therapy of autoimmune demyelination. Journal of Experimental Medicine *193*, 427–34.

Korner, H., Riminton, D. S., Strickland, D. H., Lemckert, F. A., Pollard, J. D., and Sedgwick, J. D. (1997). Critical points of tumor necrosis factor action in central nervous system autoimmune inflammation defined by gene targeting. Journal of Experimental Medicine *186*, 1585–90.

Kuchroo, V. K., Das, M. P., Brown, J. A., Ranger, A. M., Zamvil, S. S., Sobel, R. A., Weiner, H. L., Nabavi, N., and Glimcher, L. H. (1995). B7-1 and B7-2 costimulatory molecules differentially activate the TH1/TH2 developmental pathways: application to autoimmune disease therapy. Cell *80*, 707–18.

Kumar-Sinha, C., Varambally, S., Sreekumar, A., and Chinnaiyan, A. M. (2002). Molecular cross-talk between the TRAIL and interferon signaling pathways. J Biol Chem *277*, 575–85.

Ladiwala, U., Li, H., Antel, J. P., and Nalbantoglu, J. (1999). p53 induction by tumor necrosis factor-alpha and involvement of p53 in cell death of human oligodendrocytes. Journal of Neurochemistry *73*, 605–11.

Levrero, M., De Laurenzi, V., Costanzo, A., Gong, J., Wang, J. Y., and Melino, G. (2000). The p53/p63/p73 family of transcription factors: overlapping and distinct functions. Journal of Cell Science *113*, 1661–70.

Lin, Y., Devin, A., Cook, A., Keane, M. M., Kelliher, M., Lipkowitz, S., and Liu, Z. G. (2000). The death domain kinase RIP is essential for TRAIL (Apo2L)-induced activation of IkappaB kinase and c-Jun N-terminal kinase. Mol Cell Biol *20*, 6638–45.

Liu, T. S., Hilliard, B., Samoilova, E. B., and Chen, Y. (2000). Differential roles of Fas ligand in spontaneous and actively induced autoimmune encephalomyelitis. Clinical Immunology *95*, 203–11.

Lynch, D. H., Campbell, K. A., Miller, R. E., Badley, A. D., and Paya, C. V. (1996). FasL/Fas and TNF/TNFR interactions in the regulation of immune responses and disease. Behring Inst Mitt, 175–84.

Magnusson, C., and Vaux, D. L. (1999). Signalling by CD95 and TNF receptors: not only life and death. Immunol Cell Biol *77*, 41–6.

Maini, R. N., Elliott, M. J., Brennan, F. M., Williams, R. O., Chu, C. Q., Paleolog, E., Charles, P. J., Taylor, P. C., and Feldmann, M. (1995). Monoclonal anti-TNF alpha antibody as a probe of pathogenesis and therapy of rheumatoid disease. Immunological Reviews *144*, 195–223.

Malinin, N. L., Boldin, M. P., Kovalenko, A. V., and Wallach, D. (1997). MAP3K-related kinase involved in NF-kappaB induction by TNF, CD95 and IL-1. Nature *385*, 540–4.

Mariani, S. M., and Krammer, P. H. (1998). Surface expression of TRAIL/Apo-2 ligand in activated mouse T and B cells. European Journal of Immunology *28*, 1492–8.

Martin-Villalba, A., Herr, I., Jeremias, I., Hahne, M., Brandt, R., Vogel, J., Schenkel, J., Herdegen, T., and Debatin, K. M. (1999). CD95 ligand (Fas-L/APO-1L) and tumor necrosis factor-related apoptosis-inducing ligand mediate ischemia-induced apoptosis in neurons. Journal of Neuroscience *19*, 3809–17.

Meng, R. D., McDonald, E. R., 3rd, Sheikh, M. S., Fornace, A. J., Jr., and El-Deiry, W. S. (2000). The TRAIL decoy receptor TRUNDD (DcR2, TRAIL-R4) is induced by adenovirus-p53 overexpression and can delay TRAIL-, p53-, and KILLER/DR5-dependent colon cancer apoptosis. Molecular Therapy: the Journal of the American Society of Gene Therapy *1*, 130–44.

Miller, S. D., and Karpus, W. J. (1994). The immunopathogenesis and regulation of T-cell-mediated demyelinating diseases. Immunology Today *15*, 356–61.

Miller, S. D., McRae, B. L., Vanderlugt, C. L., Nikcevich, K. M., Pope, J. G., Pope, L., and Karpus, W. J. (1995). Evolution of the T-cell repertoire during the course of experimental immune-mediated demyelinating diseases. Immunological Reviews *144*, 225–44.

Miller-Blair, D. J., and Robbins, D. L. (1993). Rheumatoid arthritis: new science, new treatment. Geriatrics *48*, 28–38.

Mills, A. A., Zheng, B., Wang, X. J., Vogel, H., Roop, D. R., and Bradley, A. (1999). p63 is a p53 homologue required for limb and epidermal morphogenesis. Nature *398*, 708–13.

Miyashita, T., and Reed, J. C. (1995). Tumor suppressor p53 is a direct transcriptional activator of the human bax gene. Cell *80*, 293–9.

Moon, C., Kim, S., Wie, M., Kim, H., Cheong, J., Park, J., Jee, Y., Tanuma, N., Matsumoto, Y., and Shin, T. (2000). Increased expression of p53 and Bax in the spinal cords of rats with experimental autoimmune encephalomyelitis. Neuroscience Letters *289*, 41–4.

Morrison, R. S., and Kinoshita, Y. (2000). The role of p53 in neuronal cell death. Cell Death & Differentiation *7*, 868–79.

Mountz, J. D., Edwards, C. K., 3rd, Cheng, J., Yang, P., Wang, Z., Liu, C., Su, X., Bluethmann, H., and Zhou, T. (1996). Autoimmunity due to defective Nur77, Fas, and TNF-RI apoptosis. Adv Exp Med Biol *406*, 241–62.

Muller, M., Wilder, S., Bannasch, D., Israeli, D., Lehlbach, K., Li-Weber, M., Friedman, S. L., Galle, P. R., Stremmel, W., Oren, M., and Krammer, P. H. (1998). p53 activates the CD95 (APO-1/Fas) gene in response to DNA damage by anticancer drugs. Journal of Experimental Medicine *188*, 2033–45.

Myers, L. K. (1993). Collagen-induced arthrits. In Current protocols in immunology, J. E. Coligan, ed. (New York: Sarah Greene), pp. 15.5.1–24.

Nagata, S., and Golstein, P. (1995). The Fas death factor. Science *267*, 1449–56.

Nagata, S., and Suda, T. (1995). Fas and Fas ligand: lpr and gld mutations. Immunology Today *16*, 39–43.

Ozawa, K., Suchanek, G., Breitschopf, H., Bruck, W., Budka, H., Jellinger, K., and Lassmann, H. (1994). Patterns of oligodendroglia pathology in multiple sclerosis. Brain *117*, 1311–22.

Pan, G., Ni, J., Wei, Y. F., Yu, G., Gentz, R., and Dixit, V. M. (1997). An antagonist decoy receptor and a death domain-containing receptor for TRAIL. Science *277*, 815–8.

Pan, G., Ni, J., Yu, G., Wei, Y. F., and Dixit, V. M. (1998). TRUNDD, a new member of the TRAIL receptor family that antagonizes TRAIL signalling. FEBS Letters *424*, 41–5.

Pan, G., O'Rourke, K., Chinnaiyan, A. M., Gentz, R., Ebner, R., Ni, J., and Dixit, V. M. (1997). The receptor for the cytotoxic ligand TRAIL. Science *276*, 111–3.

Panayi, G. S. (1993). The pathogenesis of rheumatoid arthritis: from molecules to the whole patient. British Journal of Rheumatology *32*, 533–6.

Pender, M. P., McCombe, P. A., Yoong, G., and Nguyen, K. B. (1992). Apoptosis of alpha beta T lymphocytes in the nervous system in experimental autoimmune encephalomyelitis: its possible implications for recovery and acquired tolerance. Journal of Autoimmunity *5*, 401–10.

Pesch, J., Brehm, U., Staib, C., and Grummt, F. (1996). Repression of interleukin-2 and interleukin-4 promoters by tumor suppressor protein p53. J Interferon Cytokine Res *16*, 595–600.

Polyak, K., Xia, Y., Zweier, J. L., Kinzler, K. W., and Vogelstein, B. (1997). A model for p53-induced apoptosis. Nature *389*, 300–5.

Ruddle, N. H. (1992). Tumor necrosis factor (TNF-alpha) and lymphotoxin (TNF-beta). Curr Opin Immunol *4*, 327–32.

Saas, P., Walker, P. R., Hahne, M., Quiquerez, A. L., Schnuriger, V., Perrin, G., French, L., Van Meir, E. G., de Tribolet, N., Tschopp, J., and Dietrich, P. Y. (1997). Fas ligand expression by astrocytoma in vivo: maintaining immune privilege in the brain? Journal of Clinical Investigation *99*, 1173–8.

Sabelko, K. A., Kelly, K. A., Mnahm, M. H., Cross, A. H., and Russell, J. H. (1997). Fas and Fas ligand enhance the pathogenesis of experimental allergic encephalomyelitis, but are not essential for immune privilege in the central nervous system. Journal of Immunology *159*, 3096–3099.

Sabelko-Downes, K. A., Cross, A. H., and Russell, J. H. (1999). Dual Role for Fas Ligand in the Initiation of and Recovery from Experimental Allergic Encephalomyelitis. J. Exp. Med. *189*, 1195–1205.

Sakhi, S., Bruce, A., Sun, N., Tocco, G., Baudry, M., and Schreiber, S. S. (1994). p53 induction is associated with neuronal damage in the central nervous system. Proceedings of the National Academy of Sciences of the United States of America *91*, 7525–9.

Salmon, M., and Gaston, J. S. (1995). The role of T-lymphocytes in rheumatoid arthritis. British Medical Bulletin *51*, 332–45.

Santhanam, U., Ray, A., and Sehgal, P. B. (1991). Repression of the interleukin 6 gene promoter by p53 and the retinoblastoma susceptibility gene product. Proc Natl Acad Sci U S A *88*, 7605–9.

Sartor, R. B., Rath, H. C., Lichtman, S. N., and van Tol, E. A. (1996). Animal models of intestinal and joint inflammation. Baillieres Clinical Rheumatology *10*, 55–76.

Schmied, M., Breitschopf, H., Gold, R., Zischler, H., Rothe, G., Wekerle, H., and Lassmann, H. (1993). Apoptosis of T lymphocytes in experimental autoimmune encephalomyelitis. Evidence for programmed cell death as a mechanism to control inflammation in the brain. American Journal of Pathology *143*, 446–52.

Schneider, P., Bodmer, J. L., Thome, M., Hofmann, K., Holler, N., and Tschopp, J. (1997). Characterization of two receptors for TRAIL. FEBS Letters *416*, 329–34.

Schneider, P., Thome, M., Burns, K., Bodmer, J. L., Hofmann, K., Kataoka, T., Holler, N., and Tschopp, J. (1997). TRAIL receptors 1 (DR4) and 2 (DR5) signal FADD-dependent apoptosis and activate NF-kappaB. Immunity *7*, 831–6.

Schumacher, H. R., Jr., Bautista, B. B., Krauser, R. E., Mathur, A. K., and Gall, E. P. (1994). Histological appearance of the synovium in early rheumatoid arthritis. Seminars in Arthritis & Rheumatism *23*, 3–10.

Screaton, G. R., Mongkolsapaya, J., Xu, X. N., Cowper, A. E., McMichael, A. J., and Bell, J. I. (1997). TRICK2, a new alternatively spliced receptor that transduces the cytotoxic signal from TRAIL. Current Biology *7*, 693–6.

Sewell, K. L., and Trentham, D. E. (1993). Pathogenesis of rheumatoid arthritis. Lancet *341*, 283–6.

Sheikh, M. S., Burns, T. F., Huang, Y., Wu, G. S., Amundson, S., Brooks, K. S., Fornace, A. J., Jr., and el-Deiry, W. S. (1998). p53-dependent and -independent regulation of the death receptor KILLER/DR5 gene expression in response to genotoxic stress and tumor necrosis factor alpha. Cancer Research *58*, 1593–8.

Sheikh, M. S., Huang, Y., Fernandez-Salas, E. A., El-Deiry, W. S., Friess, H., Amundson, S., Yin, J., Meltzer, S. J., Holbrook, N. J., and Fornace, A. J., Jr. (1999). The antiapoptotic decoy receptor TRID/TRAIL-R3 is a p53-regulated DNA damage-inducible gene that is overexpressed in primary tumors of the gastrointestinal tract. Oncogene *18*, 4153–9.

Sheridan, J. P., Marsters, S. A., Pitti, R. M., Gurney, A., Skubatch, M., Baldwin, D., Ramakrishnan, L., Gray, C. L., Baker, K., Wood, W. I., Goddard, A. D., Godowski, P., and Ashkenazi, A. (1997). Control of TRAIL-induced apoptosis by a family of signaling and decoy receptors. Science *277*, 818–21.

Simon, A. K., Williams, O., Mongkolsapaya, J., Jin, B., Xu, X. N., Walczak, H., and Screaton, G. R. (2001). Tumor necrosis factor-related apoptosis-inducing ligand in T cell development: sensitivity of human thymocytes. Proc Natl Acad Sci U S A *98*, 5158–63.

Sneller, M. C., Wang, J., Dale, J. K., Strober, W., Middelton, L. A., Choi, Y., Fleisher, T. A., Lim, M. S., Jaffe, E. S., Puck, J. M., Lenardo, M. J., and Straus, S. E. (1997). Clincial, immunologic, and genetic features of an autoimmune lymphoproliferative syndrome associated with abnormal lymphocyte apoptosis. Blood *89*, 1341–8.

Song, K., Chen, Y., Goke, R., Wilmen, A., Seidel, C., Goke, A., Hilliard, B., and Chen, Y. (2000). Tumor necrosis factor-related apoptosis-inducing ligand (TRAIL) is an inhibitor of autoimmune inflammation and cell cycle progression. J Exp Med *191*, 1095–104.

Sprick, M. R., Weigand, M. A., Rieser, E., Rauch, C. T., Juo, P., Blenis, J., Krammer, P. H., and Walczak, H. (2000). FADD/MORT1 and caspase-8 are recruited to TRAIL receptors 1 and 2 and are essential for apoptosis mediated by TRAIL receptor 2. Immunity *12*, 599–609.

Stuart, P. M., Griffith, T. S., Usui, N., Pepose, J., Yu, X., and Ferguson, T. A. (1997). CD95 ligand (FasL)-induced apoptosis is necessary for corneal allograft survival. Journal of Clinical Investigation *99*, 396–402.

Suvannavejh, G. C., Dal Canto, M. C., Matis, L. A., and Miller, S. D. (2000). Fas-mediated apoptosis in clinical remissions of relapsing experimental autoimmune encephalomyelitis. Journal of Clinical Investigation *105*, 223–31.

Tanaka, H., Ota, K., Ikusaka, M., Ejima, M., and Maruyama, S. (1995). Expression of Fas-antigen on T cells in multiple sclerosis. Rinsho Shinkeigaku—Clinical Neurology *35*, 299–301.

Utrera, R., Collavin, L., Lazarevic, D., Delia, D., and Schneider, C. (1998). A novel p53-inducible gene coding for a microtubule-localized protein with G2-phase-specific expression. Embo J *17*, 5015–25.

Walczak, H., Degli-Esposti, M. A., Johnson, R. S., Smolak, P. J., Waugh, J. Y., Boiani, N., Timour, M. S., Gerhart, M. J., Schooley, K. A., Smith, C. A., Goodwin, R. G., and Rauch, C. T. (1997). TRAIL-R2: a novel apoptosis-mediating receptor for TRAIL. EMBO Journal *16*, 5386–97.

Walczak, H., Miller, R. E., Ariail, K., Gliniak, B., Griffith, T. S., Kubin, M., Chin, W., Jones, J., Woodward, A., Le, T., Smith, C., Smolak, P., Goodwin, R. G., Rauch, C. T., Schuh, J. C., and Lynch, D. H. (1999). Tumoricidal activity of tumor necrosis factor-related apoptosis-inducing ligand in vivo. Nature Medicine *5*, 157–63.

Waldner, H., Sobel, R. A., Howard, E., and Kuchroo, V. K. (1997). Fas- and FasL-deficient mice are resistant to induction of autoimmune encephalomyelitis. Journal of Immunology *159*, 3100–03.

Wallach, D., Boldin, M., Goncharov, T., Goltsev, Y., Mett, I., Malinin, N., Adar, R., Kovalenko, A., and Varfolomeev, E. (1996). Exploring cell death mechanisms by analyzing signaling cascades of the TNF/NGF receptor family. Behring Inst Mitt, 144–55.

Ware, C. F., VanArsdale, S., and VanArsdale, T. L. (1996). Apoptosis mediated by the TNF-related cytokine and receptor families. J Cell Biochem *60*, 47–55.

Wiley, S. R., Schooley, K., Smolak, P. J., Din, W. S., Huang, C. P., Nicholl, J. K., Sutherland, G. R., Smith, T. D., Rauch, C., Smith, C. A., and Goodwin, R. G. (1995). Identification and characterization of a new member of the TNF family that induces apoptosis. Immunity *3*, 673–82.

Wu, G. S., Burns, T. F., McDonald, E. R. r., Jiang, W., Meng, R., Krantz, I. D., Kao, G., Gan, D. D., Zhou, J. Y., Muschel, R., Hamilton, S. R., Spinner, N. B., Markowitz, S., Wu, G., and el-Deiry, W. S. (1997). KILLER/DR5 is a DNA damage-inducible p53-regulated death receptor gene. Nature Genetics *17*, 141–3.

Wu, G. S., Burns, T. F., Zhan, Y., Alnemri, E. S., and El-Deiry, W. S. (1999). Molecular cloning and functional analysis of the mouse homologue of the KILLER/DR5 tumor necrosis factor-related apoptosis-inducing ligand (TRAIL) death receptor. Cancer Research *59*, 2770–5.

Xerri, L., Devilard, E., Hassoun, J., Mawas, C., and Birg, F. (1997). Fas ligand is not only expressed in immune privileged human organs but is also coexpressed with Fas in various epithelial tissues. Molecular Pathology *50*, 87–91.

Yamanishi, Y., Boyle, D. L., Pinkoski, M. J., Mahboubi, A., Lin, T., Han, Z., Zvaifler, N. J., Green, D. R., and Firestein, G. S. (2002). Regulation of Joint Destruction and Inflammation by p53 in Collagen-Induced Arthritis. Am J Pathol *160*, 123–30.

Yamanishi, Y., Boyle, D. L., Rosengren, S., Green, D. R., Zvaifler, N. J., and Firestein, G. S. (2002). Regional analysis of p53 mutations in rheumatoid arthritis synovium. Proceedings of the National Academy of Sciences of the United States of America *99*, 10025–30.

Yang, A., Walker, N., Bronson, R., Kaghad, M., Oosterwegel, M., Bonnin, J., Vagner, C., Bonnet, H., Dikkes, P., Sharpe, A., McKeon, F., and Caput, D. (2000). p73-deficient mice have neurological, pheromonal and inflammatory defects but lack spontaneous tumours. Nature *404*, 99–103.

Yao, Q., Wang, S., Glorioso, J. C., Evans, C. H., Robbins, P. D., Ghivizzani, S. C., and Oligino, T. J. (2001). Gene transfer of p53 to arthritic joints stimulates synovial apoptosis and inhibits inflammation. Molecular Therapy *3*, 901–10.

Zamvil, S. S., and Steinman, L. (1990). The T lymphocyte in experimental allergic encephalomyelitis. Ann. Rev. Immunol. *8*, 579–621.

Zhan, Q., Bae, I., Kastan, M. B., and Fornace, A. J., Jr. (1994). The p53-dependent gamma-ray response of GADD45. Cancer Res *54*, 2755–60.

Zhang, X. D., Zhang, X. Y., Gray, C. P., Nguyen, T., and Hersey, P. (2001). Tumor necrosis factor-related apoptosis-inducing ligand-induced apoptosis of human melanoma is regulated by smac/DIABLO release from mitochondria. Cancer Res *61*, 7339–48.

Chapter 7

Oxidative Stress and Thymocyte Apoptosis

NORIKO TONOMURA,[1,2,6] RICHARD A. GOLDSBY,[3]
ERIC V. GRANOWITZ,[4,5] AND BARBARA OSBORNE[1,2]

ABSTRACT: Mitochondria play a major role in making decisions in programmed cell death. During thymocyte apoptosis, the function of mitochondrial electron transport chain becomes pro-apoptotic and produces increased levels of reactive oxygen species (ROS). The resulting oxidative stress can further aggravate apoptotic events, leading thymocytes to death. The site of ROS production during thymocyte apoptosis is most likely at complex III of the electron transport chain, and the pro-apoptotic function of electron transport chain is regulated by the proteasome.

Three Phases of Apoptosis

Programmed cell death has been extensively studied in immune cells because of their tightly regulated and rapid cell expansion and their diminution by apoptosis. Through these studies, it is now known that mitochondria play a major role in regulating apoptosis and, cell death can be divided into three phases relative to mitochondria: the initiation phase, the decision phase, and the execution or degradation phase (reviewed by Kroemer and Reed, 2000). During the initiation phase, most of the extracellular pro-apoptotic signals are delivered to mitochondria by intracellular second messengers. In the second phase, a range of proteins with apoptosis-regulatory functions that are usually confined within mitochondria are released into the cytoplasm as a result of mitochondrial outer membrane permeabilization. During this phase, the reduction of mitochondrial inner membrane potential ($\Delta\Psi_m$) is also observed, indicating the permeabilization of the inner membrane. This phase has been considered critical in the determination of cell fate. Finally, in the execution phase, the apoptosis-regulatory proteins of mitochondrial origin and the metabolic consequences of mitochondrial homeostasis disruption mediate the activation of caspases and nucleases, and the irreversible loss of cellular functions. In addition, as studies of cell death have progressed, it was also discovered that the regulatory mechanisms underlying necrosis partially overlap with those of apoptosis at mitochondrial level, implying that mitochondria may play a global role in regulating both apoptosis and necrosis (review by Kroemer and Reed, 2000).

[1]Transplantation Biology Research Center, Massachusetts General Hospital, Harvard Medical School, Boston, MA 02129; [2]The Program in Molecular and Cellular Biology, University of Massachusetts, Amherst, MA, 01003; [3]Department of Biology, Amherst College, Amherst, MA, 01002; [4]Department of Medicine, Baystate Medical Center, Springfield, MA, 01199; [5]Tufts University School of Medicine, Boston, MA, 02115; [6]*Correspondence:* tonomura@tbrc.mgh.harvard.edu

Mitochondrion as a Death Integrator

In early studies of apoptosis, the nucleus was considered to be the key regulator of the apoptotic process. However, as mentioned above, recent studies of the regulation of apoptosis have revealed that the mitochondrion is an integrator of cell death pathways for both apoptosis and necrosis (review by Kroemer and Reed, 2000). Numerous data show that while extracellular cell-death-inducing stimuli may signal through different molecular messengers, all of which appear to converge on the mitochondria to activate/release common effectors that initiate the actual killing processes. The signaling pathways upstream of the mitochondria may involve a cascade of signaling molecules, and those that reach the mitochondria affect mitochondrial membrane integrity, causing permeabilization of mitochondrial membranes. This mitochondrial membrane permeabilization is observed prior to other signs of advanced apoptosis or necrosis, irrespective of the cell type or the death-inducing stimulus, and various observations suggest that mitochondrial membrane permeabilization is involved in cell death commitment (reviewed by Green and Reed, 1998, Kroemer, 1998, Gross et al., 1999, Vander and Thompson 1999). Mitochondrial membrane permeabilization is accompanied and/or controlled by other mitochondrial cell death events such as depolarization of the mitochondrial inner membrane potential ($\Delta\Psi_m$), temporary matrix alkalinization, and release of mitochondrial proteins, including cytochrome c (Cyt c) to the cytosol, which would then mediate catabolic catastrophe in the execution phase.

Mitochondrial Membranes in Normal Cells

Mitochondria are comprised of an inner membrane and outer membrane, which forms two compartments: the matrix surrounded by the inner membrane, and the intermembrane space surrounded by the outer membrane. The inner membrane is folded into cristae, providing a large surface area for numerous intramembrane proteins, including complexes for the electron transport chain, ATP-synthesizing enzyme (ATP synthase) and adenine nucleotide translocator (ANT) that exchange ATP and ADP. In the respiring cell, electrons taken from molecules such as sugars, amino acids and lipids by various catabolic reactions are transferred from one carrier to another in the electron transport chain at the mitochondria, finally reducing oxygen to water. Overall, the transfer of electrons from one carrier to another is very favorable, with an equilibrium constant that is overwhelmingly large. At the three sites of the transport chain where the energy for ATP synthesis is generated, namely complex I, II and III, the reaction is the most favorable. These complexes contain proton pumps that force protons out of the matrix. Since the inner membrane is almost impermeable to solutes and ions under physiological conditions, a proton gradient is created at the inner membrane, keeping the matrix more alkaline. Together with other ions asymmetrically distributed across the inner membrane, the proton gradient produces electrochemical gradients, establishing a resting potential of 100–200 mV across the inner membrane (negative inside) (reviewed by Hatefi, 1985). This is the mitochondrial inner membrane potential ($\Delta\Psi_m$), which is indispensable for ATP production by ATP synthase.

In contrast to the inner membrane, the outer membrane is completely permeable to protons, presumably because of the voltage-dependent anion channel (VDAC). VDAC is the most abundant protein found in the outer membrane and is a member of the porin family of proteins. Porin is a b-sheet-type transmembrane channel protein, which transports solutes, metabolites including ADP and ATP, and proteins up to about 5,000 Da in its open configuration. Even in the closed configuration, VDAC creates a pore of approximately 1.8 Å in diameter, large enough for protons and other biologically relevant ions to pass through (Mannella et al., 1992). Therefore, the intermembrane space of mitochondria is chemically equivalent to the cytoplasm in regard to solutes with low molecular weight. With respect to proteins, about 99% of the mitochondrial proteins are encoded by the nuclear genome and translated in the cytoplasm. However, these proteins have mitochondrial targeting signals and are transported into the mitochondria, where they remain confined (reviewed by Schleiff, 2000). Therefore, mitochondria have a highly selective and unique set of proteins in the intermembrane space as well as in the matrix. As studies of the role of mitochondria in apoptosis have advanced, more than 50 different mitochondrial proteins have been found to be released into the cytoplasm, where some of them mediate catabolic catastrophe that leads to cell death (Patterson, 1999).

Mitochondrial Membrane Permeabilization

In order for these numerous proteins to enter the cytoplasm, the mitochondrial outer membrane must become permeable to larger molecules. Since these proteins mediate the downstream catabolic catastrophe, the permeabilization of the outer membrane is an imperative event in the process of apoptosis. The permeabilization of the inner membrane is also observed, however, it occurs in a more selective manner in regard to the matrix contents that are released. The inner membrane becomes permeable to solutes up to 1,500 Da, such as protons and other ions, which can equilibrate freely across the inner membrane, collapsing proton and other ion gradients, and resulting in a loss of $\Delta\Psi_m$ (reviewed by Kroemer and Reed, 2000). The depolarization of $\Delta\Psi_m$ is observed in most apoptotic cells and thus has been used as an indicator of the mitochondrial membrane permeability transition stage. Once the $\Delta\Psi_m$ collapses, the $\Delta\Psi_m$–dependent functions of mitochondria, such as import of mitochondrial proteins and metabolites and ATP production, become compromised and affect cell viability. In addition, the influx of solutes, ions, and water causes matrix expansion, stretching the inner membrane cristae, which can lead to ruptures of the outer membrane. Some sort of channel or pore, usually referred as the permeability transition pore, is thought to mediate the permeabilization of the both membranes (reviewed by Green and Reed, 1998), however, in some cases of apoptosis, it has been shown that Cyt c release occurs before or without the dissipation of $\Delta\Psi_m$, and without overt disturbance in mitochondrial morphology or functions (Kluck et al., 1997, Yang et al., 1997). Therefore, permeabilization of the outer and the inner membranes seems to be controlled and mediated separately, either by common or different mechanisms. Studies of the mechanisms of mitochondrial membrane permeabilization in different cell types have lead to several models of mitochondrial membrane permeabilization. All of the models suggest that it is not just one kind of pore that mediates permeability transition of both membranes, but instead, there are different types of channels that consist of different combinations of Bcl-2 family members, VDAC, and/or ANT. These channels regulate the permeabilization of the membrane(s) and, therefore, are most likely to control release of mitochondrial intermembrane proteins (reviewed by Kroemer and Reed, 2000, reviewed by Vieira et al., 2000). The inconsistent results regarding whether or not cells exhibit mitochondrial swelling, or exhibit susceptibility to a certain chemical treatment regulating mitochondrial membrane permeability, are most likely due to effects on components of channels that are regulated by different mechanisms.

Reactive Oxygen Species (ROS) in T Cell Apoptosis

The involvement of ROS in apoptosis has gained attention in the early 1990's. Since the discovery that phagocytes and granulocytes utilize H_2O_2 for killing of invading pathogens, involvement of H_2O_2 in causing cell death and tissue damage had been known for some time (reviewed by Klebanoff, 1980). The discovery that Bcl-2 functions in an antioxidant pathway to prevent apoptosis has lead to a hypothesis that ROS play a critical role in apoptosis, although the mechanism has not been well understood (Hockenbery et al., 1993). The requirement for generating H_2O_2 in signal transduction was made clear when Finkel et al. examined the involvement of H_2O_2 in platelet-derived growth factor-mediated signaling pathway in 1995 (Sundaresan et al., 1995). Since this discovery, the view of the role that ROS play in inducing apoptosis has been slowly shifting from one of a promiscuous cytotoxic byproduct to that of a potential signaling molecule.

In the case of T cell and thymocytes apoptosis, the involvement of the intracellular redox status and ROS was first suggested by the observations that increased levels of ROS and/or depletion of intracellular antioxidants were observed upon induction of apoptosis (Castedo et al., 1995, Zanzami et al., 1995), and the depletion of molecular oxygen and treatment of thymocyte with N-acetyle-L-cysteine (NAC) inhibited apoptosis in thymocytes *in vitro* (McLaughlin et al., 1996). However, thymocytes are far less oxygenated *in vivo*, having less than10 mmHg of oxygen tension, compared to those cultured *ex vivo* in 5% CO_2 in air (21% O_2) incubator, where the cells maintain oxygen tension above 10 mmHg at density of 1×10^7 cells/ml (Braun et al., 2001, Hale et al., 2002). We have shown a physiological relevance for NAC protection against murine thymocyte apoptosis by an *in vivo* experiment, in which 4-week-old BALB/c female mice were given two i.p. injections of NAC, followed by a injection of dexamethasone (DEX), a synthetic glucocorticoid, to induce apoptosis in the thymus. The result showed that as more death was observed in thymocytes over time, the protective effect of NAC became more apparent, strongly suggesting that ROS, most likely H_2O_2, play a critical role in DEX-induced thymocyte apoptosis *in vivo* (Figure 1).

We also find that *ex vivo*, apoptotic thymocytes experience increased levels of intracellular H_2O_2, which are reduced by the treatment with NAC (Figure 2A). The impact of the increased levels of the intracellular H_2O_2 on other apoptotic events, such as the loss of $\Delta\Psi_m$, Cyt c release, and caspase-3 activity is pro-apoptotic, which is made evident by the reduction in these events by the NAC treatment, except for the g-irradiated thymocytes (Figure 2B–D). However, lowering intracellular H_2O_2 to the untreated control level does not provide 100% protection against apoptosis, suggesting that apoptosis can occur, although at lesser extent, without increased levels of H_2O_2 in thymocytes. Therefore, the higher levels of intracellular H_2O_2 are pro-apoptotic, but do not appear to be required in thymocyte apoptosis.

In addition to our results, other studies strongly suggest the intracellular oxidative status does influence cellular susceptibility of T cells and thymocytes to apoptosis: i) depletion of molecular oxygen inhibits apoptosis (McLaughlin et al., 1996, Torres-Roca et al., 2000), ii) increased levels of ROS and/or depletion of intracellular antioxidants are observed upon induction of apoptosis (Castedo et al., 1995, Zanzami et al., 1995, Macho et al., 1997, Hildeman et al., 1999), iii) treatment of cells with ROS induces apoptosis (Ramakrishnan et al., 1998), and iv) overexpression of an antioxidant or elevated levels of intracellular antioxidants inhibits apoptosis (McLaughlin et al., 1996, Ramakrishnan et al., 1998, Hildeman et al., 1999, Tome et al., 2001, Tome and Briehl, 2001).

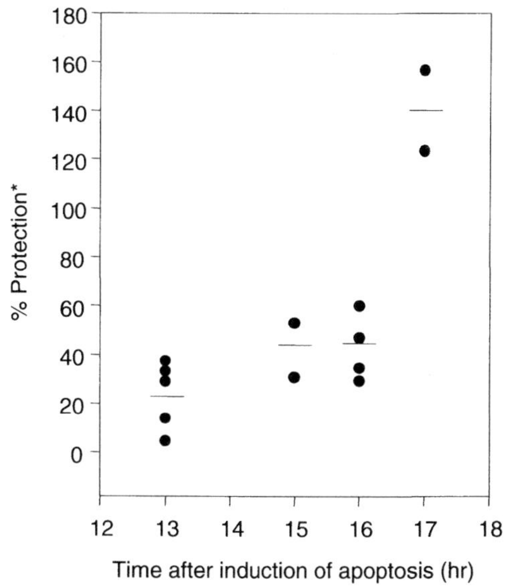

Figure 1. NAC Protects DEX-induced Thymocyte Apoptosis *in vivo*. One-month-old female BALB/c mice were given two doses of NAC or PBS (NAC control) intraperitonealy 6 hours and 1 hour before the DEX or 16% ethanol in PBS (EtHO/PBS: DEX control) injection. The four experimental groups PBS/(EtOH/PBS) and NAC/(EtOH/PBS), PBS/DEX and NAC/DEX contained a minimum of 2 mice per group. At 13, 15, 16, or 17 hours after the DEX injection, mice were sacrificed and thymocytes were isolated to measure apoptosis. The percentage of protection was calculated as below for standardized comparison.

*% Protection = 100 − [% apoptosis in NAC/DEX −% apoptosis in NAC/(EtOH/PBS)] × 100 % apoptosis in PBS/DEX − % apoptosis in PBS/(EtOH/PBS)

Reactive Oxygen Species Production in Mitochondria

For aerobic organisms, both life and death seem to depend upon oxygen. Aerobic organisms use oxygen as a terminal electron acceptor during the production of ATP in the mitochondria. This process is known as respiration or respiratory energy production. The majority of reactive oxygen species (ROS) are produced as byproducts of respiration at the inner mitochondrial membrane (Boveris and Chance, 1973). At the inner mitochondrial membrane, high-energy electrons in the electron transport chain spontaneously leak in close proximity to O_2, generating the superoxide anion (O_2^-). It has been calculated that about 2% of 10^{12} oxygen molecules consumed forms ROS, and about 2% the ROS formed damages proteins, lipids and DNA (Boveris and Chance, 1973, reviewed by Acworth and Bailey, 1995). When ROS react with macromolecules, the initial reaction generates a second radical, which then reacts with another macromolecule, creating a radical-forming chain reaction. This cascade of cytotoxic reactions that cause damage is the hallmark of oxidative stress. It is this damage that makes ROS dangerous to living cells, especially when the levels of ROS exceed the cells' natural defense capabilities. The three major types of ROS found in living cells are O_2^-, hydrogen peroxide (H_2O_2), and hydroxyl radical (OH^-). Dismutation of superoxide anion by mitochondrial manganese-superoxide dismutase (Mn-SOD) produces hydrogen peroxide, and subsequent reaction of O_2^- with H_2O_2 in the presence of copper ion (Haber-Weiss reaction), or reaction between H_2O_2 with redox active metal ions such as

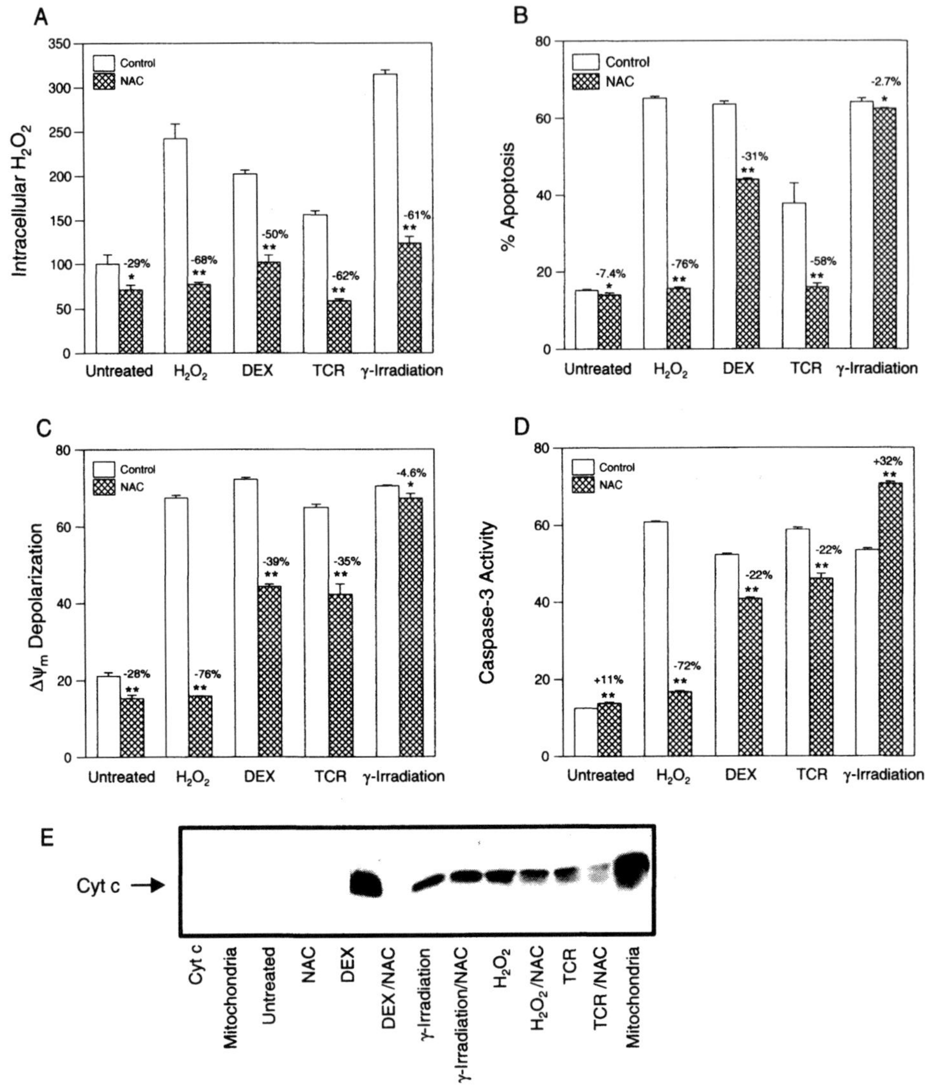

Figure 2. NAC Protects Thymocyte Apoptosis *ex vivo*. To address the influence of intercellular H_2O_2 on apoptotic events, the following parameters were measured in the presence/absence of NAC (50 mM). (A) Intercellular H_2O_2 levels measured by H_2DCFDA-di(acetoxymethyl ester). (B) Levels of apoptosis measured by Annexin V-FITC staining. (C) Mitochondrial inner membrane potential ($\Delta\Psi_m$) with $DiOC_6$. (D) Caspase-3 activity measured by PhiPhiLux$^{TM}G_2D_2$. (E) The amount of cytoplasmic Cyt c visualized by western blot analysis. Purified rat Cyt c and mitochondrial fraction were used as positive controls. Data shown is the mean ± SD of 3 samples, and representing at least 3 experiments. The numbers above asterisk(s) indicate % differences compared to each control.

* : $p < 0.05$, ** : $p < 0.01$.

Fe^{2+} ion (Fenton reaction) produces OH^- as shown in Figure 3 (reviewed by Acworth and Bailey, 1995). In most cases, O_2^- is the first ROS species generated in the living cell, and this often starts a cascade of reactions by which other ROS are generated.

The overall balance between intracellular ROS production and antioxidant defense mechanisms determines the redox status of a cell. Such antioxidant defense mechanisms

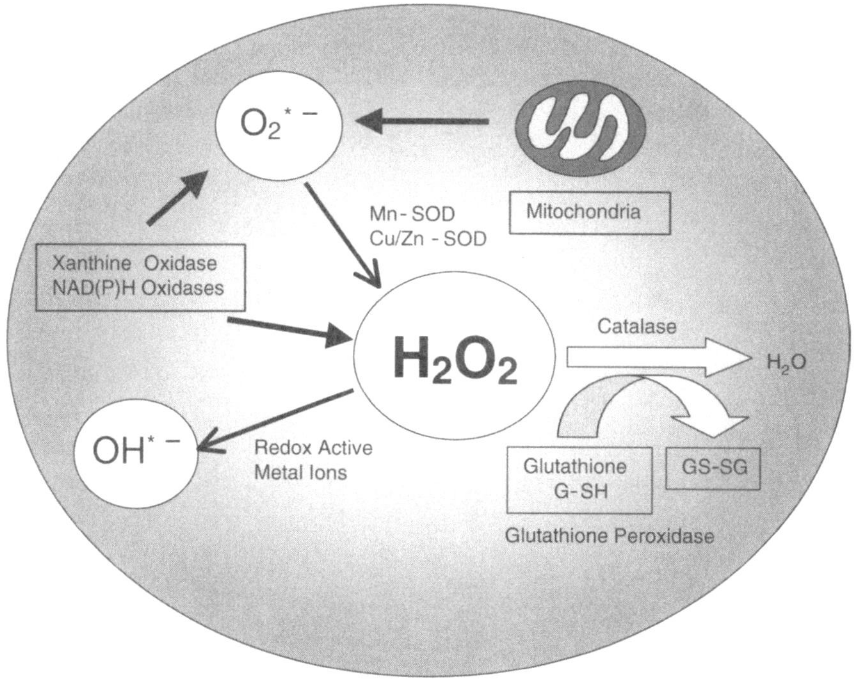

Figure 3. Intracellular Production of Reactive Oxygen Species (ROS) and Antioxidant Defense Mechanisms. See text for details.

include the enzymatic antioxidants, copper/zinc or manganese-superoxide dismutase (Cu/Zn or Mn-SOD) which convert O_2^- to H_2O_2, and catalase and peroxidases, which exhibit antioxidant activity against H_2O_2 by reducing it to water (Figure 3). The balance between the generation of ROS and the cellular antioxidant activities determines the level of oxidative stress. When the intracellular oxidative activities overwhelm the reducing equivalents, it shifts cells into a state of oxidative stress, leading to the activation of several stress signaling pathways (reviewed by Finkel and Holbrook, 2000). Although the molecular mechanisms by which ROS initially activate these pathways are not yet clear, once activation occurs, these pathways lead to various consequences, including proliferation, growth arrest, senescence and apoptosis (reviewed by Finkel and Holbrook, 2000).

Through the studies of apoptotic regulation, the functions of Bcl-2 family members and Cyt c have been highlighted in the context of cell death, but under normal circumstances, they might also play important roles in maintaining the mitochondrial homeostasis and oxidative stress levels (reviewed by Skulachev 1998, Korshunov et al., 1999, reviewed by Vander Heiden and Thompson, 1999, reviewed by Matsuyama and Reed 2000). Bcl-2 family members are involved in the regulation of VDAC and mitochondrial outer membrane permeability (reviewed by: Vander Heiden and Thompson, 1999, Harris and Thompson, 2000, Tsujimoto and Shimizu, 2000, Reed and Kroemer 2000). Skulachev et al. have shown that soluble Cyt c oxidizes O_2^- to O_2 by accepting an electron, and the release of Cyt c to the intermembrane space of mitochondria to control increased production of O_2^- has been observed *in vivo*. Furthermore, addition of Cyt c to the mitochondria strongly suppresses the formation of H_2O_2 (reviewed by Skulachev 1998, Korshunov et al., 1999). Therefore,

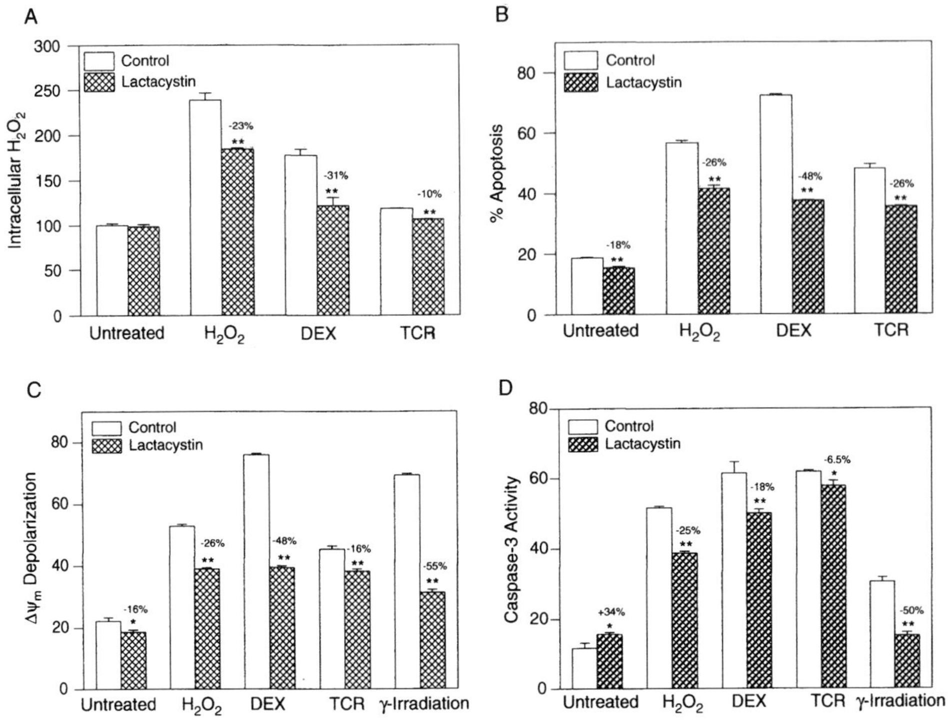

Figure 4. Involvement of Mitochondrial Electron Transport Chain in Thymocyte Apoptosis. To address the involvement of electron transport chain in thymocyte apoptosis, the following parameters were measured in the presence/absence of antimycin A (1 mM). (A) Intercellular H_2O_2 levels measured by CM-H_2DCFDA. (B) Levels of apoptosis measured by Annexin V-FITC staining. (C) Mitochondrial inner membrane potential ($\Delta\Psi_m$) with $DiOC_6$. (D) Caspase-3 activity measured by PhiPhiLux$^{TM}G_2D_2$. Data shown is the mean ± SD of 3 samples, and represents at least 3 experiments. The numbers above asterisk(s) indicate % differences compared to each control. * : $p < 0.05$, ** : $p < 0.01$.

under non-stressed condition, anti-apoptotic Bcl-2 family members and Cyt c could be part of an antioxidant defense system at the site of the mitochondria, keeping oxidative stress in check.

As described earlier, mitochondria play a major regulatory role in apoptosis induced by a variety of stimuli, and these stimuli all converge at the mitochondria through a signaling pathway unique to each stimulus in the initiation phase (reviewed by: Kroemer and Reed, 2000, Gerri and Kroemer, 2001). Accumulating evidence suggests that mitochondrial homeostasis is important in making the commitment to die, and for the progression to the execution phase of apoptosis (reviewed by: Vander Heiden and Thompson, 1999, Matsuyama and Reed, 2000, Von Ahsen et al., 2000). Mitochondrial homeostasis is maintained through the regulation of permeability of the mitochondrial inner and outer membranes, energy production, oxygen consumption and proton redistribution by the electron transport chain, and trans-membrane transportation of ions and molecules including Cyt c (reviewed by: Vander Heiden and Thompson, 1999, Kroemer and Reed, 2000, Matsuyama and Reed, 2000, Von Ahsen et al., 2000).

Since mitochondria are the major site for ROS production in healthy cells and mitochondrial contribution to the progression of apoptosis indicates some disturbances in normal mitochondrial functions, it is reasonable to suggest that the mitochondria might be involved

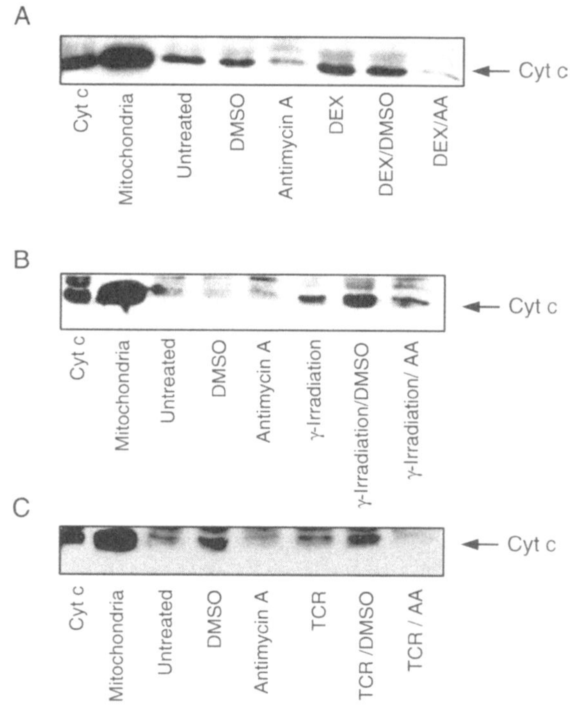

Figure 5. Antimycin A Inhibits Cyt c Release in Thymocyte Apoptosis. The inhibitory effect of antimycin A on Cyt c release in (A) DEX treated, (B) g -irradiated, and (C) TCR-stimulated thymocytes was assessed by western blot analysis of cytoplasmic Cyt c in each condition. Purified rat Cyt c and mitochondrial fraction were used as positive controls.

in causing the increased oxidative stress during apoptosis. We tested this hypothesis by using mitochondrial electron transport chain inhibitors, which block the flow of electrons at specific sites. Blocking the electron flow upstream of the site of O_2^- formation caused by electron leakage would also reduce the rate of H_2O_2 production at the mitochondria. The increased levels of H_2O_2, as well as levels of apoptosis, loss of $\Delta\Psi_m$, caspase-3 activation and Cyt c release in apoptotic thymocytes were significantly reduced by rotenone, a NADH dehydrgenase inhibitor (complex I) and antimycin A, ubiquinol-Cyt c oxidoreductase (complex III) inhibitor (Figure 4 and 5), but not by azide, a complex IV inhibitor (data not shown). Our data suggest that intracellular H_2O_2 in apoptotic thymocytes is generated by the electron transport chain at the complex III. The function(s) of the electron transport chain might be necessary to control the oxidative stress, since given with the inhibitors, exogenously added H_2O_2 aggravated all the apoptotic events measured.

Antimycin A inhibits electron transfer within the ubiquinone cycle of the complex III at a specific site. Ubiquinone, also called coenzyme Q is a two-electron carrier, which has three redox states. Ubiquinone with one electron, called ubisemiquinone anion, is generated as an intermediate at Q_o site and Q_i site, the ubiquinone binding sites located at the complex III. In living cells, the majority of ROS are produced by ubisemiquinone anions (reviewed by Finkel and Holbrook, 2000). Antimycin A inhibits electron transfer from Cyt b_{566} to ubiquinone at the Q_i site (Kim et al., 1998), blocking the formation of ubisemiquinone anion at the Q_i site. Therefore, our data strongly suggest that the increased production of H_2O_2

by death-inducing signals is most likely mediated by the formation of O_2^- at the ubiquinone cycle via formation of ubisemiquinone anions at the Q_i site.

Proteasomal Contribution to Apoptosis

The proteasome is a large, multi-catalytic complex consisting of over twenty distinct peptides that has multiple enzymatic proteolytic activities (reviewed by Grimm and Osborne, 1999). An active 26S proteasome is composed of a 20S catalytic core and a 19S regulator/activator cap, which rapidly degrades various target proteins in highly regulated and selective manner. The substrates are commonly targeted by the addition of a chain of ubiquitin proteins via ATP-dependent reactions mediated by a series of enzymes (ubiquitinases) known as E1 (activating enzyme), E2 (conjugating enzyme), and E3 (ligating enzyme). The polyubiquitin chain on a substrate then signals to the active proteasome for the degradation of the substrate. The substrates for proteasome-dependent degradation comprise a wide variety of proteins ranging from damaged and abnormal proteins to functional proteins that play crucial roles in cell functions.

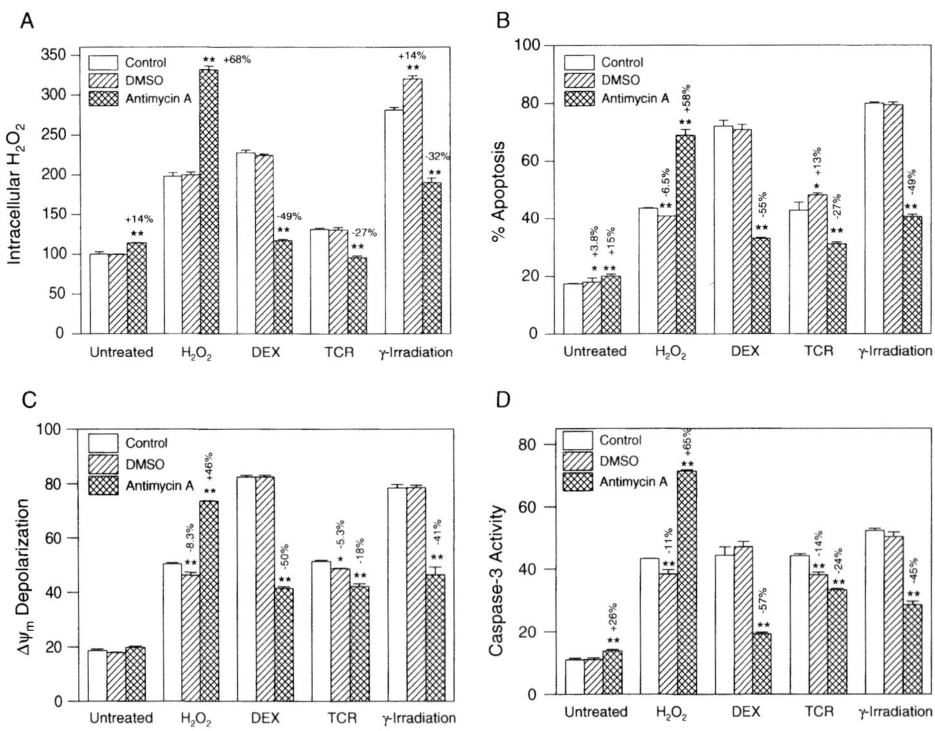

Figure 6. Proteasome Regulates Apoptotic Events in Thymocyte Apoptosis. Proteasomal regulation of apoptotic events was addressed by measuring the following parameters in the thymocytes in the presence/absence of lactacystin (10 mM). (A) Intercellular H_2O_2 levels measured by CM-H_2DCFDA. (B) Levels of apoptosis measured by Annexin V-FITC staining. (C) Mitochondrial inner membrane potential ($\Delta\Psi_m$) with $DiOC_6$. (D) Caspase-3 activity measured by PhiPhiLux$^{TM}G_2D_2$. Data shown is the mean ± SD of 3 samples, and represents at least 3 experiments. The numbers above asterisk(s) indicate % differences compared to each control.
* : $p < 0.05$, ** : $p < 0.01$.

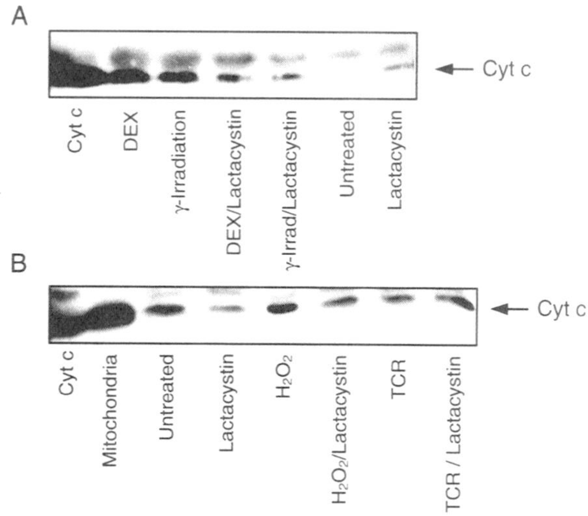

Figure 7. Proteasome Regulates Cyt c Release during Thymocyte Apoptosis. The inhibitory effect of lactacystin on Cyt c release in (A) DEX-treated or g -irradiated, and (B) H_2O_2-treated or TCR-stimulated thymocytes were assessed by western blot analysis of cytoplasmic Cyt c in each condition. Purified rat Cyt c and mitochondrial fraction were used as positive controls.

The discovery of caspases and their role in apoptosis has raised interest in examining the involvement of other proteolytic pathways in apoptosis. Grimm et al. have shown that the activities of the proteasome are required during DEX, phorbol myristate acetate (PMA), a phorbol ester, and g -irradiation-induced thymocyte apoptosis (Grimm et al., 1996). Inhibition of proteasome activities during thymocyte apoptosis results in reduced loss of $\Delta\Psi_m$. Furthermore, Grimm et al. found that the inhibiting proteasome functions prevented the cleavage of poly (ADP-ribose) polymerase (PARP), a caspase-3 substrate, suggested that the inhibition of proteasome activities might reduce the activation of caspase-3. Therefore, it was presumed that proteasomes play regulatory role upstream of caspase activation, possibly at or above the mitochondria in the thymocyte apoptotic pathways. We addressed the placement of proteasomes relative to mitochondria by using lactacystin, a peptide inhibitor specific to the protease function of proteasomes. Our data demonstrate that inhibition of proteasome activity reduced mitochondrial cell death events such as the depolarization of $\Delta\Psi_m$, H_2O_2 generation, Cyt c release, caspase-3 activation and apoptosis (Figure 6 and 7). These results demonstrate that in order for mitochondria to mediate cell death events, proteasome activity is necessary. Therefore, we conclude that the proteasome, at least in part, regulates death-inducing signals upstream of mitochondria in the apoptotic signaling pathway in thymocytes.

Conclusions

Our studies of oxidative stress in thymocyte apoptosis are summarized in Figure 8. Although more detailed molecular mechanisms are yet to be discovered, currently available data reviewed here establish the mitochondrial electron transport chain as a key mediator of increased oxidative stress and apoptosis in thymocyte apoptosis. When thymocytes are

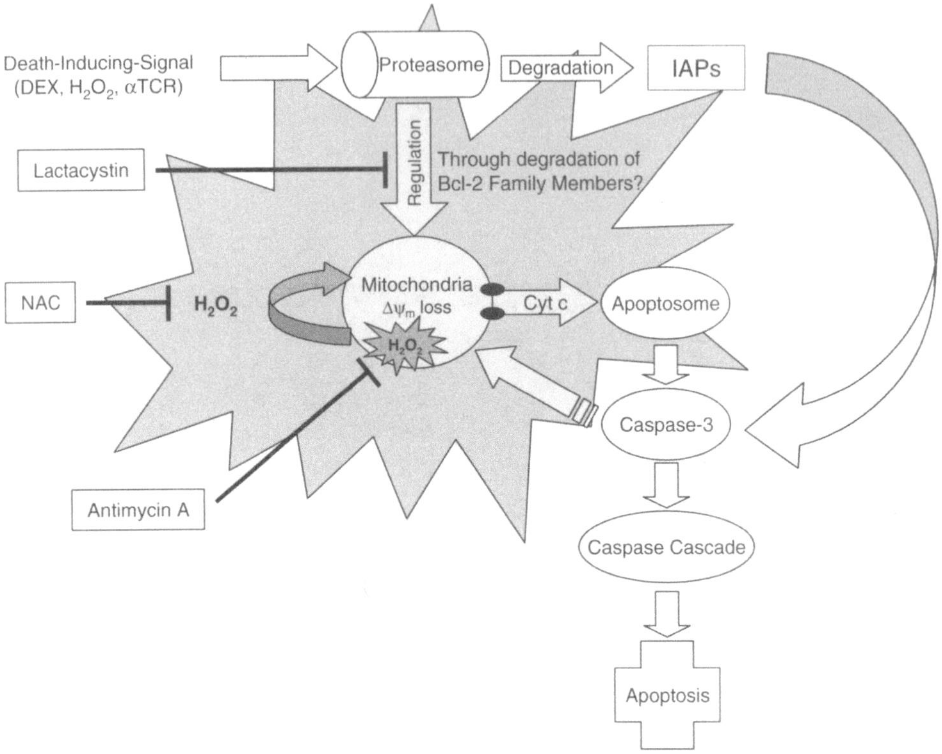

Figure 8. Hypothetical Model of Apoptosis Regulation in Thymocytes. A schematic drawing summarizing the data reviewed in this article and introducing hypothetical mechanisms by which proteasomes, mitochondria and oxidative stress may regulate thymocyte apoptosis. See text for details.

induced to die by apoptosis by various stimuli including DEX, g -irradiation, TCR engagement, and exogenous H_2O_2, thymocytes experience increased levels of intracellular H_2O_2. Therefore, as with in many different cell types, apoptotic thymocytes suffer from increased oxidative stress. Due to conflicting evidence showing apoptosis could proceed in the absence of oxygen, or in the presence of antioxidants (Hug et al., 1994, Sidoti-de Fraisse et al., 1998), scientists have debated for years whether or not oxygen and ROS are involved in the execution of T cell or thymocyte apoptosis. However, further studies conducted to determine the molecular mechanisms involved in apoptosis in T cells and thymocytes, revealed oxygen/ROS-independent apoptotic pathways that are also independent of mitochondrial control (reviewed by Krammer, 2000). This type of apoptosis is induced through death receptors, such as Fas and tumor necrosis factor (TNF) receptor (TNFR) or by granzyme B. All these apoptosis inducing pathways lead to the direct activation of procaspase-8, and activated caspase-8 activate the caspase cascade in a mitochondria-independent manner. However, thymocyte apoptosis induced by DEX, addition of exogenous H_2O_2, TCR-engagement or g -irradiation is all mitochondria-dependent, and all of these stimuli lead to the increased production of ROS at mitochondrial electron transport chain during apoptosis. The enhanced production of ROS in mitochondria results in the increased levels of intracellular H_2O_2, which could promote other apoptotic events downstream of mitochondrial events.

Although the elevated levels of intracellular H_2O_2 could be pro-apoptotic, it does not seem to be a requirement for the progression of apoptotic events in thymocytes. However, the function(s) of the electron transport chain, which is regulated by proteasome, is essential. One possible mechanism by which proteasomes regulate mitochondria is through the degradation of Bcl-2 family member(s). Evidence shows that proteasomes modulate the balance among pro- and anti-apoptotic Bcl-2 family members through selective degradation (Dimmeler et al., 1999, Li and Dou, 2000, Marshansky et al., 2001). Since most of the channels/pores that are thought to mediate mitochondrial membrane permeabilization contain Bcl-2 family member(s) as constituent(s), the modulation of the balance among pro- and anti-apoptotic Bcl-2 family members could lead to the compromised function of the electron transport chain (Marshansky et al., 2001). We cannot rule out other effects of the proteasome which could occur following the contribution of mitochondria. It is possible and indeed suggested by the work of Ashwell and colleagues that proteasomal degradation of inhibitor of apoptosis proteins (IAPs) is an important feature of thymocyte apoptosis (Yang et al., 2000). Since IAPs are known to bind and prevent activation of caspases, it is likely that the proteasome is also regulating caspase-3 activation downstream of mitochondrial events. Slee and colleagues have shown that in Jurkat cells, caspase-3 establishes a pro-apoptotic feedback loop to mitochondria through direct cleavage of BID, causing further permeabilization of the outer membrane (Slee et al., 2000). Thus, it is possible that the proteasome may play a critical role in regulating apoptosis in T cells/thymocytes at several points in the cell death cascade.

The detailed molecular mechanism(s) that leads to the production of H_2O_2 in the mitochondria during apoptosis is not yet elucidated. The production of H_2O_2 could be one of the consequences of a necessary process in mediating apoptosis. Once the permeabilization of mitochondrial inner membrane, indicated by the loss of $\Delta\Psi_m$, occurs in the course of apoptotic actions in the mitochondria, the electron transport chain may pump more protons through uncoupled-oxidation of substrates in an effort to restore the collapsed electrochemical gradients, resulting in the over-production of ROS. Our experimental data revealed that exogenously added H_2O_2 causes the production of H_2O_2 at the mitochondria (Figure 2A), so once the production of H_2O_2 occurs at the mitochondrial electron transport chain, H_2O_2 could create an amplifying loop at the mitochondria to promote catastrophe.

ACKNOWLEDGEMENT: We acknowledge Kathie Curnick and Becky Lawlor for their excellent technical assistance.

References

Boveris, A., and Chance, B. (1973). The mitochondrial generation of hydrogen peroxide. General properties and effect of hyperbaric oxygen. Biochem J *134*, 707–716.

Braun, R. D., Lanzen, J. L., Snyder, S. A., and Dewhirst, M. W. (2001). Comparison of tumor and normal tissue oxygen tension measurements using OxyLite or microelectrodes in rodents. Am J Physiol Heart Circ Physiol *280*, H2533–2544.

Dimmeler, S., Breitschopf, K., Haendeler, J., and Zeiher, A. M. (1999). Dephosphorylation targets Bcl-2 for ubiquitin-dependent degradation: a link between the apoptosome and the proteasome pathway. J Exp Med *189*, 1815–1822.

Finkel, T., and Holbrook, N. J. (2000). Oxidants, oxidative stress and the biology of ageing. Nature *408*, 239–247.

Ferri, K. F., and Kroemer, G. (2001). Mitochondria—the suicide organelles. Bioessays *23*, 111–115.

Green, D. R., and Reed, J. C. (1998). Mitochondria and apoptosis. Science *281*, 1309–1312.

Grimm, L. M., Goldberg, A. L., Poirier, G. G., Schwartz, L. M., and Osborne, B. A. (1996). Proteasomes play an essential role in thymocyte apoptosis. Embo J *15*, 3835–3844.

Grimm, L. M., and Osborne, B. A. (1999). Apoptosis and the proteasome. Results Probl Cell Differ *23*, 209–228.

Gross, A., McDonnell, J. M., and Korsmeyer, S. J. (1999). BCL-2 family members and the mitochondria in apoptosis. Genes Dev *13*, 1899–1911.

Hale, L. P., Braun, R. D., Gwinn, W. M., Greer, P. K., and Dewhirst, M. W. (2002). Hypoxia in the thymus: role of oxygen tension in thymocyte survival. Am J Physiol Heart Circ Physiol *282*, H1467–1477.

Harris, M. H., and Thompson, C. B. (2000). The role of the Bcl-2 family in the regulation of outer mitochondrial membrane permeability. Cell Death Differ *7*, 1182–1191.

Hatefi, Y. (1985). The mitochondrial electron transport and oxidative phosphorylation system. Annu Rev Biochem *54*, 1015–1069.

Hockenbery, D. M., Oltvai, Z. N., Yin, X. M., Milliman, C. L., and Korsmeyer, S. J. (1993). Bcl-2 functions in an antioxidant pathway to prevent apoptosis. Cell *75*, 241–251.

Hug, H., Enari, M., and Nagata, S. (1994). No requirement of reactive oxygen intermediates in Fas-mediated apoptosis. FEBS Lett *351*, 311–313.

Klebanoff, S. J. (1980). Oxygen metabolism and the toxic properties of phagocytes. Ann Intern Med *93*, 480–489.

Kluck, R. M., Bossy-Wetzel, E., Green, D. R., and Newmeyer, D. D. (1997). The release of cytochrome c from mitochondria: a primary site for Bcl-2 regulation of apoptosis. Science *275*, 1132–1136.

Korshunov, S. S., Krasnikov, B. F., Pereverzev, M. O., and Skulachev, V. P. (1999). The antioxidant functions of cytochrome c. FEBS Lett *462*, 192–198.

Kroemer, G., Dallaporta, B., and Resche-Rigon, M. (1998). The mitochondrial death/life regulator in apoptosis and necrosis. Annu Rev Physiol *60*, 619–642.

Kroemer, G., and Reed, J. C. (2000). Mitochondrial control of cell death. Nat Med *6*, 513–519.

Li, B., and Dou, Q. P. (2000). Bax degradation by the ubiquitin/proteasome-dependent pathway: involvement in tumor survival and progression. Proc Natl Acad Sci U S A *97*, 3850–3855.

Macho, A., Hirsch, T., Marzo, I., Marchetti, P., Dallaporta, B., Susin, S. A., Zamzami, N., and Kroemer, G. (1997). Glutathione depletion is an early and calcium elevation is a late event of thymocyte apoptosis. J Immunol *158*, 4612–4619.

Mannella, C. A., Forte, M., and Colombini, M. (1992). Toward the molecular structure of the mitochondrial channel, VDAC. J Bioenerg Biomembr *24*, 7–19.

Marshansky, V., Wang, X., Bertrand, R., Luo, H., Duguid, W., Chinnadurai, G., Kanaan, N., Vu, M. D., and Wu, J. (2001). Proteasomes modulate balance among proapoptotic and antiapoptotic Bcl-2 family members and compromise functioning of the electron transport chain in leukemic cells. J Immunol *166*, 3130–3142.

Matsuyama, S., and Reed, J. C. (2000). Mitochondria-dependent apoptosis and cellular pH regulation. Cell Death Differ *7*, 1155–1165.

McLaughlin, K. A., Osborne, B. A., and Goldsby, R. A. (1996). The role of oxygen in thymocyte apoptosis. Eur J Immunol *26*, 1170–1174.

Patterson, S. D., Spahr, C. S., Daugas, E., Susin, S. A., Irinopoulou, T., Koehler, C., and Kroemer, G. (2000). Mass spectrometric identification of proteins released from mitochondria undergoing permeability transition. Cell Death Differ *7*, 137–144.

Ramakrishnan, N., Chen, R., McClain, D. E., and Bunger, R. (1998). Pyruvate prevents hydrogen peroxide-induced apoptosis. Free Radic Res *29*, 283–295.

Schleiff, E. (2000). Signals and receptors—the translocation machinery on the mitochondrial surface. J Bioenerg Biomembr *32*, 55–66.

Sidoti-de Fraisse, C., Rincheval, V., Risler, Y., Mignotte, B., and Vayssiere, J. L. (1998). TNF-alpha activates at least two apoptotic signaling cascades. Oncogene *17*, 1639–1651.

Skulachev, V. P. (1998). Cytochrome c in the apoptotic and antioxidant cascades. FEBS Lett *423*, 275–280.

Sundaresan, M., Yu, Z. X., Ferrans, V. J., Irani, K., and Finkel, T. (1995). Requirement for generation of H2O2 for platelet-derived growth factor signal transduction. Science *270*, 296–299.

Tome, M. E., Baker, A. F., Powis, G., Payne, C. M., and Briehl, M. M. (2001). Catalase-overexpressing thymocytes are resistant to glucocorticoid-induced apoptosis and exhibit increased net tumor growth. Cancer Res *61*, 2766–2773.

Tome, M. E., and Briehl, M. M. (2001). Thymocytes selected for resistance to hydrogen peroxide show altered antioxidant enzyme profiles and resistance to dexamethasone-induced apoptosis. Cell Death Differ *8*, 953–961.

Torres-Roca, J. F., Tung, J. W., Greenwald, D. R., Brown, J. M., Herzenberg, L. A., and Katsikis, P. D. (2000). An early oxygen-dependent step is required for dexamethasone-induced apoptosis of immature mouse thymocytes. J Immunol *165*, 4822–4830.

Tsujimoto, Y., and Shimizu, S. (2000). VDAC regulation by the Bcl-2 family of proteins. Cell Death Differ *7*, 1174–1181.

Vander Heiden, M. G., and Thompson, C. B. (1999). Bcl-2 proteins: regulators of apoptosis or of mitochondrial homeostasis? Nat Cell Biol *1*, E209–216.

Vieira, H. L., Haouzi, D., El Hamel, C., Jacotot, E., Belzacq, A. S., Brenner, C., and Kroemer, G. (2000). Permeabilization of the mitochondrial inner membrane during apoptosis: impact of the adenine nucleotide translocator. Cell Death Differ *7*, 1146–1154.

Von Ahsen, O., Waterhouse, N. J., Kuwana, T., Newmeyer, D. D., and Green, D. R. (2000). The 'harmless' release of cytochrome c. Cell Death Differ *7*, 1192–1199.

Yang, J., Liu, X., Bhalla, K., Kim, C. N., Ibrado, A. M., Cai, J., Peng, T. I., Jones, D. P., and Wang, X. (1997). Prevention of apoptosis by Bcl-2: release of cytochrome c from mitochondria blocked. Science *275*, 1129–1132.

Zamzami, N., Marchetti, P., Castedo, M., Decaudin, D., Macho, A., Hirsch, T., Susin, S. A., Petit, P. X., Mignotte, B., and Kroemer, G. (1995). Sequential reduction of mitochondrial transmembrane potential and generation of reactive oxygen species in early programmed cell death. J Exp Med *182*, 367–377.

Chapter 8

Activation-Induced Cell Death and T Helper Subset Differentiation

YUFANG SHI, SATISH DEVADAS, XIAOREN ZHANG, LIYING ZHANG, ACHSAH KEEGAN, KRISTY GREENELTCH, JENNIFER SOLOMON, ZENGRONG YUAN, ERWEI SUN, CATHERINE LIU, JYOTI DAS, MEGHA THAYYIL SATISH, LIXIN WEI, JIAN-NIAN ZHOU, AND ARTHUR I. ROBERTS*

ABSTRACT: Activation-induced cell death (AICD) has been demonstrated to occur in T cell hybridomas, immature thymocytes, and activated mature T cells. However, the molecular mechanisms and the physiological role of AICD in the differentiation of T helper cell subpopulations remain elusive. We have recently shown that activation-induced cell death in Th1 and Th2 cells is executed via distinct mechanisms. Our results suggest that cytokine signals initiate the differentiation program, but it is the selective action of death effectors that determines the endpoint balance of differentiating T helper subsets. Activation-induced expression of TNF-related apoptosis-inducing ligand (TRAIL) is observed exclusively in Th2 clones and primary T helper cells differentiated under Th2 conditions, while the expression of CD95L (FasL) is mainly in Th1 cells. Furthermore, Th2 cells are more resistant to either TRAIL- or CD95L-induced apoptosis than are Th1 cells. Both Th1 and Th2 cells can induce apoptosis in labeled Th1 but not Th2 cells, and caspase inhibitor z-VAD inhibits AICD in Th1 but not Th2 cells, implicating different mechanisms of AICD in these subpopulations. Blocking TRAIL CD95L significantly enhances IFN-g production in vitro. Likewise, young CD95-defective transgenic (MRL/MpJ-lpr/lpr) mice show a greater Th1 response to ovalbumin immunization compared to control mice. Therefore, apoptosis mediated by CD95L and TRAIL is critical in determining the fate of differentiating T helper cells.

Introduction

Programmed cell death through apoptosis plays a fundamental role in various physiological processes and its deregulation is a crucial part of a variety of diseases such as autoimmunity, cancer, stroke, degenerative diseases, and AIDS (Duke et al., 1996; Reed &

*Department of Molecular Genetics, Microbiology and Immunology, University of Medicine and Dentistry of New Jersey-Robert Wood Johnson Medical School, 661 Hoes Lane, Piscataway, New Jersey 08854.

Tomaselli, 2000; Schmitz et al., 2000). In the last 15 years, apoptosis has received a great deal of attention in the scientific community and intensive research in this exciting field has led to the identification of key mechanisms of this vital biological phenomenon. It is now known that apoptosis is an active cell suicide process carried out by a cascade of molecular events involving a number of membrane receptors and cytoplasmic proteins (Habibovic et al., 2000; Sharma et al., 2000; Zimmermann et al., 2001). In the immune system, apoptosis appears to occur physiologically during T and B lymphocyte repertoire selection (Ashwell et al., 2000; Green et al., 1992; McConkey et al., 1992) and during cytolytic T cell-mediated killing (Henkart et al., 1997; Trapani et al., 1999). Apoptosis is also involved in the removal of excess T and B cells as part of immune homeostasis (Chan et al., 2000; Green & Scott, 1994; Scott, 1993; Scott et al., 1996). Defects in this process could thus lead to the accumulation of potentially autoreactive T cells and B cells.

Activation-Induced Cell Death

One of the best-characterized systems in which apoptosis can be recognized is activation-induced cell death (AICD) in T cells. We demonstrated that AICD could be induced specifically in immature thymocytes *in vivo* by administration of antibodies against the TCR complex (Shi et al., 1991; Shi et al., 1989). This was also observed in fetal organ culture, using either antibody to CD3 (Smith et al., 1989) or the superantigen, staphylococcal enterotoxin B (Jenkinson et al., 1989). *In vivo* administration of specific antigen also induced apoptosis in the thymus of OVA-specific TCR transgenic mice (Murphy et al., 1990). Although it was suspected that AICD of immature thymocytes in this system is likely due to TNFa produced by TCR-activated mature T cells, it is now known that antigen activation-promoted AICD of thymocytes in TCR transgenic mice cannot be inhibited by anti-TNFa (Sytwu et al., 1996). While the mechanism of AICD in immature thymocytes is unknown, it is likely the means of negative selection of autoreactive T cells during their development in the thymus (Hamad & Schneck, 2001).

Immature T cells do not proliferate *in vitro* and are difficult to maintain, thus elucidation of the molecular mechanisms controlling AICD in these cells has been challenging. The first model system in which AICD was demonstrated was a T cell hybridoma. It was demonstrated that activation of T cell hybridoma cells often leads to cessation of cell proliferation and cytolysis (Mercep et al., 1988). Further studies revealed that these activation-induced changes have characteristics of apoptosis, showing cell membrane blebbing, nuclear condensation, and genomic DNA fragmentation, thus this phenomenon was termed AICD (Shi et al., 1990). The induction of growth arrest and apoptosis by activation is not restricted to immature thymocytes and T cell hybridomas, but can be similarly triggered in activated, mature, non-transformed T lymphocytes (Boehme & Lenardo, 1993a; Lenardo, 1991).

AICD in Mature T Cells

Before encountering specific antigens, T cells remain at a resting stage. These cells are normally long-lived, with a life span of months in rodents and years in humans (Sprent & Tough, 1994; Tough & Sprent, 1995). When triggered via their antigen-specific T-cell receptor, these resting mature T lymphocytes are activated to elicit an appropriate immune response. The activation signals are often robust and lead to massive proliferation: when appropriate costimulation is supplied, continuous TCR activation can lead to an increase in cell number of 10^9 to 10^{11} fold in 60 days (Levine, et al., 1997). Considering that the human body is composed of approximately 10^{14} cells in total, this expansion is extraordinary. Without proper control, the accumulation of activated T cells after antigenic stimulation would be devastating. As a result, the mechanism of AICD of previously-activated T cells has evolved as a means to specifically eliminate useless or potentially harmful cells in order to preserve cellular homeostasis, terminate cellular immune responses, and ensure peripheral tolerance.

Early studies showed that TCR activation could inhibit IL-2-driven proliferation of activated murine and human mature T lymphocytes (Breitmeyer et al., 1987; Nau et al., 1987; Webb et al., 1985). This effect was confirmed to occur by AICD by several groups (Bensussan, et al., 1990; Janssen, et al., 2000; Janssen, et al., 1992; Kabelitz & Janssen, 1997; Russell, 1995). Such observations of AICD in normal T lymphocytes stirred a great deal of interest in how AICD might contribute to the regulation of cellular homeostasis during immune responses (Baumann et al., 2002; Donjerkovic & Scott, 2000; Mountz et al., 1996). Detailed elucidation of the mechanisms controlling the induction or prevention of AICD could lead to new ways to modulate the immune system.

Mechanisms of AICD

It is generally believed that negative selection of T-cells in the thymus occurs activation-induced apoptosis, but the exact molecular mechanisms remain largely unknown. Recent studies with knockout technology indicate that the pro-apoptotic Bcl-2 family protein, Bim, might play a important role (Bouillet et al., 2002). It was found that activation leads to increased Bim expression in thymocytes. Thymocytes from mice lacking Bim are refractory to apoptosis induced by TCR stimulation. Moreover, in mice bearing autoreactive TCRs, the deletion of autoreactive thymocytes is severely impaired by Bim deficiency. The deletion of autoreactive cells is not completely inhibited in Bim-deficient mice, however, so Bim may not be the only mediator of AICD in immature thymocytes.

Early work from the Sprent lab has shown that apoptosis of double-positive immature T cells induced by high-dose staphylococcal enterotoxin B (SEB) is mediated by the interaction of CD95 and CD95L (Kishimoto et al., 1998). Yet overexpression of CD95 leads to an increase in the size of the thymus (Zhou et al., 1995). Similarly, it has been reported that activation of thymocytes leads to an increase in apoptosis-inducing receptors for TRAIL (tumor necrosis factor-related apoptosis-inducing ligand) and increased sensitivity to TRAIL-mediated apoptosis, but TRAIL does not play a major role in antigen-induced deletion of autoreactive thymocytes (Simon et al., 2001). A recent study showed that mice deficient in TRAIL have a severe defect in thymocyte apoptosis induced by TCR-ligation. TRAIL-deficient mice are also hypersensitive to collagen-induced arthritis and streptozotocin-induced diabetes. Thus, TRAIL might play a role in mediating AICD during negative selection (Lamhamedi-Cherradi et al., 2003). Another TNF family member, RANKL (ligand for receptor activator of nuclear factor kappa B), has been shown to be highly expressed in the thymus and we have recently shown that its expression can be up-regulated by TCR activation and dexamethasone treatment. Its role in AICD of thymocytes has yet to be established.

In recently-generated DR3 (Apo3/LARD/TRAMP)-deficient mice, negative selection of thymocytes and anti-CD3-induced apoptosis are both somewhat affected, while superantigen-induced negative selection and positive selection are normal (Wang et al., 2001). Among transgenic mice created so far, the one with the most significant change in the thymus is an overexpressor of LIGHT (a tumor necrosis factor superfamily member with homology to lymphotoxin that is expressed by activated T cells). These mice show dramatic increases in negative selection and much reduced thymus size. Yet LIGHT knockout mice have lymphoid organs of normal sizes and microarchitectures, and there are no differences in total cell numbers and composition of lymphocyte subsets in thymus, spleen, and lymph nodes (Tamada et al., 2002). Therefore, the changes in LIGHT-overexpressing mice may be a result of excessive costimulation provided by LIGHT ligand. In summary, the molecules that mediate AICD during negative selection still remain to be identified.

Mice with a mutation in CD95 (*lymphoproliferation*, *lpr*), or CD95L (*generalized lymphoproliferative disease*, *gld*) develop severe lymphadenopathy, splenomegaly, and autoimmune diseases, even though there are no other obvious abnormalities. Clearly, CD95 and CD95L must play critical roles in maintaining homeostasis of peripheral lymphocytes. This is further supported by the observation that AICD in activated mature T cells and in a T cell hybridoma proceeds via ligation of the T-cell receptor and signal transduction events culminating in expression of the CD95 and CD95L genes (Van Parijs & Abbas, 1996; Van Parijs et al., 1996). Moreover, CD95L antagonists such as soluble CD95 and antibody against CD95L are able to block AICD (Brunner et al., 1995; Ettinger, et al., 1995; Ju, et al., 1995; Rothstein, et al., 1995).

Naive ($CD45RA^+$) T cells express little or no cell-surface Fas whereas previously activated ($CD45RO^+$) T cells express relatively high amounts (Pawelec et al., 1996). Primary activation of naïve T cells through engagement of TCR/CD3 enables these cells to undergo proliferation and cytokine production *in vitro* (Orchansky & Teh, 1994). However, restimulation results in expression of FasL and leads to cell death. Fas and FasL expression are critical to AICD as activated T-cells of *lpr* and *gld* mice are defective in AICD. Moreover, addition of Fas-Fc fusion protein to block FasL prevents AICD in restimulated T cells (Ettinger et al., 1995). Therefore, AICD in peripheral T cells is dependent mainly on the interaction of Fas and FasL, though TNF-a may also play a role in this process (Sytwu et al., 1996).

AICD is generally believed to proceed as follows: upon stimulation of activated T cells via the CD3/TCR complex, the expression of FasL is rapidly induced. FasL binds to the surface-expressed CD95 molecule on the same or on neighboring cells triggering Fas-dependent apoptosis. Due to the existence of microvilli on the cell surface, or, alternatively, to FasL cleavage by metalloproteases, Fas/FasL interactions result in both autocrine "suicide" and paracrine "fratricide." Although the role of Fas/FasL interaction in the induction of T-cell apoptosis is unambiguous, it is evident that other molecules are perhaps equally important. As mentioned above, under certain circumstances, TNF-a can act as a mediator of apoptosis in mature T lymphocytes. We have recently shown that TRAIL, another well-known membrane-bound inducer of apoptosis, can also contribute to cell death in activated T cells (Zhang, 2003).

The induction of AICD sensitivity of mature T cells is also influenced by several factors. The T-cell growth factor IL-2 has been implicated in the priming of mature T lymphocytes for AICD (Lenardo, 1991; Wang et al., 1997). It has been suggested that the role of growth factors in the induction of apoptosis is to drive T cells into the S phase of the cell cycle, where they are sensitive to TCR-triggered AICD (Boehme & Lenardo, 1993b). We have reported that c-myc is required for the activation-induced expression of FasL (Wang et al., 1998). Importantly, functionally distinct subpopulations of $CD4^+$ T cells seem to vary in AICD sensitivity, with Th1 cells being much more AICD-sensitive than Th2 cells (Oberg et al., 1997; Ramsdell et al., 1994).

Differential Regulation of AICD in T-Helper Subsets

In response to activation signals, T helper cells differentiate into at least two distinct subsets, Th1 and Th2, as defined by the immune functions they mediate (Kim et al., 1985) and cytokines they secrete (Mosmann et al., 1986). Th1 cells secrete IL-2, IFN-g, and lymphotoxin and are important in promoting cellular immunity, while Th2 cells secrete IL-4, IL-5, IL-10, and IL-13 and are crucial in supporting humoral immune responses (Murphy, 1998). T helper precursors commit to either a Th1 or Th2 phenotype within a few days of activation (Miner & Croft, 1998). Although several factors, such as dose and route of antigen administration and engagement of costimulatory molecules, have been shown to influence the differentiation of T helper subsets, the most potent regulators of T helper differentiation are undoubtedly cytokines.

However, it is not known whether a naïve precursor T helper cell is pre-programmed to the Th1 or Th2 lineage before activation. There are two competing and mutually-exclusive models of T cell differentiation orientation. The stochastic model predicts that differentiation orientation is genetically pre-determined. The selection signals will determine only which subset in a mixed T helper population will grow out (Abbas et al., 1996; Constant & Bottomly, 1997; Paul & Seder, 1994). The instructive model proposes that a naïve T cell is malleable and fully capable of becoming either Th1 or Th2 cells. The differentiation process is completely dependent upon instructive signals such as the cytokine milieu and costimulation molecules (Farrar et al., 2001). Although more evidence supports the stochastic model (Kamogawa et al., 1993; Naramura et al., 1998; Riviere et al., 1998), the resolution of this paradox awaits single cell-based experiments. Regardless, after the initiation of differentiation, the establishment of the appropriate type of immune response is likely maintained by regulating cell survival under specific conditions. Several reports have shown that Th1 and Th2 cells have different propensities to undergo activation-induced cell death (AICD) (Accornero et al., 1997; Oberg et al., 1997; Ramsdell et al., 1994; Varadhachary et al., 1997). It has also been shown that activation-induced FasL surface expression varies between Th1 and Th2 cells (Oberg et al., 1997; Ramsdell et al., 1994). Following TCR ligation, AICD-sensitive Th1 clones express significantly higher levels of FasL than do Th2 cells. This difference has also been reported in Th1 and Th2 cell lines. All Th1 and Th0 cell lines, but only two out of five tested Th2 clones, express FasL mRNA in response to activation signals (Suda et al., 1995). In contrast, it was reported that completely differentiated murine Th1 and Th2 cells express equal levels of FasL and are also equally susceptible to Fas-mediated AICD (Watanabe et al., 1997). We recently reported that TRAIL and CD95L, which are critical for the regulation of lymphocyte apoptosis, are expressed in distinct patterns in both cloned and *in vitro* differentiated Th1 and Th2 cells upon activation through the T cell antigen receptor (TCR). TRAIL is observed exclusively in Th2 cells, while CD95L is detected mainly in Th1 cells. Furthermore, Th2 cells are more resistant to either TRAIL- or CD95L-induced apoptosis than are Th1 cells. Blocking TRAIL or CD95L significantly enhanced IFN-g secretion and slightly decreased IL-4 in activated T helper cells (Zhang, 2003). Additionally, we found that young Fas-deficient

MRL/MpJ-lpr/lpr mice demonstrate a stronger Th1 response than MRL/MpJ-+/+ in response to ovalbumin immunization (Zhang, 2003). Therefore, apoptosis through CD95L and TRAIL plays a critical role in the development of Th1 and Th2 cells from the activated T cell population.

TRAIL and AICD in Th2 Cells

TRAIL is a well-known apoptosis-inducing ligand. Apoptosis via its receptors, however, involves several unique features that point to a very complex regulation of cell death pathways. First, while all five known receptors bind TRAIL with similar affinities, only the DR4 and DR5 death receptors contain functional death domains and are able to mediate apoptosis. Thus, the cytoplasmic domains of decoy receptors, DcR1 and DcR2, may interfere with trimer formation by DR4 or DR5 and actually inhibit the death receptors (Zhou, et al., 2002). So far the role of TRAIL in AICD is not well-documented. Some studies have shown that human Jurkat leukemia cells are sensitive to TRAIL-induced apoptosis (Secchiero, et al., 2001). A neutralizing antibody to TRAIL partially suppressed apoptosis induced by supernatants from phytohemagglutinin (PHA)-prestimulated Jurkat cells (Martinez-Lorenzo, et al., 1998). We showed that TRAIL is induced only in Th2, not Th1 cells. Membrane-bound TRAIL on Th2 cells can kill Th1 cells as determined by cytotoxicity assay (Zhang, 2003). It should be noted that although Th2 cells are resistant to CD95L- and TRAIL-mediated apoptosis, they can still undergo AICD in a TRAIL- and CD95-independent manner. AICD in Th2 cells is not inhibited by caspase inhibitor Z-VAD (Devadas and Shi, unpublished observation), suggesting that AICD is mediated by different pathways in Th1 and Th2 cells. Nevertheless, unless there is a sustained Th1 cytokine environment to support continuous Th1 differentiation, the immune response will be pushed towards the Th2 type by apoptosis mediated through CD95L and TRAIL. The best evidence for this is that the absence of exogenous cytokines or the inhibition of apoptosis by anti-CD95L and soluble DR5 both lead to an increase in Th1 cells. Thus the process of Th1 and Th2 differentiation is a result of continuous cell proliferation and apoptosis; cytokine signals initiate the differentiation program while apoptosis determines the destiny of T helper subsets, supporting the stochastic model of T-helper cell differentiation.

Concluding Remarks

AICD plays an important role in T cell development and peripheral lymphocyte homeostasis. AICD is a complex process regulated by multiple signals, only some of which have been discussed here. In this review, we have focused on AICD of peripheral T-cell subpopulations. Although it is clear that the cytokine milieu is critical in Th1 and Th2 differentiation, it is not known which factors are involved in the establishment and stability of each type of immune response. Our findings of the differential expression of CD95L and TRAIL in Th1 and Th2 cells and the role of apoptosis mediated by these molecules in the balance between Th1 and Th2 cell development demonstrate that after the direction of differentiation is set by cytokine signals, the destiny of T helper subset differentiation is determined by apoptosis.

References

Abbas, A. K., Murphy, K. M., and Sher, A. (1996). Functional diversity of helper T lymphocytes. *Nature* **383**, 787–93.

Accornero, P., Radrizzani, M., Delia, D., Gerosa, F., Kurrle, R., and Colombo, M. P. (1997). Differential susceptibility to HIV-GP120-sensitized apoptosis in CD4+ T-cell clones with different T-helper phenotypes: role of CD95/CD95L interactions. *Blood* **89**, 558–69.

Ashwell, J. D., Lu, F. W., and Vacchio, M. S. (2000). Glucocorticoids in T cell development and function*. *Annu Rev Immunol* **18**, 309–45.

Baumann, S., Krueger, A., Kirchhoff, S., and Krammer, P. H. (2002). Regulation of T cell apoptosis during the immune response. *Curr Mol Med* **2**, 257–72.

Bensussan, A., Leca, G., Corvaia, N., and Boumsell, L. (1990). Selective induction of autocytotoxic activity through the CD3 molecule. *Eur J Immunol* **20**, 2615–9.

Boehme, S. A., and Lenardo, M. J. (1993a). Ligand-induced apoptosis of mature T lymphocytes (propriocidal regulation) occurs at distinct stages of the cell cycle. *Leukemia* **7 Suppl 2**, S45–9.

Boehme, S. A., and Lenardo, M. J. (1993b). Propriocidal apoptosis of mature T lymphocytes occurs at S phase of the cell cycle. *Eur J Immunol* **23**, 1552–60.

Bouillet, P., Purton, J. F., Godfrey, D. I., Zhang, L. C., Coultas, L., Puthalakath, H., Pellegrini, M., Cory, S., Adams, J. M., and Strasser, A. (2002). BH3-only Bcl-2 family member Bim is required for apoptosis of autoreactive thymocytes. *Nature* **415**, 922–6.

Breitmeyer, J. B., Oppenheim, S. O., Daley, J. F., Levine, H. B., and Schlossman, S. F. (1987). Growth inhibition of human T cells by antibodies recognizing the T cell antigen receptor complex. *J Immunol* **138**, 726–31.

Brunner, T., Mogil, R. J., LaFace, D., Yoo, N. J., Mahboubi, A., Echeverri, F., Martin, S. J., Force, W. R., Lynch, D. H., Ware, C. F., and et al. (1995). Cell-autonomous Fas (CD95)/Fas-ligand interaction mediates activation-induced apoptosis in T-cell hybridomas. *Nature* **373**, 441–4.

Chan, K. F., Siegel, M. R., and Lenardo, J. M. (2000). Signaling by the TNF receptor superfamily and T cell homeostasis. *Immunity* **13**, 419–22.

Constant, S. L., and Bottomly, K. (1997). Induction of Th1 and Th2 CD4+ T cell responses: the alternative approaches. *Annu Rev Immunol* **15**, 297–322.

Donjerkovic, D., and Scott, D. W. (2000). Activation-induced cell death in B lymphocytes. *Cell Res* **10**, 179–92.

Duke, R. C., Ojcius, D. M., and Young, J. D. (1996). Cell suicide in health and disease. *Sci Am* **275**, 80–7.

Ettinger, R., Panka, D. J., Wang, J. K., Stanger, B. Z., Ju, S. T., and Marshak-Rothstein, A. (1995). Fas ligand-mediated cytotoxicity is directly responsible for apoptosis of normal CD4+ T cells responding to a bacterial superantigen. *J Immunol* **154**, 4302–8.

Farrar, J. D., Ouyang, W., Lohning, M., Assenmacher, M., Radbruch, A., Kanagawa, O., and Murphy, K. M. (2001). An instructive component in T helper cell type 2 (Th2) development mediated by GATA-3. *J Exp Med* **193**, 643–50.

Green, D. R., Bissonnette, R. P., Glynn, J. M., and Shi, Y. (1992). Activation-induced apoptosis in lymphoid systems. *Semin Immunol* **4**, 379–88.

Green, D. R., and Scott, D. W. (1994). Activation-induced apoptosis in lymphocytes. *Curr Opin Immunol* **6**, 476–87.

Habibovic, S., Hrgovic, Z., Bukvic, I., and Hrgovic, I. (2000). [Molecular mechanisms in apoptosis]. *Med Arh* **54**, 33–40.

Hamad, A. R., and Schneck, J. P. (2001). Antigen-induced T cell death is regulated by CD4 expression. *Int Rev Immunol* **20**, 535–46.

Henkart, P. A., Williams, M. S., Zacharchuk, C. M., and Sarin, A. (1997). Do CTL kill target cells by inducing apoptosis? *Semin Immunol* **9**, 135–44.

Janssen, O., Sanzenbacher, R., and Kabelitz, D. (2000). Regulation of activation-induced cell death of mature T-lymphocyte populations. *Cell Tissue Res* **301**, 85–99.

Janssen, O., Wesselborg, S., and Kabelitz, D. (1992). Immunosuppression by OKT3—induction of programmed cell death (apoptosis) as a possible mechanism of action. *Transplantation* **53**, 233–4.

Jenkinson, E. J., Kingston, R., Smith, C. A., Williams, G. T., and Owen, J. J. (1989). Antigen-induced apoptosis in developing T cells: a mechanism for negative selection of the T cell receptor repertoire. *Eur J Immunol* **19**, 2175–7.

Ju, S. T., Panka, D. J., Cui, H., Ettinger, R., el-Khatib, M., Sherr, D. H., Stanger, B. Z., and Marshak-Rothstein, A. (1995). Fas(CD95)/FasL interactions required for programmed cell death after T-cell activation. *Nature* **373**, 444–8.

Kabelitz, D., and Janssen, O. (1997). Antigen-induced death of T-Lymphocytes. *Front Biosci* **2**, d61–77.

Kamogawa, Y., Minasi, L. A., Carding, S. R., Bottomly, K., and Flavell, R. A. (1993). The relationship of IL-4- and IFN gamma-producing T cells studied by lineage ablation of IL-4-producing cells. *Cell* **75**, 985–95.

Kim, J., Woods, A., Becker-Dunn, E., and Bottomly, K. (1985). Distinct functional phenotypes of cloned Ia-restricted helper T cells. *J Exp Med* **162**, 188–201.

Kishimoto, H., Surh, C. D., and Sprent, J. (1998). A role for Fas in negative selection of thymocytes in vivo. *J Exp Med* **187**, 1427–38.

Lamhamedi-Cherradi, S. E., Zheng, S. J., Maguschak, K. A., Peschon, J., and Chen, Y. H. (2003). Defective thymocyte apoptosis and accelerated autoimmune diseases in TRAIL(−/−) mice. *Nat Immunol* **10**, 10.

Lenardo, M. J. (1991). Interleukin-2 programs mouse alpha beta T lymphocytes for apoptosis. *Nature* **353**, 858–61.

Levine, B. L., Bernstein, W. B., Connors, M., Craighead, N., Lindsten, T., Thompson, C. B., and June, C. H. (1997). Effects of CD28 costimulation on long-term proliferation of CD4+ T cells in the absence of exogenous feeder cells. *J Immunol* **159**, 5921–30.

Martinez-Lorenzo, M. J., Alava, M. A., Gamen, S., Kim, K. J., Chuntharapai, A., Pineiro, A., Naval, J., and Anel, A. (1998). Involvement of APO2 ligand/TRAIL in activation-induced death of Jurkat and human peripheral blood T cells. *Eur J Immunol* **28**, 2714–25.

McConkey, D. J., Jondal, M., and Orrenius, S. (1992). Cellular signaling in thymocyte apoptosis. *Semin Immunol* **4**, 371–7.

Mercep, M., Bluestone, J. A., Noguchi, P. D., and Ashwell, J. D. (1988). Inhibition of transformed T cell growth in vitro by monoclonal antibodies directed against distinct activating molecules. *J Immunol* **140**, 324–35.

Miner, K. T., and Croft, M. (1998). Generation, persistence, and modulation of Th0 effector cells: role of autocrine IL-4 and IFN-gamma. *J Immunol* **160**, 5280–7.

Mosmann, T. R., Cherwinski, H., Bond, M. W., Giedlin, M. A., and Coffman, R. L. (1986). Two types of murine helper T cell clone. I. Definition according to profiles of lymphokine activities and secreted proteins. *J Immunol* **136**, 2348–57.

Mountz, J. D., Zhou, T., Su, X., Cheng, J., Pierson, M., Bluethmann, H., and Edwards, C. K., 3rd (1996). Autoimmune disease results from multiple interactive defects in apoptosis induction molecules and signaling pathways. *Behring Inst Mitt*, 200–19.

Murphy, K. M. (1998). T lymphocyte differentiation in the periphery. *Curr Opin Immunol* **10**, 226–32.

Murphy, K. M., Heimberger, A. B., and Loh, D. Y. (1990). Induction by antigen of intrathymic apoptosis of CD4+CD8+TCRlo thymocytes in vivo. *Science* **250**, 1720–3.

Naramura, M., Kole, H. K., Hu, R. J., and Gu, H. (1998). Altered thymic positive selection and intracellular signals in Cbl-deficient mice. *Proc Natl Acad Sci U S A* **95**, 15547–52.

Nau, G. J., Moldwin, R. L., Lancki, D. W., Kim, D. K., and Fitch, F. W. (1987). Inhibition of IL 2-driven proliferation of murine T lymphocyte clones by supraoptimal levels of immobilized anti-T cell receptor monoclonal antibody. *J Immunol* **139**, 114–22.

Oberg, H. H., Lengl-Janssen, B., Kabelitz, D., and Janssen, O. (1997). Activation-induced T cell death: resistance or susceptibility correlate with cell surface fas ligand expression and T helper phenotype. *Cell Immunol* **181**, 93–100.

Orchansky, P. L., and Teh, H. S. (1994). Activation-induced cell death in proliferating T cells is associated with altered tyrosine phosphorylation of TCR/CD3 subunits. *J Immunol* **153**, 615–22.

Paul, W. E., and Seder, R. A. (1994). Lymphocyte responses and cytokines. *Cell* **76**, 241–51.

Pawelec, G., Sansom, D., Rehbein, A., Adibzadeh, M., and Beckman, I. (1996). Decreased proliferative capacity and increased susceptibility to activation-induced cell death in late-passage human CD4+ TCR2+ cultured T cell clones. *Exp Gerontol* **31**, 655–68.

Ramsdell, F., Seaman, M. S., Miller, R. E., Picha, K. S., Kennedy, M. K., and Lynch, D. H. (1994). Differential ability of Th1 and Th2 T cells to express Fas ligand and to undergo activation-induced cell death. *Int Immunol* **6**, 1545–53.

Reed, J. C., and Tomaselli, K. J. (2000). Drug discovery opportunities from apoptosis research. *Curr Opin Biotechnol* **11**, 586–92.

Riviere, I., Sunshine, M. J., and Littman, D. R. (1998). Regulation of IL-4 expression by activation of individual alleles. *Immunity* **9**, 217–28.

Rothstein, T. L., Wang, J. K., Panka, D. J., Foote, L. C., Wang, Z., Stanger, B., Cui, H., Ju, S. T., and Marshak-Rothstein, A. (1995). Protection against Fas-dependent Th1-mediated apoptosis by antigen receptor engagement in B cells. *Nature* **374**, 163–5.

Russell, J. H. (1995). Activation-induced death of mature T cells in the regulation of immune responses. *Curr Opin Immunol* **7**, 382–8.

Schmitz, I., Kirchhoff, S., and Krammer, P. H. (2000). Regulation of death receptor-mediated apoptosis pathways. *Int J Biochem Cell Biol* **32**, 1123–36.

Scott, D. W. (1993). Analysis of B cell tolerance in vitro. *Adv Immunol* **54**, 393–425.

Scott, D. W., Grdina, T., and Shi, Y. (1996). T cells commit suicide, but B cells are murdered! *J Immunol* **156**, 2352–6.

Secchiero, P., Gonelli, A., Celeghini, C., Mirandola, P., Guidotti, L., Visani, G., Capitani, S., and Zauli, G. (2001). Activation of the nitric oxide synthase pathway represents a key component of tumor necrosis factor-related apoptosis-inducing ligand-mediated cytotoxicity on hematologic malignancies. *Blood* **98**, 2220–8.

Sharma, K., Wang, R. X., Zhang, L. Y., Yin, D. L., Luo, X. Y., Solomon, J. C., Jiang, R. F., Markos, K., Davidson, W., Scott, D. W., and Shi, Y. F. (2000). Death the Fas way: regulation and pathophysiology of CD95 and its ligand. *Pharmacol Ther* **88**, 333–47.

Shi, Y. F., Bissonnette, R. P., Parfrey, N., Szalay, M., Kubo, R. T., and Green, D. R. (1991). In vivo administration of monoclonal antibodies to the CD3 T cell receptor complex induces cell death (apoptosis) in immature thymocytes. *J Immunol* **146**, 3340–6.

Shi, Y. F., Sahai, B. M., and Green, D. R. (1989). Cyclosporin A inhibits activation-induced cell death in T-cell hybridomas and thymocytes. *Nature* **339**, 625–6.

Shi, Y. F., Szalay, M. G., Paskar, L., Sahai, B. M., Boyer, M., Singh, B., and Green, D. R. (1990). Activation-induced cell death in T cell hybridomas is due to apoptosis. Morphologic aspects and DNA fragmentation. *J Immunol* **144**, 3326–33.

Simon, A. K., Williams, O., Mongkolsapaya, J., Jin, B., Xu, X. N., Walczak, H., and Screaton, G. R. (2001). Tumor necrosis factor-related apoptosis-inducing ligand in T cell development: sensitivity of human thymocytes. *Proc Natl Acad Sci U S A* **98**, 5158–63.

Smith, C. A., Williams, G. T., Kingston, R., Jenkinson, E. J., and Owen, J. J. (1989). Antibodies to CD3/T-cell receptor complex induce death by apoptosis in immature T cells in thymic cultures. *Nature* **337**, 181–4.

Sprent, J., and Tough, D. F. (1994). Lymphocyte life-span and memory. *Science* **265**, 1395–400.

Suda, T., Okazaki, T., Naito, Y., Yokota, T., Arai, N., Ozaki, S., Nakao, K., and Nagata, S. (1995). Expression of the Fas ligand in cells of T cell lineage. *J Immunol* **154**, 3806–13.

Sytwu, H. K., Liblau, R. S., and McDevitt, H. O. (1996). The roles of Fas/APO-1 (CD95) and TNF in antigen-induced programmed cell death in T cell receptor transgenic mice. *Immunity* **5**, 17–30.

Tamada, K., Ni, J., Zhu, G., Fiscella, M., Teng, B., van Deursen, J. M., and Chen, L. (2002). Cutting edge: selective impairment of CD8+ T cell function in mice lacking the TNF superfamily member LIGHT. *J Immunol* **168**, 4832–5.

Tough, D. F., and Sprent, J. (1995). Life span of naive and memory T cells. *Stem Cells* **13**, 242–9.

Trapani, J. A., Sutton, V. R., and Smyth, M. J. (1999). CTL granules: evolution of vesicles essential for combating virus infections. *Immunol Today* **20**, 351–6.

Van Parijs, L., and Abbas, A. K. (1996). Role of Fas-mediated cell death in the regulation of immune responses. *Curr Opin Immunol* **8**, 355–61.

Van Parijs, L., Ibraghimov, A., and Abbas, A. K. (1996). The roles of costimulation and Fas in T cell apoptosis and peripheral tolerance. *Immunity* **4**, 321–8.

Varadhachary, A. S., Perdow, S. N., Hu, C., Ramanarayanan, M., and Salgame, P. (1997). Differential ability of T cell subsets to undergo activation-induced cell death. *Proc Natl Acad Sci U S A* **94**, 5778–83.

Wang, E. C., Thern, A., Denzel, A., Kitson, J., Farrow, S. N., and Owen, M. J. (2001). DR3 regulates negative selection during thymocyte development. *Mol Cell Biol* **21**, 3451–61.

Wang, R., Brunner, T., Zhang, L., and Shi, Y. (1998). Fungal metabolite FR901228 inhibits c-Myc and Fas ligand expression. *Oncogene* **17**, 1503–8.

Wang, R., Ciardelli, T. L., and Russell, J. H. (1997). Partial signaling by cytokines: cytokine regulation of cell cycle and Fas-dependent, activation-induced death in CD4+ subsets. *Cell Immunol* **182**, 152–60.

Watanabe, N., Arase, H., Kurasawa, K., Iwamoto, I., Kayagaki, N., Yagita, H., Okumura, K., Miyatake, S., and Saito, T. (1997). Th1 and Th2 subsets equally undergo Fas-dependent and -independent activation-induced cell death. *Eur J Immunol* **27**, 1858–64.

Webb, S. R., Li, J. H., MacNeil, I., Marrack, P., Sprent, J., and Wilson, D. B. (1985). T cell receptors for responses to Mls determinants and allo-H-2 determinants appear to be encoded on different chromosomes. *J Exp Med* **161**, 269–74.

Zhang, X. R., L. Y. Zhang, L. Li, L. H. Glimcher, A. D. Keegan, and Y. F. Shi (2003). Reciprocal Expression of TRAIL and CD95L in Th1 and Th2 Cells: Role of Apoptosis in T Helper Subset Differentiation. *Cell Death and Differentiation* **In Press**.

Zhou, T., Edwards, C. K., 3rd, and Mountz, J. D. (1995). Prevention of age-related T cell apoptosis defect in CD2-fas-transgenic mice. *J Exp Med* **182**, 129–37.

Zhou, T., Mountz, J. D., and Kimberly, R. P. (2002). Immunobiology of tumor necrosis factor receptor superfamily. *Immunol Res* **26**, 323–36.

Zimmermann, K. C., Bonzon, C., and Green, D. R. (2001). The machinery of programmed cell death. *Pharmacol Ther* **92**, 57–70.

Chapter 9

The $Bax^{-/-}Bak^{-/-}$ Mouse: a Model for Apoptosis

WEI-XING ZONG, JEFFREY C. RATHMELL, JEFFREY A. GOLDEN, AND TULLIA LINDSTEN*

ABSTRACT: The *bax*$^{-/-}$*bak*$^{-/-}$ double-deficient mouse has provided a unique tool to study apoptosis. These mice display severe defects in both developmental apoptosis and regulation of tissue homeostasis. Developmental defects include persistent interdigital webs in the adult mouse. A defect in tissue homeostasis is illustrated in the hematopoietic system where lymphocytosis, enlarged lymphoid organs and lymphocytic infiltration in parenchymal organs is seen. In addition, defects in thymocyte development, negative selection and T cell homeostasis is seen. The central nervous system of the *bax*$^{-/-}$*bak*$^{-/-}$ mice show an excess of cells in areas where neural progenitor cells are known to reside. Primary cells as well as cell lines derived from the *bax*$^{-/-}$*bak*$^{-/-}$ mouse display resistance to a multitude of apoptotic stimuli and has helped to provide a model for the regulation of the cell intrinsic apoptotic pathway.

Gene Targeting Models of Bcl-2 Family Members

It is well established that programmed cell death or apoptosis plays a crucial role in regulation of tissue homeostasis as well as during development. There are two major apoptotic pathways (Strasser et al., 1995; Green, 2000). The cell extrinsic pathway is regulated by receptors such as Fas and TNF-R1, which when bound to ligand will initiate cell death via recruitment and activation of caspases. The cell intrinisic apoptotic pathway is initiated when the outer mitochondrial membrane loses its integrity (Green, 2000). This leads to the release of cytochrome c and other factors that will lead to the activation of caspases and cell death. The Bcl-2 family of proteins regulates the cell intrinsic apoptotic pathway. This family includes anti-apoptotic members such as Bcl-2 and Bcl-xL, pro-apoptotic members such as Bax and Bak and a group of proteins termed BH3-only proteins (Table 1). It is thought that the Bcl-2 family of proteins exerts their effect on the mitochondrial membrane such that Bcl-2 and Bcl-xL will preserve the integrity of the outer mitochondrial membrane whereas Bak and Bax have the opposite effect promoting the release of cytochrome c.

bcl-xL deficient mice die during embryogenesis due to massive apoptosis in multiple organ systems (Motoyama et al., 1995). *bcl-2* deficient mice show defects in renal development and develop immunodeficiency (Nakayama et al., 1993). Thus, loss of these two major anti-apoptotic genes will lead to severe deficiencies in the apoptotic pathway. *bax* and

*Departments of Pathology and Laboratory Medicine, Cancer Biology, Abramson Family Cancer Research Institute, University of Pennsylvania, Philadelphia, PA 19104. *Correspondence:* lindsten@mail.med.upenn.edu

Table 1. Listing of the members of the Bcl-2 family of proteins.

The Bcl-2 Family of Proteins		
Anti-apoptotic	Pro-apoptotic	BH3-only
Bcl-2	Bax	Bad
Bcl-xL	Bak	Bim
Mcl-1	Bok	Bid
Bfl-1/A1		Bik
Boo/DIVA		Noxa
Bcl-w		PUMA
NR-13		EGL-1
Ced-9		Bmf
E1B-19K		BNIP3
		NIX
		HRK

bak are ubiquitously expressed genes whereas *bok* expression is limited to the reproductive system. It was surprising to find that when *bax* deficient mice were generated no major defects in the apoptotic pathway was seen (Knudson et al., 1995). The only phenotype these mice exhibit is male sterility due to a developmental arrest in spermatogenesis that leads to increased cell death. In addition, *bak*-deficient mice exhibit no discernable phenotype (Lindsten et al., 2000). These results could indicate that there could be other more important pro-apoptotic genes not yet identified, or that *bax* and *bak* are redundant in their function. The latter possibility was found to be correct when $bax^{-/-}bak^{-/-}$ mice were generated (Lindsten et al., 2000).

Physical Signs

$bax^{-/-}bak^{-/-}$ mice show very high perinatal mortality. Only about 10% of pups born survive to adulthood. It was noted that in litters where $bax^{-/-}bak^{-/-}$ mice were expected, pups would be found pushed away from the nest and neglected. These pups died and were confirmed to be $bax^{-/-}bak^{-/-}$. This could indicate that part of the perinatal mortality could be attributed to defects in social behavior. When examining adult $bax^{-/-}bak^{-/-}$ mice severe defects in developmental apoptosis were found which is illustrated by the persistence of interdigital webs on their paws. Interestingly, the presence of a single allele of either *bax* or *bak* completely reversed the phenotype. In addition to interdigital webs, female $bax^{-/-}bak^{-/-}$ mice showed imperforate vaginas. The $bax^{-/-}bak^{-/-}$ mice were unresponsive to auditory stimulation and they exhibited circling behavior, particularly when exposed to stress.

CNS Defects

The neural phenotype exhibited by the $bax^{-/-}bak^{-/-}$ mice prompted an investigation of the central nervous system. When coronal sections of brains from the $bax^{-/-}bak^{-/-}$ mice were examined masses of small, densely staining cells were noted first in the subventricular zone. Upon further investigation, similar masses of cells were found in the hippocampus, olfactory bulb, corpus callosum, basal ganglia, white matter, the dorsal midline of the mesencephalon and cerebellum. These are areas where during embryogenesis neural progenitor cells are known to reside. Neural progenitor cells can also be recovered from the adult mouse from the subventricular zones, the hippocampus, the olfactory bulbs and the cerebellum but they are not present in such excess numbers as in the adult *bax*$^{-/-}$*bak*$^{-/-}$ mice. Confirming the nature of these cells, preliminary data indicate that the excess cells in the adult *bax*$^{-/-}$*bak*$^{-/-}$ brain stain positive for nestin, a marker specific for neural progenitor cells as well as for NeuN, an early postmitotic neural marker.

Defects in Lymphoid Homeostasis

Peripheral Lymphoid Defects

Upon dissection it was noted that the spleens and lymph nodes of the *bax*$^{-/-}$*bak*$^{-/-}$ mice were greatly enlarged. Histological analysis of the spleens showed an expanded red pulp that contained a large number of plasma cells and histiocytes. In addition to lymphoid accumulation in the peripheral lymphoid organs, lymphocytic infiltration was also found in the liver and in the kidney. Some older mice also showed glomerulonephropathy.

The white blood cell counts in the *bax*$^{-/-}$*bak*$^{-/-}$ mice were elevated. The largest increase was in the lymphocyte count with as much as a 10-fold increase in some mice. The number of myeloid colony forming units was also increased when examined in hematopoietic colony assays. Peripheral lymphocytes from lymph nodes were examined by flow cytometry to determine the phenotypic characteristics of the lymphocytes that accumulate in the *bax*$^{-/-}$*bak*$^{-/-}$ mice. The ratio of CD4 and CD8 positive cells were normal, however the T cells were small as measured by forward angle light scatter and they displayed a memory phenotype in that there was an increased expression of CD44 and a decreased expression of CD62L. The lymphocytes lacked expression of activation markers such as CD25. The number of B cells as measured by B220 staining was normal, however also B cells were skewed towards cells with a memory phenotype as an increase in cells that were B220^{bright}IgD^{-} was seen.

Defects in T Cell Development

Despite the splenomegaly and enlarged lymph nodes the thymus of *bax*$^{-/-}$*bak*$^{-/-}$ mice was normal in size (Rathmell et al., 2002). However, when the distribution of CD4 and CD8 was determined, profound defects were found. The CD4^{-}CD8^{-} double-negative, CD4^{+}CD8^{-} and CD4^{-}CD8^{+} single positive populations were increased whereas there was fewer of the CD4^{+}CD8^{+} double positive population of cells. This loss of double-positive cells was not due to stress-induced apoptosis as thymocytes from *bax*$^{-/-}$*bak*$^{-/-}$ mice were completely resistant to treatment with dexamethasone at doses that killed normal thymocytes. Thymocytes from *bax*$^{-/-}$*bak*$^{-/-}$ mice, when cultured without stimulation stayed alive whereas normal thymocytes die rapidly, indicating that Bax and Bak regulate death by neglect. Failure to die by neglect may explain the high number of double-negative thymocytes. To confirm that the altered thymocyte subpopulations was intrinsic to the *bax*$^{-/-}$*bak*$^{-/-}$ hematopoietic system, lethally irradiated RAG1$-/-$ mice were reconstituted with bone marrow from *bax*$^{-/-}$*bak*$^{-/-}$ and control mice.

It was found that the defects in the distribution of CD4+ and CD8+ subpopulations was recapitulated in the bone marrow chimeras derived from $bax^{-/-}bak^{-/-}$ mice. The $bax^{-/-}bak^{-/-}$ bone marrow chimeras proved a useful system to examine the role of apoptosis in the regulation of negative selection, thymopoiesis and T cell homeostasis by allowing analysis of hematopoietic intrinsic defects. To study negative selection, both an in vitro and in vivo systems were used. It was shown that when $bax^{-/-}bak^{-/-}$ thymocytes were cultured on anti-CD3 coated plastic dishes, in the presence of anti-CD28 antibodies to mimic T cell receptor (TCR) signaling, they did not undergo cell death as compared to normal thymocytes. To test whether negative selection was impaired in vivo, the following experimental design was used. It has been shown that for a given H-2 background there is a preferential elimination of certain TCR b-chain variable regions when double-positive thymocytes are compared to single $CD4^+$ T cells, as a consequence of stimulation with endogenous retroviral superantigens. In the case of the experimental system above, about 60% of $CD4^+$ TCR V_b5^+ T cells should be eliminated relative to the frequency of double-positive TCR V_b5^+ thymocytes, whereas the frequency of TCR V_b6^+cells are unaffected. When mice reconstituted with $bax^{-/-}bak^{-/-}$ bone marrow were examined, the number of V_b5^+ cells were the same for double positive thymocytes as compared to $CD4^+$ T cells, indicating that negative selection is dependent on apoptosis regulated by Bax and Bak and could explain the accumulation of double-negative thymocytes in the $bax^{-/-}bak^{-/-}$ mice. It was noted that as bone marrow chimeras from $bax^{-/-}bak^{-/-}$ mice aged the cellularity of the thymus decreased. To study thymopoiesis, mice that had been reconstituted with bone marrow from $bax^{-/-}bak^{-/-}$ and normal mice for 10 weeks, were injected with the nucleotide analog bromodeoxyuridine (Brdu). After seven daily injections the mice were sacrificed and the thymocytes analyzed for the expression of CD4, CD8 and Brdu. It was found that the Brdu incorporation was much lower in the $bax^{-/-}bak^{-/-}$ mice, in particular in the double-negative subset where a 90% reduction of Brdu incorporation was noted as compared to normal mice.

While the thymus showed decreased cellularity with age, the peripheral T cells in the $bax^{-/-}bak^{-/-}$ mice gradually increased manifested in splenomegaly and enlarged lymph nodes as mentioned previously. The T cells that accumulated in the periphery had a memory cell phenotype. One interpretation of this phenomenon is that in the absence of *bax* and *bak* lymphocytes are not eliminated after antigen clearance in immune responses. This was tested experimentally by immunizing mice reconstituted with $bax^{-/-}bak^{-/-}$ or control bone marrow with the staphylococcal enterotoxin B (SEB). The mice were tested at different time points after immunization for the presence of T cells with TCR V_b8, in the blood that are specifically activated by SEB. It was found that in normal mice the number of V_b8 $CD4^+$ T cells showed an initial increase followed by a decrease to a basal level just above background by day 16. However, in the $bax^{-/-}bak^{-/-}$ mice the number of V_b8 $CD4^+$ T cells remained at high levels throughout the test period. This elevated level of V_b8 $CD4^+$ T cells was also observed in the spleen where at day 20 after immunization there were elevated levels of $CD4^+$ T cells bearing V_b8 as compared to control mice. Thus, it appears that indeed Bax and Bak are required for elimination of antigen activated T cells. This failure to eliminate peripheral T cells after an immune response may act to inhibit thymopoesis found in aged $bax^{-/-}bak^{-/-}$ mice indicating a role for Bax and Bak in the normal turnover of T cells.

In conclusion, these findings show that Bax and Bak play an important role in thymopoiesis, T cell development, negative selection and maintenance of T cell homeostasis in the immune response.

Table 2. Resistance to apoptotic stimuli in cell types of the $bax^{-/-}bak^{-/-}$ mouse.

Cell Type	Apoptotic Stimulus	Apoptotic Resistance
Thymocytes	Dexamethasone	+
	Neglect	+
	g irradiation	+
	Etoposide	+
	anti-CD3+anti-CD28	+
	anti-Fas	−
Splenocytes	Neglect	+
MEF	Neglect (serum withdrawal)	+
	Etoposide	+
	g irradiation	+
	Anoikis	+
	Staurosporine	+
	Thapsigargin	+
	Tunicamycin	+
	Brefeldin A	+
	Oxygen deprivation	+
	JNK signaling	+
	TNF-a	−
BMK	TNF-a	+
Primary hepatocytes	anti-Fas	+

$Bax^{-/-}Bak^{-/-}$ Cells Show Resistance to Apoptotic Stimuli

As mentioned in the previous section thymocytes obtained either from $bax^{-/-}bak^{-/-}$ mice or mice reconstituted with $bax^{-/-}bak^{-/-}$ bone marrow showed resistance to cell death induced by dexamethasone or neglect. To extend these findings several cellular systems and different apoptotic stimuli were investigated as outlined in Table 2.

It was found that also splenocytes from $bax^{-/-}bak^{-/-}$ mice were resistant to cell death when cultured without stimulus for several days, also referred to as death by neglect. In addition, thymocytes were also found to be resistant to g irradiation and to treatment with etoposide. Considering the difficulty in raising $bax^{-/-}bak^{-/-}$ mice to adulthood, it proved useful to establish cell lines from tissues of $bax^{-/-}bak^{-/-}$ mice. The first such lines to be established were mouse embryonic fibroblast (MEF) lines. These cell lines have been used to test a large number of apoptotic stimuli as well as provide a means to dissect the cell intrinsic apoptotic pathway.

Initially, it was shown that MEFs from $bax^{-/-}bak^{-/-}$ mice but not those from the single *bax* or *bak* knock-out or control mice were resistant to multiple signals directed at multiple sites of the cell machinery including g irradiation, serum withdrawal (death by neglect), anoikis (loss of adherence), treatments with staurosporine (a kinase inhibitor), etoposide (a topoisomerse II inhibitor), and endoplasmic reticulum stress signals: thapsigargin, tunicamycin, and brefeldin A (Wei et al., 2001). $bax^{-/-}bak^{-/-}$ MEFs were also resistant to oxygen deprivation-induced cell death (McClintock et al., 2002). The c-Jun NH2-terminal kinase (JNK) signaling pathway is required for stress-induced release of mitochondrial cytochrome c and apoptosis (Lei et al., 2002). Also in this experimental system it was shown that $bax^{-/-}bak^{-/-}$ MEFs were resistant to apoptosis.

When activation of the cell extrinsic pathway was investigated in primary thymocytes by treatment with anti-Fas antibodies, $bax^{-/-}bak^{-/-}$ thymocytes were as susceptible to cell death as thymocytes from normal mice, indicating the separation of the cell intrinsic vs. the cell extrinsic pathway (Lindsten et al., 2000). Similar results were obtained when MEFs were treated with tumor necrosis factor-a (TNF-a) in the presence of actinomycin D (Wei et al., 2001). The response to TNF-a stimulation is known to activate two separate signaling cascades depending on the cell type investigated (Scaffidi et al., 1998). Thus, type I cells have been defined as cells where caspase-8 is activated within seconds by the death-inducing signaling complex (DISC) followed by activation of caspase 3. In type II cells there is a much slower response such that cleavage of caspase 8 and 3 does not occur until about 60 minutes following receptor engagement. Overexpression of Bcl-xL or Bcl-2 was shown to block caspase 8 and caspase 3 activation as well as apoptosis in type II but not in type I cells.

Thus, when baby mouse kidney (BMK) cells derived from $bax^{-/-}bak^{-/-}$ mice were transformed with E1A and dominant-negative p53 and treated with TNF-a, they were found to be resistant to cell death compared to control cell lines (Degenhardt et al., 2002). Thus, primary MEFs are type I cells whereas BMK cell lines are type II cells. Similarly, when the response to anti-Fas injection in vivo, was studied, $bax^{-/-}bak^{-/-}$ mice were shown to be resistant to the severe hepatocellular apoptosis and rapid death, seen in single knock-out and normal mice. Thus, liver cells appear to be an example of type II cells.

The importance of Bax and Bak are evident from these studies including both primary and transformed cells from $bax^{-/-}bak^{-/-}$ mice in the response to a multitude of apoptotic stimuli directed at multiple points of the cell machinery. The MEFs would also prove to be useful in dissecting the apoptotic pathway.

Model for the Regulation of Apoptosis by the Bcl-2 Family of Proteins

In addition to the multi-domain pro- and anti-apoptotic sub-groups of the Bcl-2 family proteins, another group of proteins namely “BH3-only proteins” have been the focus of recent studies. They share homology with the BH3 domain of the multi-domain Bcl-2 family members. Up to date more than 10 BH3-only proteins have been identified, which include Bid, Bad, Bik/Nbk, Bim, Bmf, Noxa, Puma, and Bnip3. BH3-only proteins have been proposed to be allosteric regulators of the Bcl-2 family of proteins, and serve as central effectors of apoptotic signaling pathways. All BH3-only proteins identified so far are proapoptotic. Divergent mechanisms regulate the proapoptotic activities of numerous BH3-only proteins, which show the complexity involved in the control of apoptosis in response to different death-inducing stimuli. For example, Bid is truncated into tBid during death receptor-induced apoptosis (Li et al., 1998; Luo et al., 1998), Bim and Bmf are released from microtubule-associated dynein light chain or myosin V actin complex during growth factor withdrawal- and anoikois-induced apoptosis (Puthalakath et al., 1999; Puthalakath et al., 2001). Puma and Noxa are transcriptionally up regulated in p53-mediated apoptosis (Oda et al., 2000; Nakano and Vousden, 2001; Yu et al., 2001). Puma and Bim have been found to be up regulated by growth factor withdrawal (Han et al., 2001; Shinjyo et al., 2001; Dijkers et al., 2000).

When activated, the BH3-only proteins are localized to mitochondria. Thus these proteins may function as effectors, or as regulators to bind the multi-domain Bcl-2 family proteins, at the mitochondria to initiate apoptosis. The $bax^{-/-}bak^{-/-}$ MEFs provided an important tool to test this hypothesis. When tBid, Bim, Bad, Noxa, and Puma were

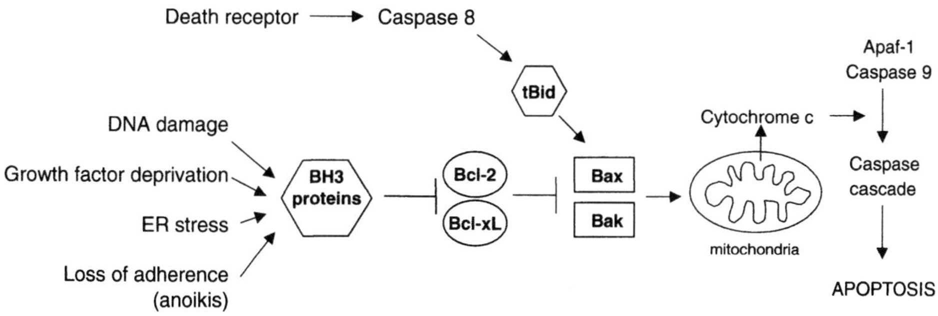

Figure 1. A model for the cell intrinisic apoptotic pathway.

introduced by retroviral transduction in wild type cells they are potent inducers of apoptosis, but do not induce apoptosis in $bax^{-/-}bak^{-/-}$ MEFs. Re-introducing Bax or Bak back into $bax^{-/-}bak^{-/-}$ cells restored their sensitivity to BH3-only proteins (Wei et al., 2001; Zong et al., 2001; Cheng et al., 2001).

The BH3-only proteins have been demonstrated to have selective binding to the multi-domain Bcl-2 family proteins: such that tBid binds proapoptotic Bax and Bak, or Bad and Bim that bind anti-apoptotic Bcl-2 and Bcl-xL. The selective binding activities of these BH3-only proteins are determined by their distinct BH3 domains (Letai et al., 2002). It has been long established that the cell's fate to live or to die is determined by the balance between the anti-apoptotic members such as Bcl-2 and Bcl-xL, and the proapoptotic members such as Bax and Bak. The fact that both kinds of BH3-only proteins fail to induce apoptosis in the absence of Bax and Bak indicates that the major role of Bcl-2 and Bcl-xL is to antagonize the apoptotic activity of Bax and Bak. In the absence of Bax and Bak, modulating Bcl-2 activity with BH3-only proteins does not affect the cell's fate to live or to die.

The data support a model in which the default status of a cell, in the absence of both pro- and anti-apoptotic Bcl-2 family members is survival (Figure 1). MEF cell death requires the effector function of either Bax or Bak. In a healthy cell, pro-survival Bcl-2 proteins antagonize this effect. The BH3-only proteins can disrupt the balance between these two groups. In response to cell death signals, BH3-only proteins are activated and trigger the activation of Bax/Bak through binding and neutralizing pro-survival Bcl-2 members, or activating directly the proapoptotic Bax and Bak, ultimately leading to apoptosis.

References

Cheng, E. H., Wei, M. C., Weiler, S., Flavell, R. A., Mak, T. W., Lindsten, T., and Korsmeyer, S. J. (2001). BCL-2, BCL-X(L) sequester BH3 domain-only molecules preventing BAX- and BAK-mediated mitochondrial apoptosis. Mol Cell *8*, 705–711.

Degenhardt, K., Sundararajan, R., Lindsten, T., Thompson, C., and White, E. (2002). Bax and Bak independently promote cytochrome C release from mitochondria. J Biol Chem *277*, 14127–14134.

Dijkers, P. F., Medema, R. H., Lammers, J. W., Koenderman, L., and Coffer, P. J. (2000). Expression of the pro-apoptotic Bcl-2 family member Bim is regulated by the forkhead transcription factor FKHR-L1. Curr Biol *10*, 1201–1204.

Green, D. R. (2000). Apoptotic pathways: paper wraps stone blunts scissors. Cell *102*, 1–4.

Han, J., Flemington, C., Houghton, A. B., Gu, Z., Zambetti, G. P., Lutz, R. J., Zhu, L., and Chittenden, T. (2001). Expression of bbc3, a pro-apoptotic BH3-only gene, is regulated by diverse cell death and survival signals. Proc Natl Acad Sci U S A *98*, 11318–11323.

Knudson, C. M., Tung, K. S., Tourtellotte, W. G., Brown, G. A., and Korsmeyer, S. J. (1995). Bax-deficient mice with lymphoid hyperplasia and male germ cell death. Science *270*, 96–99.

Lei, K., Nimnual, A., Zong, W. X., Kennedy, N. J., Flavell, R. A., Thompson, C. B., Bar-Sagi, D., and Davis, R. J. (2002). The Bax subfamily of Bcl2-related proteins is essential for apoptotic signal transduction by c-Jun NH(2)-terminal kinase. Mol Cell Biol *22*, 4929–4942.

Letai, A., Bassik, M. C., Walensky, L. D., M.D., S., Wiler, S., and Korsmeyer, S. K. (2002). Distinct BH3 domains either senitize or activate mitochondrial apoptosis serving as prototype cancer therapeutics. Cancer Cell *2*, 183–192.

Li, H., Zhu, H., Xu, C. J., and Yuan, J. (1998). Cleavage of BID by caspase 8 mediates the mitochondrial damage in the Fas pathway of apoptosis. Cell *94*, 491–501.

Lindsten, T., Ross, A. J., King, A., Zong, W. X., Rathmell, J. C., Shiels, H. A., Ulrich, E., Waymire, K. G., Mahar, P., Frauwirth, K., *et al.* (2000). The combined functions of proapoptotic Bcl-2 family members bak and bax are essential for normal development of multiple tissues. Mol Cell *6*, 1389–1399.

Luo, X., Budihardjo, I., Zou, H., Slaughter, C., and Wang, X. (1998). Bid, a Bcl2 interacting protein, mediates cytochrome c release from mitochondria in response to activation of cell surface death receptors. Cell *94*, 481–490.

McClintock, D. S., Santore, M. T., Lee, V. Y., Brunelle, J., Budinger, G. R., Zong, W. X., Thompson, C. B., Hay, N., and Chandel, N. S. (2002). Bcl-2 family members and functional electron transport chain regulate oxygen deprivation-induced cell death. Mol Cell Biol *22*, 94–104.

Motoyama, N., Wang, F., Roth, K. A., Sawa, H., Nakayama, K., Negishi, I., Senju, S., Zhang, Q., Fujii, S., and et al. (1995). Massive cell death of immature hematopoietic cells and neurons in Bcl-x-deficient mice. Science *267*, 1506–1510.

Nakano, K., and Vousden, K. H. (2001). PUMA, a novel proapoptotic gene, is induced by p53. Mol Cell *7*, 683–694.

Nakayama, K., Negishi, I., Kuida, K., Shinkai, Y., Louie, M. C., Fields, L. E., Lucas, P. J., Stewart, V., Alt, F. W., and et al. (1993). Disappearance of the lymphoid system in Bcl-2 homozygous mutant chimeric mice. Science *261*, 1584–1588.

Oda, E., Ohki, R., Murasawa, H., Nemoto, J., Shibue, T., Yamashita, T., Tokino, T., Taniguchi, T., and Tanaka, N. (2000). Noxa, a BH3-only member of the Bcl-2 family and candidate mediator of p53-induced apoptosis. Science *288*, 1053–1058.

Puthalakath, H., Huang, D. C., O'Reilly, L. A., King, S. M., and Strasser, A. (1999). The proapoptotic activity of the Bcl-2 family member Bim is regulated by interaction with the dynein motor complex. Mol Cell *3*, 287–296.

Puthalakath, H., Villunger, A., O'Reilly, L. A., Beaumont, J. G., Coultas, L., Cheney, R. E., Huang, D. C., and Strasser, A. (2001). Bmf: a proapoptotic BH3-only protein regulated by interaction with the myosin V actin motor complex, activated by anoikis. Science *293*, 1829–1832.

Rathmell, J. C., Lindsten, T., Zong, W.-X., Cinalli, R. M., and Thompson, C. B. (2002). Deficiency in Bak and Bax perturbs thymic selection and lympoid homeostasis. Nature Immunology *3*, 932–939.

Scaffidi, C., Fulda, S., Srinivasan, A., Friesen, C., Li, F., Tomaselli, K. J., Debatin, K. M., Krammer, P. H., and Peter, M. E. (1998). Two CD95 (APO-1/Fas) signaling pathways. Embo J *17*, 1675–1687.

Shinjyo, T., Kuribara, R., Inukai, T., Hosoi, H., Kinoshita, T., Miyajima, A., Houghton, P. J., Look, A. T., Ozawa, K., and Inaba, T. (2001). Downregulation of Bim, a proapoptotic relative of Bcl-2, is a pivotal step in cytokine-initiated survival signaling in murine hematopoietic progenitors. Mol Cell Biol *21*, 854–864.

Strasser, A., Harris, A. W., Huang, D. C., Krammer, P. H., and Cory, S. (1995). Bcl-2 and Fas/APO-1 regulate distinct pathways to lymphocyte apoptosis. Embo J *14*, 6136–6147.

Wei, M. C., Zong, W. X., Cheng, E. H., Lindsten, T., Panoutsakopoulou, V., Ross, A. J., Roth, K. A., MacGregor, G. R., Thompson, C. B., and Korsmeyer, S. J. (2001). Proapoptotic BAX and BAK: a requisite gateway to mitochondrial dysfunction and death. Science *292*, 727–730.

Yu, J., Zhang, L., Hwang, P. M., Kinzler, K. W., and Vogelstein, B. (2001). PUMA induces the rapid apoptosis of colorectal cancer cells. Mol Cell *7*, 673–682.

Zong, W. X., Lindsten, T., Ross, A. J., MacGregor, G. R., and Thompson, C. B. (2001). BH3-only proteins that bind pro-survival Bcl-2 family members fail to induce apoptosis in the absence of Bax and Bak. Genes Dev *15*, 1481–1486.

Chapter **10**

Novel Transcriptional Regulatory Pathways of IL-3-Dependent Survival Responses

JEFFREY J.Y. YEN[1], YUNG-LUEN YU[1], WANNHSIN CHEN[1,2], AND YUN-JUNG CHIANG[1]

ABSTRACT: In the past several years, we devoted in the identification of cellular transcriptional factors that could recognize the CES2/E2A-HLF binding element (CBE) and were involved in apoptotic regulation. We firstly demonstrated the existence of multiple binding complexes of CBE in various cell types and tissues, and identified cAMP-responsive element binding protein (CREB) as a component in one major CBE-complex of hematopoietic cell lines. Stimulation of hematopoietic cells with IL-3 promptly induced phosphorylation of CREB at serine 133 partially via a PKA-dependent pathway. Alteration of function of PKA or CREB strongly correlated with the survival capacity of hematopoietic cells. Secondly, we explored the IL-3-dependent transactivation mechanism of another CBE-binding protein, E4BP4. We demonstrated that *E4bp4* was regulated by IL-3 mainly at the transcriptional level. Promoter mapping and binding analyses revealed that GATA-1 and GATA-2 proteins were responsible for *E4bp4* transcription via a conserved GATA site. Functional assays also suggested that GATA-1 not only modulated the expression of the E4bp4 gene but also controlled apoptosis. The discovery of involvement of CREB and GATA transcriptional factors in IL-3's survival responses extends our knowledge on the anti-apoptotic pathways in hematopoietic cells and may contribute to our further understanding of the process of hematopoiesis and leukemogenesis.

Introduction

As one of the most characterized cytokines, interleukin 3 (IL-3) is well known for its survival effect on both hematopoietic stem and progenitors cells. Although with the extensive studies, the signaling pathway and underlying mechanism leading to survival responses of IL-3 still are not completely understood. Recently, an apoptotic genetic pathway of *Caenorhabditis elegans* was suggested to be evolutionally conserved in the cytokine-dependent anti-apoptotic responses in mammalian hematopoietic cell lineages. In this pathway, *Ces-2* is required for a cell-type specific apoptotic program and encodes a basic region-leucine zipper (bZIP) family transcriptional factor recognizing the same DNA motif

[1]Institute of Biomedical Sciences, Academia Sinica, Taipei, Taiwan; [2]Current address: Division of Biomaterials and Tissue Engineering, Biomedical Engineering Center, Industrial Technology Research Institute, Hsinchu, Taiwan. *Correspondence:* Dr. Jeffrey J.Y. Yen, Institute of Biomedical Sciences, Academia Sinica, Taipei, 11529 Taiwan. Phone: 886 2 789 90001; Fax: 886 2 785 3569; E-mail: bmjyen@ibms.sinica.edu.tw

as an oncoprotein E2A-HLF does, created by chromosome translocation in certain human acute lymphoblastic leukemia. In the past several years, we devoted in the identification and characterization of two of these cellular transcriptional factors that recognize this CES-2/E2A-HLF binding element (CBE) and are involved in apoptotic regulation. Our works clearly established that at least two novel transcriptional regulatory pathways are involved in transducing IL-3 survival signal. The discovery of involvement of PKA as well as GATA factors in IL-3 signal transduction extends our knowledge on the apoptotic pathway in hematopoietic cells, and may contribute to our further understanding of the processes of hematopoiesis and leukemogenesis.

The Role of the *Ces-2* Gene in Apoptosis Pathway of *C. elegans*

During the development of *C. elegans* embryo, 131 out of the 1,090 cells undergo programmed cell death (1,2). Genetic studies have identified mutations in many genes that specifically affect this process, and recent findings suggest that these genes define an evolutionarily conserved genetic pathway for programmed cell death from *C. elegans* to the humans. Two genes that only affect the death of a small number of cells were suggested involving in developmentally regulated death program in a cell-lineage specific manner, meanwhile the other genes affecting all 131 dying cells are mutations of the basic apoptotic machinery. In early embryogenesis, the sister neurosecretory motor (NSM) neurons in the pharynx undergo programmed cell death depending on the expression of two cell death specification genes, *Ces-1* and *Ces-2* (3-5). Dominant gain-of-function mutations in the *Ces-1* gene and recessive loss-of-function mutations in the *Ces-2* gene allow these two neuron cells to survive, while *Ces-1* loss-of-function mutation suppresses the phenotype of *Ces-2* loss-of-function mutation and adopts the wild-type phenotype in NSM sister neurons. Therefore, the *Ces-2* gene was suggested to encode a pro-apoptotic protein that in turn represses the function of the antiapoptotic *Ces-1* gene. Recently, the *Ces-2* gene was molecularly cloned, and the predicted protein possessed the characteristic functional domains of bZIP proteins (6). Not only did the amino acid sequence of CES-2 share significant homology with E4BP4 and PAR family proteins, including HLF and TEF, in the basic DNA-binding domains, but also the DNA sequence recognized by the CES-2 protein was identical to that of the PAR family and E4BP4.

The E2A-HLF Oncoprotein in Human Acute B-Lineage Leukemia

On the other hand, molecular analysis of chromosomal translocation t(17;19)(q22;p13) in human acute lymphoblastic leukemia reveals that this translocation results in B-cell-specific expression of a chimeric protein that fused HLF to the transcription factor E2A (7,8). This fusion protein, E2A-HLF, retains the DNA binding specificity of its parental HLF protein but has greater trans-activation and transformation potential (9,10). The close homology of the basic DNA-binding domain of HLF to that of the Ces-2 protein of *C. elegans* further suggests the involvement of E2A-HLF in the conserved cell death pathway. Recently, anti-apoptotic activity was suggested to underlie the mechanism of transformation of the E2A-HLF gene (11). Surprisingly, structural study reveals that disabling mutations of the HLF bZIP domain had little effect on the antiapoptotic activity of the chimeric protein (12). But in the context of an intact HLF bZIP domain, the AD1 but not the AD2 transcription

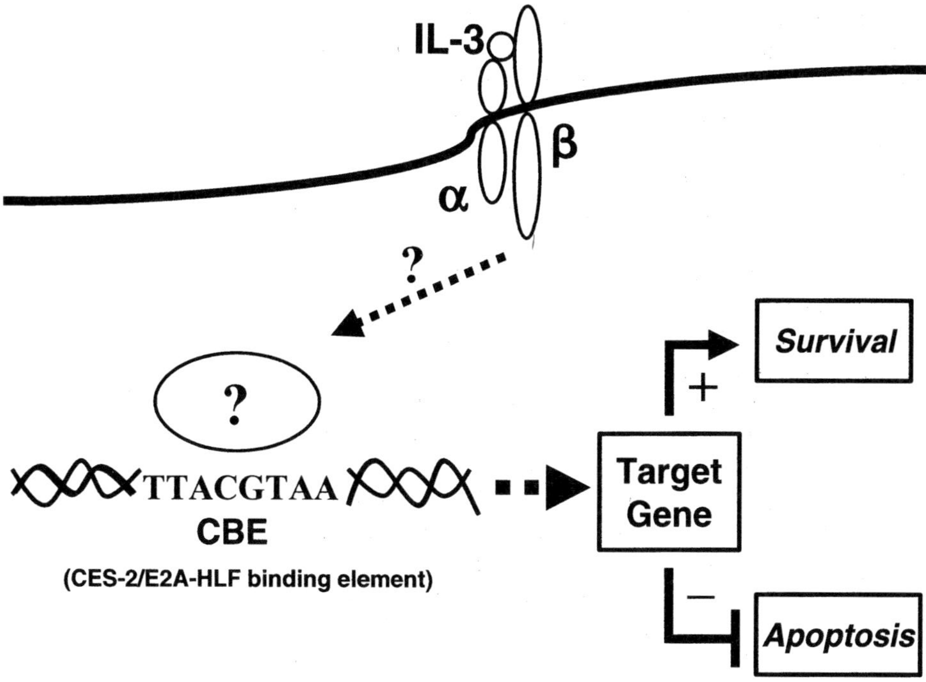

Figure 1. Identification and characterization of cellular CES-2/E2A-HLF binding element (CBE) binding proteins that involve in IL-3-dependent survival responses. In hematopoietic cell lineages, cytokine-withdrawal induced-apoptosis can be suppressed by oncoprotein E2A-HLF which binds to CBE and activates CBE-containing survival genes. Hypothetically, E2A-HLF will transactivate the CBE-containing target genes by competing out the unknown cellular CBE modulators (the "?" circle) which are constantly under the control of cytokines via an un-identified pathway.

domain of E2A was required for antiapoptotic activity. However, either AD1 or AD2 could promote cell survival after cytokine deprivation in conjunction with a defective bZIP domain. Therefore, a dual mechanism model has been raised that in one hand AD1 and bZIP domains act cooperatively to block apoptosis and on the other hand interaction of HLF with the amino-terminal region of E2A act to block the expression of genes that normally control the apoptotic machinery of pro-B cells (12). Since overexpression of E2A-HLF in a factor-dependent cell line resulted in prolonged survival, the targets of the E2A-HLF protein may be involved in antiapoptotic functions in responding to stimulation of cytokines (Figure 1). These target genes may be normally under the control of a set of cellular regulators that recognize the conserved DNA sequences on promoters as E2A-HLF does.

CES-2/E2A-HLF Binding Element (CBE) Binding Proteins

To identify the evolutionarily conserved CES-2/E2A-HLF homologues in mammalian cells, we have set out two means to identify these factors (13). One is a biochemical approach by binding the nuclear extracts with the radioactive ^{32}P-labeled consensus binding oligonucleotide probe, also known as electrophoretic mobility shift assay (EMSA). Another approach is a molecular biological means by screening the E. coli expression library with tandem-ligated ^{32}P-labeled consensus binding oligonucleotide probe. By the EMSA method, we found that the nuclear extract from a murine factor-dependent pro-B cell line Ba/F3 bound to the CBE oligonucleotide probe and formed multiple binding complexes in the native polyacrylamide gel. A complete inhibition was observed with an unlabeled CBE probe but not with a 4-bp mismatched M4 probe, thus verifying the sequence specificity of these complexes. Most of these complexes were present in all extracts tested, including extracts from human GM-CSF-dependent TF-1, murine IL-2-dependent HT-2, human cervical carcinoma HeLa, human hepatocarcinoma HepG2, murine fibroblast NIH 3T3, and Chinese hamster ovary carcinoma CHOP cell lines, and in extracts from several tissues, including thymocytes and splenocytes of BALB/c mice and thymocytes of porcine origin. The levels of slow-migrating group of complexes were more variable while that of fast-migrating group were more constant.

In ion-free buffer, very little binding activity was detected with Ba/F3 and TF-1 nuclear extracts. A drastic increase of binding activity was observed in buffers containing KCl from 50 to 200 mM, with a peak at around 133 mM. All of these complexes seemed to require an optimal ionic concentration similar to that recommended for the binding of E2A-HLF to CBE, i.e., 133 mM (7). None of these binding complexes were heat resistant, and they rapidly diminished when the temperature exceeded 55°C. When nuclear extract was prepared from cells depleted of cytokine, the longer the cells were starved, the more the binding capacity of nuclear extract decreased. Western blot analysis revealed that the nuclear content of CREB also decreased and this could be caused either by the decrease of nuclear localization, modification, or synthesis of the transcriptional factors in responding to cytokine deprivation.

Alternatively, we used a ^{32}P-labeled CBE oligonucleotide to screen three expression libraries, including that of human leukocytes, mouse embryonic stem cells, and mouse 15-day embryo, in E. coli. E4BP4 and C/EBPb were very abundant and readily picked up from the leukocyte library by CBE probe. Thyroid embryonic factor (TEF), a PAR family member, and ATF2, a CREB family member, were found in the embryonic stem cell library. During early embryonic development (at day 15), the expression of three different transcriptional factor families could be detected, including the TEF and HLF family, CREB and ATF2 family, and C/EBPb family. These data are consistent with the EMSA study by that multiple transcriptional factors were able to bind to the CBE motif in a given cell type, and that the composition in the CBE complexes of different tissues might vary accordingly (Figure 2). By no means our data suggest all of these genes are involved in survival responses of the target cells, however, it is certainly worth of careful re-examination of the antiapoptotic potential of these genes during embryogenesis.

CREB and IL-3's Function

In the attempt to identify the cellular CBE-binding protein(s) that may be involved in apoptosis regulation in mammals, cyclic AMP-responsive element (CRE)-binding protein (CREB) was identified as one major CBE complex of Ba/F3 and TF-1 cells (13). Initially, we were attempted to explore the molecular identities of these CBE-binding proteins by applying UV to cross-link nuclear proteins to ^{32}P-labeled CBE and then fractionated the proteins by SDS-PAGE to determine their molecular weights. From the sizes of these CBE-binding components and the sequence homology between CBE and the CRE, we postulated that one of the 46-kDa proteins was CREB. We went further to demonstrate the presence of

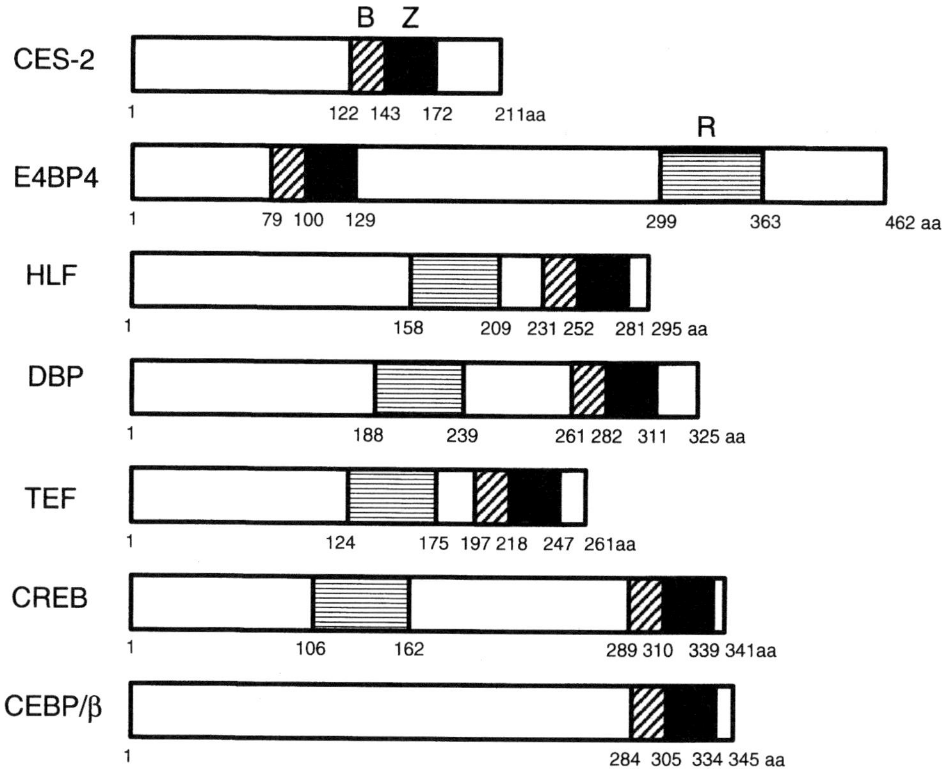

Figure 2. Schematic representation of tentative cellular CBE-binding transcriptional factors. The basic and leucine Zipper (bLIZ) domain of each protein is indicated as a hatched and solid boxes; the transcriptional regulatory domains are indicated as a shaded "R" box.

CREB in the fast-migrating group of CBE-binding complexes by supershifting the specific complexes with anti-CREB antibody. However, CREB bound to both sequences equally well, suggesting that CREB binds to both elements with similar affinities. Activation of CREB by pharmacological agents or the catalytic subunit of protein kinase A (PKAc) resulted in induction of the CBE-driven reporter gene. Intriguingly, stimulation of Ba/F3 cells with interleukin-3 (IL-3) promptly induced phosphorylation of CREB at serine133 partially via an unknown PKA-dependent pathway (unpublished data). Consistently, Ba/F3 cell survival in the absence of IL-3 was prolonged by activation of PKA. Conversely, treatment of cells with a PKA inhibitor or expression of the dominant negative forms of PKA and CREB overrode the survival activity of IL-3. Last, a functional CBE motif was identified in the *bcl-2* promoter in responding to PKA and IL-3 stimulation, suggesting an underlying mechanism for the anti-apoptotic activity of PKA/CREB/CBE transcriptional pathway (Figure 3).

GATA Factors and IL-3's Function

GATA family proteins are a group of transcription factors containing two related zinc fingers that mediate DNA binding. Among the six known members, GATA-1, GATA-2,

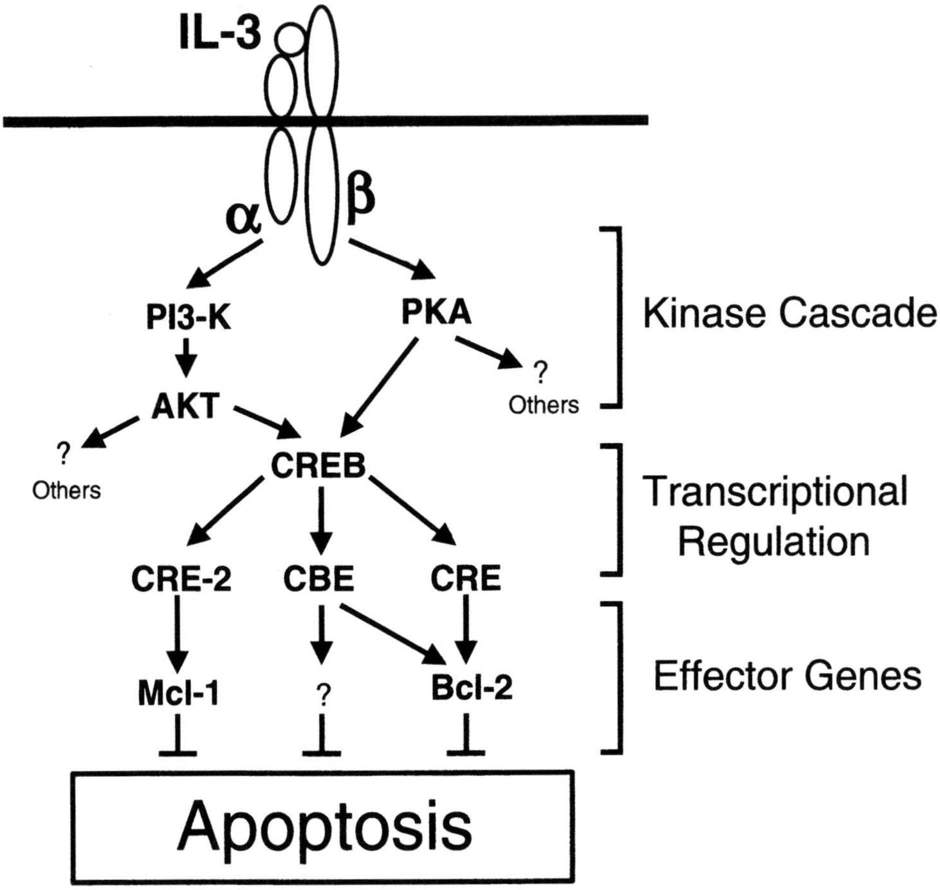

Figure 3. The CREB protein can work on various target promoter elements once it is activated by interleukin 3 via either PI-3K/Akt or PKA pathway.

and GATA-3 are preferentially expressed in hematopoietic cells, whereas the other GATA factors are expressed exclusively in nonhematopoietic tissues. The GATA factors were suggested to play an important role in the differentiation of the erythropoietic lineages due to the fact that the functionally important GATA motifs were identified in virtually all erythroid-specific genes. Not until recently they are suggested to be also critical in maintaining the viability of their target cells, by studying the GATA-1 and GATA-2-null embryos (14,15). We independently discovered the anti-apoptotic function of GATA-1 during the study of mechanism underlying the IL-3-dependent regulation of antiapoptotic transcriptional factor E4BP4 (16). *E4bp4*, a member of bZIP transcriptional factor subfamily, has been shown to be up-regulated by the IL-3 and plays an important role in IL-3's anti-apoptotic response. In this study, we demonstrated that *E4bp4* is regulated by IL-3 mainly at the transcriptional level. Promoter mapping and binding analyses revealed that GATA-1 and GATA-2 proteins bind to a critical GATA motif in the *E4bp4* promoter in vitro and chromatin immunoprecipitation (ChIP) assay further confirmed the in vivo binding of GATA-1 to the *E4bp4* promoter. In IL-3-dependent cell line, overexpression of GATA factors transactivated the *E4bp4* reporter, increased the expression of endogenous *E4bp4* gene, and postponed cell death, while

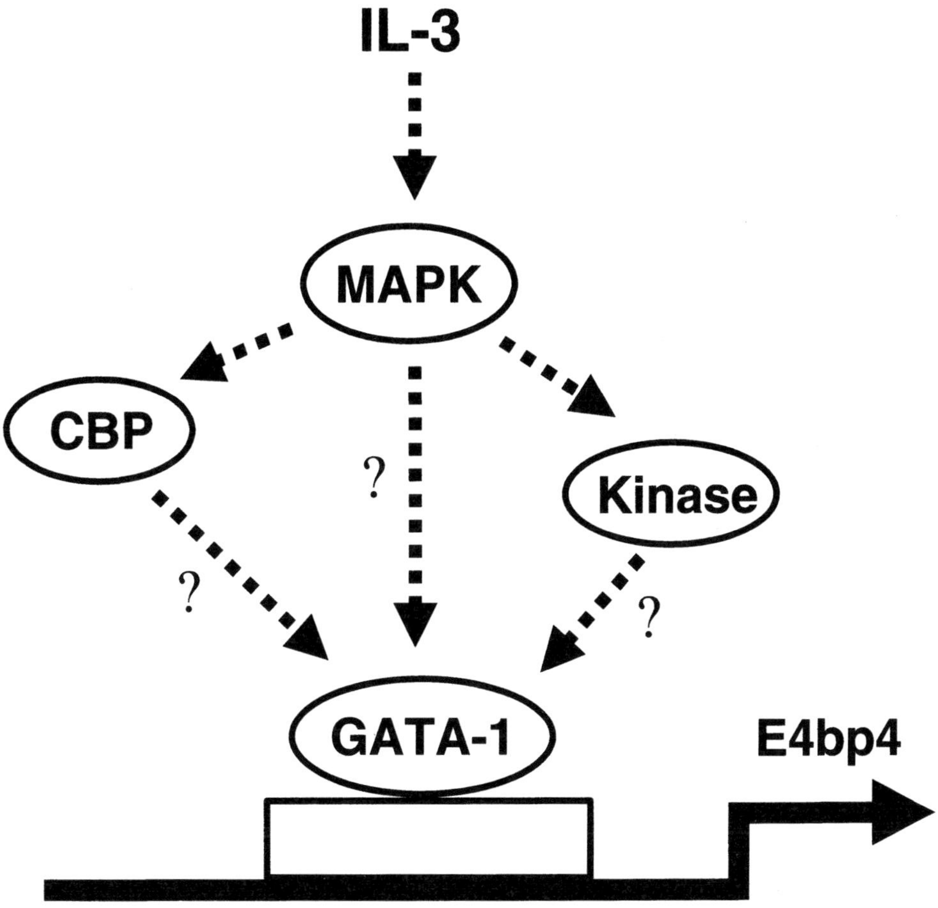

Figure 4. The possible mechanisms that GATA-1 is regulated by IL-3 in hematopoietic cells. There are several potential mechanisms are indicated; The GATA proteins can be activated via phosphorylation either directly by MAPK, or indirectly by an indermediate kinase, or via acetylation by the CREB-binding protein (CBP).

interference of GATA function by dominant negative GATA-1 mutant abolished all the effects, suggesting a critical role of GATA factors in survival responses of IL-3-dependent Ba/F3 cells. Intriguingly, the binding capability of GATA-1 to the promoter DNA seems to be regulated by cytokine, and GATA-1 becomes hyper-phosphorylated in responding to IL-3 stimulation via a yet to-be-identified RAS/RAF/MAPK pathway (unpublished data). Further investigation on the mechanism of in vivo chromatin unwinding and assembly of the transcriptional activators on this GATA-dependent *E4bp4* gene will unravel some molecular insights on how a gene is regulated in development (Figure 4).

A Novel Transcriptional Regulation Pathway in Development

There are two clues suggest that *E4bp4* may be a very unique target of the GATA factor family members. One is in the promoter analysis experiments; the clustered point mutations at the GATA site in the 1139-bp promoter fragment almost completely abolished both the

Table 1. Expression on *E4bp4* and *GATA Factor* mRNA in Various Tissues

Tissue	E4BP4	GATA-1	GATA-2	GATA-3	GATA-4	GATA-5	GATA-6
Leukocyte	+++	+++	+++				
Skeletal muscle	+++				+		+
Testis	+++	++	++		+		+
Placenta	++		++	++	+		+
Heart	++		+		+++	+++	+++
Lung	++		+	+	++	++	++
Ovary	++	+	+		++		++
Prostate	++		+	+			
Spleen	++	++	++	+++			
Pancreas	+				+	+	+
Kidney	+		+	+			
Colon	+				+	+	+
Small intestine	+				+	+	+
Thymus	+			++			
Liver	+		+	+	+	+	+
Brain	+		+	+			
Smooth muscle	+		+		+		+

Note: +++ = Strong expression; ++ = moderate expression; + = weak expression.

basal and inducible promoter activity of *E4bp4*, suggesting the presence of GATA factor is very crucial for the basic transcriptional capability of the *E4bp4* gene. The other one is in the in vitro EMSA analyses; the *E4bp4* GATA site could not distinguish among the GATA family proteins, binding with similar affinity at least to GATA-1, GATA-2, and GATA-3. It would be conceivable that in tissues wherever with high levels of expression of any GATA factors, *E4bp4* may also be highly inducible by these GATA factors. Indeed a recent observation in T cell differentiation supports the idea. The *E4bp4* mRNA was highly expressed during differentiation of both CD4+ and CD8+ type 2 T cells where GATA-3 was strongly expressed (17). These observations were consistent with the possibility that *E4bp4* may contribute to the GATA-dependent survival effect during B cell and T cell development (18–21). Furthermore, *E4bp4* transcripts have a wide tissue distribution with a particularly high expression level in peripheral blood leukocytes, testis, skeletal muscle, placenta, heart, prostate, ovary, and lung (22) (Table 1). Interestingly, high level expression of GATA-1 or GATA-2 has been reported in peripheral blood leukocytes and testis (23–25, 26,27), and other GATA family members are preferentially expressed in tissues like heart, prostate, ovary, and lung (28). Thus, it would be of high interest to further explore whether the GATA/E4bp4 transcriptional regulatory pathway represents a novel survival response commonly utilized during the development of many hematopoietic and nonhematopoietic tissues.

References

1. Sulston, J.E., and Horvitz, H.R. (1977). Post-embryonic cell lineages of the nematode, Caenorhabditis elegans. Dev. Biol. *56*, 110–156.
2. Sulston, J.E., Schierenberg, E., White, J.G., and Thomson, J.N. (1983). The embryonic cell lineage of the nematode Caenorhabditis elegans. Dev. Biol. *100*, 64–119.
3. Ellis, H.M., and Horvitz, H.R. (1986). Genetic control of programmed cell death in the nematode C. elegans. Cell. *44*, 817–829.

4. Ellis, R.E., and Horvitz, H.R. (1991). Two C. elegans genes control the programmed deaths of specific cells in the pharynx. Development *112*, 591–603.
5. Trent, C., Tsuing, N., and Horvitz, H.R. (1983). Egg-laying defective mutants of the nematode Caenorhabditis elegans. Genetics *104*, 619–647.
6. Metzstein, M.M., Hengartner, M.O., Tsung, N., Ellis, R.E., and Horvitz, H.R. (1996). Transcriptional regulator of programmed cell death encoded by Caenorhabditiselegans gene ces-2. Nature *382*, 545–547.
7. Hunger, S.P., Ohyashiki, K., Toyama, K., and Cleary, M.L. (1992). Hlf, a novel hepatic bZIP protein, shows altered DNA-binding properties following fusion to E2A in t(17;19) acute lymphoblastic leukemia. Genes Dev. *6*, 1608–1620.
8. Inaba, T., Roberts, W.M., Shapiro, L.H., Jolly, K.W., Raimondi, S.C., Smith, S.D., and Look, A.T. (1992). Fusion of the leucine zipper gene HLF to the E2A gene in human acute B-lineage leukemia. Science *257*, 531–534.
9. Inaba, T., Shapiro, L.H., Funabiki, T., Sinclair, A.E., Jones, B.G., Ashmun, R.A., and Look, A.T. (1994). DNA-binding specificity and trans-activating potential of the leukemia-associated E2A-hepatic leukemia factor fusion protein. Mol. Cell. Biol. *14*, 3403–3413.
10. Yoshihara, T., Inaba, T., Shapiro, L.H., Kato, J.Y., and Look, A.T. (1995). E2A-HLF-mediated cell transformation requires both the trans-activation domains of E2A and the leucine zipper dimerization domain of HLF. Mol. Cell. Biol. *15*, 3247–3255.
11. Inaba, T., Inukai, T., Yoshihara, T., Seyschab, H., Ashmun, R.A., Canman, C.E., Laken, S.J., Kastan, M.B., and Look, A.T. (1996). Reversal of apoptosis by the leukaemia-associated E2A-HLF chimaeric transcription factor. Nature *382*, 541–544.
12. Inukai, T., Inaba, T., Ikushima, S., and Look, A.T. (1998). The AD1 and AD2 transactivation domains of E2A are essential for the antiapoptotic activity of the chimeric oncoprotein E2A-HLF. Mol. Cell. Biol. *18*, 6035–6043.
13. Chen, W., Yu, Y.L., Lee, S.F., Chiang, Y.J., Chao, J.R., Huang, J.H., Chiong, J.H., Huang, C.J., Lai, M.Z., Yang-Yen, H.F., and Yen, J.J. (2001). CREB is one component of the binding complex of the Ces-2/E2A-HLF binding element and is an integral part of the interleukin-3 survival signal. Mol. Cell. Biol. *21*, 4636–4646.
14. Pevny, L., Simon, M.C., Robertson, E., Klein, W.H., Tsai, S.F., D'Agati, V., Orkin, S.H., and Costantini, F. (1991). Erythroid differentiation in chimaeric mice blocked by a targeted mutation in the gene for transcription factor GATA-1. Nature *349*, 257–260.
15. Tsai, F.Y., Keller, G., Kuo, F.C., Weiss, M., Chen, J., Rosenblatt, M., Alt, F.W., and Orkin, S.H. (1994). An early haematopoietic defect in mice lacking the transcription factor GATA-2. Nature *371*, 221–226.
16. Yu, Y.L., Chiang, Y.J., and Yen, J.J. (2002). GATA factors are essential for transcription of the survival gene E4bp4 and the viability response of interleukin-3 in Ba/F3 hematopoietic cells. J. Biol. Chem. *277*, 27144–27153.
17. Chtanova, T., Kemp, R.A., Sutherland, A.P., Ronchese, F., and Mackay, C.R. (2001). Gene microarrays reveal extensive differential gene expression in both CD4(+) and CD8(+) type 1 and type 2 T cells. J. Immunol. *167*, 3057–3063.
18. Weiss, M.J., and Orkin, S.H. (1995). Transcription factor GATA-1 permits survival and maturation of erythroid precursors by preventing apoptosis. Proc. Natl. Acad. Sci. USA. *92*, 9623–9627.
19. Tsai, F.Y., and Orkin, S.H. (1997). Transcription factor GATA-2 is required for proliferation/survival of early hematopoietic cells and mast cell formation, but not for erythroid and myeloid terminal differentiation. Blood *89*, 3636–3643.
20. Pandolfi, P.P., Roth, M.E., Karis, A., Leonard, M.W., Dzierzak, E., Grosveld, F.G., Engel, J.D., and Lindenbaum, M.H. (1995). Targeted disruption of the GATA3 gene causes severe abnormalities in the nervous system and in fetal liver haematopoiesis. Nat. Genet. *11*, 40–44.
21. Hendriks, R.W., Nawijn, M.C., Engel, J.D., van Doorninck, H., Grosveld, F., and Karis, A. (1999). Expression of the transcription factor GATA-3 is required for the development of the earliest T cell progenitors and correlates with stages of cellular proliferation in the thymus. Eur. J. Immunol. *29*, 1912–1918.
22. Lai, C.K., and Ting, L.P. (1999). Transcriptional repression of human hepatitis B virus genes by a bZIP family member, E4BP4. J. Virol. *73*, 3197–3209.
23. Yamamoto, M., Ko, L.J., Leonard, M.W., Beug, H., Orkin, S.H., and Engel, J.D. (1990). Activity and tissue-specific expression of the transcription factor NF-E1 multigene family. Genes Dev. *4*, 1650–1662.
24. Martin, D.I., Zon, L.I., Mutter, G., and Orkin, S.H. (1990). Expression of an erythroid transcription factor in megakaryocytic and mast cell lineages. Nature *344*, 444–447.

25. Romeo, P.H., Prandini, M.H., Joulin, V., Mignotte, V., Prenant, M., Vainchenker, W., Marguerie, G., and Uzan, G. (1990). Megakaryocytic and erythrocytic lineages share specific transcription factors. Nature *344*, 447–449.
26. Ito, E., Toki, T., Ishihara, H., Ohtani, H., Gu, L., Yokoyama, M., Engel, J.D., and Yamamoto, M. (1993). Erythroid transcription factor GATA-1 is abundantly transcribed in mouse testis. Nature *362*, 466–468.
27. Yomogida, K., Ohtani, H., Harigae, H., Ito, E., Nishimune, Y., Engel, J.D., and Yamamoto, M. (1994). Developmental stage- and spermatogenic cycle-specific expression of transcription factor GATA-1 in mouse Sertoli cells. Development. *120*, 1759–1766.
28. Weiss, M.J., and Orkin, S.H. (1995). GATA transcription factors: key regulators of hematopoiesis. Exp. Hematol. *23*, 99–107.

Chapter 11

MAP-1 Is a Putative Ligand for the Multidomain Proapoptotic Protein Bax

KUAN ONN TAN, SHING-LENG CHAN, NAIYANG FU, AND VICTOR C. YU*

ABSTRACT: Recent data on gene deletion analyses revealed that cells derived from animals that are deficient in both *Bax* and *Bak* (two key members of the "multidomain" proapoptotic Bcl-2 subfamily), but not cells lacking one of the two genes, are completely resistant to apoptotic death triggered by diverse stimuli. These data suggest that engagement of a "multidomain" proapoptotic member, BAX or BAK, is essential for apoptotic signaling. To gain further understanding of the molecular functions of the "multidomain" proapoptotic molecules, we used yeast two hybrid screen to identify protein partners of BAX. A proapoptotic protein, termed MAP-1 (Modulator of Apoptosis), was identified as a BAX-interacting protein from a human brain cDNA library. MAP-1 is a member of a growing family of proteins that were initially identified as onconeural antigens. MAP-1 contains a BH3-like motif and mediates caspase-dependent death in mammalian cells when overexpressed. Mutagenesis analyses revealed that the BH3-like domain of MAP-1 is required for association with BAX and for mediating apoptosis. Interestingly, in contrast to all other previously identified BAX-associating proteins that require only one of the three BH domains of BAX for binding, all the three BH (BH1, BH2 and BH3) domains of BAX are necessary for binding to MAP-1. Taken together, these data suggest that MAP-1 may mediate its proapoptotic function by engaging its BH3-like domain in binding the Bcl-X_L-like hydrophobic pocket of BAX.

Key Words: BCL-2; BAX; BAK; multidomain; BH domain; apoptosis; paraneoplastic syndromes; onconeural antigen

Introduction

Bcl-2 family of proteins is central regulators of cell death and survival that can be grouped into sub-families of pro-survival and pro-apoptotic molecules (Adams and Cory, 1998). They are characterized by the presence of several conserved motifs, known as the Bcl-2 homology (BH) domains. While the N-terminal BH4 domain is restricted to some pro-survival members, BH1, BH2, and BH3 can generally be found among members of both the pro-survival and pro-apoptotic sub-families (Adams and Cory, 1998). Mutagenesis

*Institute of Molecular and Cell Biology, 30 Medical Drive, Singapore 117609, Republic of Singapore. *Correspondence:* Victor C. Yu, Ph.D. Institute of Molecular and Cell Biology, 30 Medical Drive, Singapore. Tel: 65 6874-3740; Fax: 65 6779-1117; E-mail: mcbyuck@imcb.nus.edu.sg and structural studies preformed in pro-survival members BCL-2 and Bcl-X_L revealed that the BH domains are important functional domains that are also required for dimerization functions (Zha et al., 1996). Interestingly, association of the pro-survival member Bcl-X_L with the pro-apoptotic member Bax requires the BH1, BH2, and BH3 domains of the former but only the BH3 domain of the latter (Zha et al., 1996, Sattler et al., 1997).

In addition to its role as a protein-protein interaction domain, the BH3 domain of pro-apoptotic members appears to be important for mediating their pro-apoptotic function (Adams and Cory, 1998). This notion is further supported by the recent discovery of a new group of cell death agonists containing only the BH3 domain (Kelekar and Thompson, 1998). Similar to other pro-apoptotic members of the Bcl-2 family, the BH3 domain in these molecules plays an important role in mediating pro-apoptotic function as well as association with the pro-survival members of the Bcl-2 family.

The three-dimensional structure of Bcl-X_L demonstrates that the BH1, BH2, and BH3 regions form an elongated hydrophobic cleft to which a BH3 amphipathic alpha helix can bind (Muchmore et al., 1996, Sattler wt al., 1997). It was thus postulated the multidomain as well as the BH3-only subfamily of proapoptotic molecules mediate cell killing by engaging its BH3 domain as death ligand through binding to the receptor domain of the prosurvival members of the BCL-2 family. Interestingly, recent data from gene knock-out study demonstrated that cells lacking both Bax and Bak are completely resistant to diverse apoptotic stimuli, including those triggered by the "BH3-only" molecules (Lindsten et al., 2000, Wei et al., 2001). These data suggest that BAX and BAK can act downstream of BH3-only molecules and activation of a "multidomain" proapoptotic member, BAX or BAK, is pre-requisite for apoptosis signaling.

Proapoptotic members of the Bcl-2 family that contains multiple BH domains such as Bax are thus proposed to exist in two different conformations: one that is similar to Bcl-X_L and the other with the BH3 domain rotated outside to allow its insertion into the hydrophobic cleft of a pro-survival protein (Adams and Cory, 1998). Recent data from NMR analyses confirms the existence of a Bcl-X_L-like conformation in BAX (Suzuki et al., 2000). This structural feature of BAX suggests that the elongated Bcl-X_L-like hydrophobic binding pocket could potentially serve as a receptor for ligands with structural similarity to the BH3 domain.

Results

Identification of a BAX-Associating Protein, MAP-1, That Contains a Novel BH3-Like Motif

In our effort to gain further understanding of the functions of the elongated hydrophobic pocket formed by the BH domains of Bax, we used the yeast two-hybrid screen to identify Bax-associating proteins. Screens were carried out in Human Brain, HeLa and Human B-cell libraries (Clontech) by using mouse cDNAs encoding amino acids 1-171 of Bax (BaxΔC21) as baits. From approximately six million transformants, more than one hundred clones from the primary screen were identified, five of which were found to interact specifically with BaxΔC21 but not with other unrelated Gal-4 fusion proteins tested. Interestingly, one of the clones, B26, encodes a protein fragment containing a BH3-like motif (Fig. 1). Upon sequencing, the B26 clone was found to contain a partial complementary DNA (cDNAs). Using B26 cDNA fragment as probe, we screened a human brain (cerebellum) cDNA library and obtained several cross-hybridizing cDNA clones. The longest clone, B9, contained a

Figure 1. Primary protein sequence and sequences alignment of MAP-1 family of proteins. Predicted amino acid sequence of human MAP-1 is aligned with that of the human Ma1, Ma2 and Ma3. Amino acid residues that are identical to human MAP-1 are boxed. The Bax-interacting clone isolated from the yeast two-hybrid screen was fused to Gal-4 activation domain at the positions indicated by B26. The putative BH3-like domains are in boldface and underlined.

2.2 kb cDNA insert with a 1056 nucleotide open reading frame beginning with a translational initiation consensus sequence and ending with an in-frame stop codon that encoded a protein of 351 amino acids with a predicted molecular mass of 39 kDa (Fig. 1). We named this protein MAP-1 for modulator of apoptosis (Tan et al., 2001).

A region of eight amino acids (amino acids 120–127) in MAP-1 was found to be highly similar to the BH3 domains present among members of the Bcl-2 family (Fig. 1). The BH3-like motif identified appears to be most similar to the recently identified BH3-B domain present in the N-terminal region of BID (Tan et al., 1999). Interestingly, the BH3-only molecule, BID, is unique among other BH3-only molecules because of its ability in binding both Bcl-X_L and BAX (Wang et al., 1996). It was proposed that the binding of BID to BAX might be a critical step in the activation of BAX in apoptosis signaling. The BH3-B motif of BID is required only for mediating intramolecular interaction, but is dispensable for binding to BAX or Bcl-X_L (Tan et al., 1999).

MAP-1 Is a Member of a Protein Family of Onconeural Antigens

Data base searches revealed that the predicted amino acid sequence of the human MAP-1 protein shares extensive similarity with three other human proteins (Fig. 1), Ma1 (58%) (Dalmau et al., 1999), Ma2 (47%) (Voltz et al., 1999), and MA3 (40%) (Rosenfeld et al., 2001) suggesting that MAP-1 is a member of a gene family. Ma1, Ma2 and MA3 were all initially identified as antigens recognized by autoantibodies present in the sera of patients with paraneoplastic neurological syndromes (PNS). PNS are neurologic degenerative disorders that occur in patients with neoplasms outside the nervous system. PNS usually precede detection of the tumor and it may affect any part of the nervous system and are often more debilitating than the cancer itself. Using sera obtained from patients with PNS afflicted with underlying malignancy to undertake expression cloning, several onconeural antigens, termed MA1, MA2 and MA3, with highly homologous protein sequences to each other, were cloned. Ma1 and Ma2 have a highly restricted expression pattern. Ma1 protein is detected only in brain and testis (Dalmau et al., 1999) while Ma2 protein is found exclusively in brain (Voltz et al., 1999). At present, no function has been described for the Ma1, Ma2 and MA3 proteins. Interestingly, despite high degree of amino acid sequence similarity of MAP-1 to the MA proteins, MAP-1 has not been reported as an onconeural antigen.

MAP-1 Associates with Only a Subset of BCL-2 Family of Proteins

The ability of MAP-1 to associate with Bax was further demonstrated *in vivo* in transient co-transfection experiments. HA-tagged Bax, Bid, BimL, Bik, Bcl-2 and Bcl-X_L were transiently co-expressed with myc-tagged MAP-1 in 293T cells and lysates were prepared for immunoprecipatation as previously described (Chan et al., 1999, 2000). MAP-1 immunoprecipitated specifically with Bax, Bcl-2, and Bcl-X_L but not other members of BCL-2 family tested (Fig. 2). These interactions observed in the immunoprecipitation

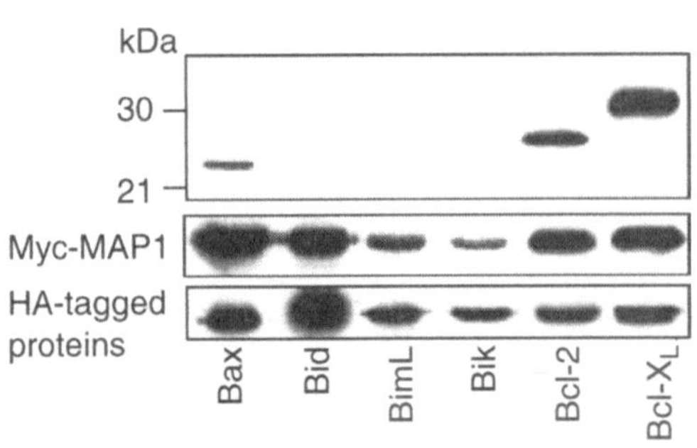

Figure 2. MAP-1 associates with a subset of BCL-2 family of proteins. 293T cells (100 mm plates at 70% confluency) were transiently co-transfected with 10 mg each of the indicated expression plasmids. Myc-MAP-1 was immunoprecipitated with the polyclonal anti-Myc antibody (A14). Co-precipitating HA-tagged proteins were detected by Western blot analysis using the monoclonal HA antibody (F7). Expressions of the Myc- and HA-tagged proteins were determined by Western blot analysis of an aliquot (1%) of the total extract (1 ml) with monoclonal antibodies (lower panels). The bands representing the levels of expressed HA-tagged proteins were aligned (lowest panel). Position of molecular weight standards are shown.

assays were further demonstrated *in vitro* by using the GST-pulldown assay. GST-MAP-1 was able to pulldown in vitro translated ^{35}S-labelled MAP-1, Bax, Bcl-2 and Bcl-X_L, but not the ^{35}S-labelled Bid (Tan et al., 2001).

MAP-1 Mediates Caspase-Dependent Death When Overexpressed

We next examined the possible role of MAP-1 in modulating cell death by transient transfection experiments. SH-SY5Y human neuroblastoma cells (Ronca et al., 1997) were transfected with Myc-tagged MAP-1 expression plasmid with or without the addition of the broad spectrum caspase inhibitor, ZVAD-fmk (10 uM) to the culture media 6 h post transfection. At 24 h post-transfection, a significant percentage of Myc-MAP-1 positive cells displayed apoptotic morphology as evidenced by the appearance of shrunken and rounded cells (left upper panel, Fig. 3). DNA strand breaks were detected at 24 h post-transfection by using fluorescent TUNEL (Pharmingen) that labeled DNA strand break with FITC-dUTP (right upper panel, Fig. 3). Caspase inhibitor, ZVAD-fmk, was found to block MAP-1-induced apoptosis efficiently as evidenced by the absence of round cell morphology (left lower panel, Fig. 3) and TUNEL staining (right lower panel, Fig. 3).

The BH3-Like Motif of MAP-1 Is Required for Pro-Apoptotic Function and Is Also Involved in Binding BAX Through Its BH Domains

The conserved leucine residue in the BH3 domains of Bax and Bad was reported to be critical for mediating interaction with Bcl-2 or Bcl-X_L (Zha et al., 1997, Wang et al., 1998) as well as for the N-terminal BH3-B domain of Bid to interact with its C-terminus (Tan et al., 1999). Deletion of the entire BH3-like domain (Table 1) or substitution of this conserved leucine of the BH3 domain in MAP-1 was found to be sufficient to abolish its interaction with Bax (Tan et al., 2001). Interestingly, in contrast to most proapoptotic members of the Bcl-2 family, the BH3-like domain in MAP-1 did not appear to be required for interaction with Bcl-X_L as deletion of the entire BH3-like domain of MAP-1 did not affect its binding to Bcl-X_L (Table 1). These data provide evidence that MAP-1 is not a classical BH3-only member of BCL-2 family as the BH3-like domain in this case is required for association with BAX but not Bcl-X_L.

Table 1. Summary of Data from Immunoprecipitation Experiments for Evaluating the Protein-Protein Interation Function of the BH3-Like Domain in MAP-1 and the Three BH Domains in Bax

Protein Partners	Bax	Bax (G108A) (BH1)	Bax (W151A) (BH2)	Bax (L63E) (BH3)	Bcl-X_L
BCl-X_L	+	+	+	−	N.T.
MAP-1	+	−	−	−	+
MAP-1(Δ120–127)	−	N.T.	N.T.	N.T.	+

Note: 293T cells were transiently co-transfected with the indicated Myc-tagged (column) and HA-tagged (row) expression constructs. Transfection and co-immunoprecipitation procedures were carrried out essentially the same as the experiment described in Figure 2. (+), strong to medium interaction; (−), no detectable interaction; N.T., not tested.

Similar to many BH3-domain containing molecules (Kelekar and Thompson, 1998), the BH3-like domain in MAP-1 appeared to be required for its cell death function as deletion of this domain or mutations of the highly conserved amino acid in this domain resulted in non-apoptotic protein (Tan et al., 2001).

Several highly conserved amino acid residues of the BH1, BH2, and BH3 domains among Bcl-2 family members have been demonstrated to be critical for hetero-dimerization functions (Yin et al., 1994, Wang et al., 1996, Wang et al., 1998). To evaluate the requirement of the BH domains of Bax in mediating interaction with MAP-1, substitution point mutations of these critical amino acid residues in the BH1, BH2, and BH3 domains were made. In agreement with previous data, mutation of the BH3 (L63E), but not the BH1 (G108A) or BH2 (W151A) domains of Bax abolished its binding to Bcl-X_L (Tan et al., 2001, and Table 1). However, none of the point mutants were able to bind MAP-1 (Tan et al., 2001 and Table 1), suggesting that the three BH domains (BH1, BH2, and BH3) of Bax are all required for mediating protein-protein interaction with MAP-1.

Discussion

MAP-1 contains a putative BH3 domain and associates with Bax, Bcl-X_L and Bcl-2 *in vitro* and *in vivo* in mammalian cells. It mediates caspase-dependent apoptosis when overexpressed. The BH3-like domain in MAP-1 is required for binding to Bax and for mediating apoptosis. Interestingly, in contrast to other known Bax-associating proteins, the binding of Bax to MAP-1 requires all the BH (BH1, BH2, and BH3) domains of Bax as point mutations affecting any one of the BH domains abolished its binding to MAP-1. MAP-1 thus represents the first protein partner of Bax identified that may potentially bind BAX through its elongated hydrophobic pocket as defined by the three BH domains.

The crystal structure of Bcl-X_L suggested that it shares similarity to the pore-forming domains of bacterial toxins such as colicins A1 and E1 and Diphtheria toxin (Muchmore et al., 1996). It has been reported that regions encompassing part of the BH1 and BH2 domains may have pore-forming function (Adams and Cory, 1998). In addition, Bax was shown to form ion channel in *vitro* (Tsujimoto and Shimizu, 2000). However, it still remains to be determined whether Bcl-2 family of proteins actually forms channels in *vivo* and whether these proteins regulate apoptosis via the creation of ion channels (Tsujimoto and Shimizu, 2000).

Instead of having the BH1, BH2, and BH3 domains forming a receptor structure as in the case of Bcl-X_L (Adams and Cory, 1998), the BH domains in Bax have been suggested to serve independent functions. The BH3 domains of Bax has been proposed to be involved in binding the permeability transition pore and inducing permeability transition and cytochrome c release from the mitochondria (Green and Reed, 1998, Tsujimoto and Shimizu, 2000). Recently, the BH3-domain only molecule Bid was found to associate with Bax through its BH1 domain (Wang et al., 1996). Animals with two key members of multidomain pro-apoptotic genes, BAX and BAK, deleted exhibited profound phenotype. Furthermore, cells derived from these animals failed to process apoptotic signals originated from diverse stimuli provided further evidence to support the argument that multidomain molecules, BAX and BAK, act as central integrators for apoptosis signals. Identification of MAP-1 may thus provide further opportunities for investigating the complex mechanisms operated by Bax.

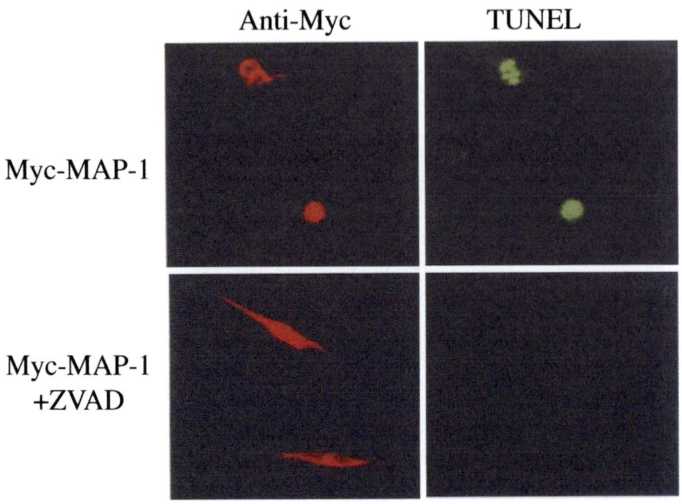

Fig.3. MAP-1 mediates caspase-dependent apoptosis in mammalian cells. MAP-1 induces caspase-dependent cell shrinkage and DNA strand breaks in SH-SY5Y cells. SH-SY5Y cells were transiently transfected with 1.5 μg each of Myc-tagged MAP-1 plasmid. 6h post-transfection, media was removed and fresh media was added with or without broad spectrum caspase inhibitor ZVAD-fmk (10uM) where indicated. Myc-MAP-1 expressing cells were detected by anti-Myc antibody, followed by immunofluoresence staining with anti-mouse Cy3. DNA strand breaks were detected at 24h post-treatment by fluorescent terminal deoxynucleotidyltransferase dUTP nick end labeling (TUNEL). DNA strand breaks were labeled with FITC-dUTP.

ACKNOWLEDGMENTS: We are grateful to Drs. Suzanne Cory and David Huang for providing us with many of the cDNA clones employed in this study. This work was supported by grants from the Agency for Science, Technology and Research (A*STAR) of Singapore.

References

Adams, J.M. and Cory, S. (1998). The Bcl-2 protein family: arbiters of cell survival. Science *281*: 1322–1326.

Chan, S.L., Tan, K.O., Zhang, L., Yee, K.S., Ronca, F., Chan, M.Y., and Yu, V.C. (1999). F1Aalpha, a death receptor-binding protein homologous to the Caenorhabditis elegans sex-determining protein, FEM-1, is a caspase substrate that mediates apoptosis. J. Biol. Chem. *274*: 32461–32468.

Chan, S.L., Yee, K.S., Tan, K.M., and Yu, V.C. (2000). The Caenorhabditis elegans sex determination protein FEM-1 is a CED-3 substrate that associates with CED-4 and mediates apoptosis in mammalian cells. J. Biol. Chem. 275: 17925–17928.

Dalmau, J., Gultekin, S.H., Voltz, R., Hoard, R., DesChamps, T., Balmaceda, C., Batchelor, T., Gerstner, E., Eichen, J., Frennier, J., Posner, J.B., and Rosenfeld, M.R. (1999). Ma1, a novel neuron- and testis-specific protein, is recognized by the serum of patients with paraneoplastic neurological disorders. Brain *122*: 27–39.

Green, D.R., and Reed, J.C. (1998). Mitochondria and apoptosis. Science *281*: 1309–1312.

Kelekar, A., and Thompson, C.B. (1998). Bcl-2-family proteins: the role of the BH3 domain in apoptosis. Trends Cell Biol. *8*: 324–330.

Lindsten, T., Ross, A.J., King, A., Zong, W.X., Rathmell, J.C., Shiels, H.A., Ulrich, E., Waymire, K.G., Mahar, P., Frauwirth, K., Chen, Y., Wei, M., Eng, V.M., Adelman, D.M., Simon, M.C., Ma, A., Golden, J.A., Evan, G., Korsmeyer, S.J., MacGregor, G.R., and Thompson, C.B. (2000). The combined functions of proapoptotic Bcl-2 family members bak and bax are essential for normal development of multiple tissues. Mol. Cell *6*: 1389–1399.

Muchmore, S.W., Sattler, M., Liang, H., Meadows, R.P., Harlan, J.E., Yoon, H.S., Nettesheim, D., Chang, B.S., Thompson, C.B., Wong, S.L., Ng, S.L., and Fesik, S.W. (1996). X-ray and NMR structure of human Bcl-xL, an inhibitor of programmed cell death. Nature *381*: 335–341.

Ronca, F., Chan, S.L., and Yu, V.C. (1997). 1-(5-Isoquinolinesulfonyl)-2-methylpiperazine induces apoptosis in human neuroblastoma cells, SH-SY5Y, through a p53-dependent pathway. J. Biol. Chem. *272*: 4252–4260.

Rosenfeld, M.R., Eichen, J.G., Wade, D.F., Posner, J.B., and Dalmau, J. (2001). Molecular and clinical diversity in paraneoplastic immunity to Ma proteins. Ann. Neurol. *50*: 339–348.

Sattler, M., Liang, H., Nettesheim, D., Meadows, R.P., Harlan, J.E., Eberstadt, M., Yoon, H.S., Shuker, S.B., Chang, B.S., Minn, A.J., Thompson, C.B., and Fesik, S.W. (1997). Structure of Bcl-xL-Bak peptide complex: recognition between regulators of apoptosis. Science *275*: 983–986.

Suzuki, M., Youle, R.J., and Tjandra, N. (2000). Structure of Bax: coregulation of dimer formation and intracellular localization. Cell *103*: 645–654.

Tan, K.O., Tan, K.M., and Yu, V.C. (1999). A novel BH3-like domain in BID is required for intramolecular interaction and autoinhibition of pro-apoptotic activity. J. Biol. Chem. *274*: 23687–23690.

Tan, K.O., Tan, K.M., Chan, S.L., Yee, K.S., Bevort, M., Ang, K.C., and Yu, V.C. (2001). MAP-1, a novel proapoptotic protein containing a BH3-like motif that associates with Bax through its Bcl-2 homology domains. J. Biol Chem. *276*: 2802–2807.

Tsujimoto, Y. and Shimizu, S. (2000). Bcl-2 family: life-or-death switch. FEBS Lett. *466*: 6–10.

Voltz, R., Gultekin, S.H., Rosenfeld, M.R., Gerstner, E., Eichen, J., Posner, J.B., and Dalmau, J. (1999). A serologic marker of paraneoplastic limbic and brain-stem encephalitis in patients with testicular cancer. N. Engl. J. Med. *340*: 1788–1795.

Wang, K., Yin, X.M., Chao, D.T., Milliman, C.L., and Korsmeyer, S.J. (1996). BID: a novel BH3 domain-only death agonist. Genes Dev. *10*: 2859–2869.

Wang, K., Gross, A., Waksman, G., and Korsmeyer, S.J. (1998). Mutagenesis of the BH3 domain of BAX identifies residues critical for dimerization and killing. Mol. Cell Biol. *18*: 6083–6089.

Wei, M.C., Zong, W.X., Cheng, E.H., Lindsten, T., Panoutsakopoulou, V., Ross, A.J., Roth, K.A., MacGregor, G.R., Thompson, C.B., and Korsmeyer, S.J. (2001). Proapoptotic BAX and BAK: a requisite gateway to mitochondrial dysfunction and death. Science *292*: 727–730.

Yin, X.M., Oltvai, Z.N., and Korsmeyer, S.J. (1994). BH1 and BH2 domains of Bcl-2 are required for inhibition of apoptosis and heterodimerization with Bax. Nature *369*: 321–323.

Zha, H., Fisk, H.A., Yaffe, M.P., Mahajan, N., Herman, B., and Reed, J.C. (1996). Structure-function comparisons of the proapoptotic protein Bax in yeast and mammalian cells. Mol. Cell Biol. *16*: 6494–6508.

Zha, J., Harada, H., Osipov, K., Jockel, J., Waksman, G., and Korsmeyer, S.J. (1997). BH3 domain of BAD is required for heterodimerization with BCL-XL andproapoptotic activity. J. Biol. Chem. *272*: 24101–24104.

Chapter 12

The Mechanisms and Significance of Apoptotic Cell-Mediated Immune Regulation

ERWEI SUN[1,2] AND YUFANG SHI[2]

ABSTRACT: The discovery of apoptosis has prompted scientists in biology and medicine to explore the mechanisms and clinical significance of this fascinating phenomenon. In the last decade, accumulating evidence has revealed that apoptosis plays a pivotal role in almost every aspect of growth, differentiation and development. Although it is known that apoptotic cells are rapidly scavenged as a means to prevent inflammation, it remains to be determined whether the clearance of apoptotic cells is related to immune regulation. A review of the recent literature combined with our own work shows that apoptosis is not only a programmed and conserved process of cell death, but also exerts unique effects on the immune system, i.e. apoptotic cell-mediated immune regulation (AMIR). It seems that phagocytes, along with their cytokines released upon binding, phagocytosing and processing apoptotic cells, lie at the heart of AMIR. Investigation of AMIR will not only bring about a better understanding of many important physiological and pathological mechanisms, but also provide new hope for patients of transplantation and autoimmune diseases.

Apoptosis is a normal process in growth, development and aging while deregulation of the apoptotic process or clearance of apoptotic cells may result in a variety of diseases (Thompson, 1995). Studies in the last decade have revealed that apoptosis is an active, programmed, and energy-dependent process that is characterized by sequential activation of a series of suicide enzymes to eliminate unnecessary, aged, or diseased cells. The morphological characteristics of apoptosis include extracellular exposure of phosphatidylserine, blebbing of the cell membrane, and nuclear condensation, and fragmentation. Apoptotic cells are rapidly and discreetly scavenged and digested by phagocytes so that the contents of apoptotic cells are not released directly into the interstitial space. In contrast, necrosis generally takes place in pathological situations such as infections, physical injury or ischemia. Differing from apoptosis, necrosis is an incidental, non-programmed, and passive process that results in cell swelling and membrane rupture. Cell debris is released directly into surrounding tissue space and stimulates local inflammation (Barrington et al., 2001).

Once apoptotic cells are quietly phagocytosed, do they evoke biological responses? This question has been ignored until the recent discovery that recognition, phagocytosis,

[1]Organ Transplantation Department, Zhujiang Hospital, Guangzhou 510282, China. [2]Molecular Genetics, Microbiology and Immunology, University of Medicine and Dentistry of New Jersey, Piscataway, NJ 08854, USA. *Correspondence:* Erwei Sun, 661 Hoes Lane, UMDNJ-RWJMS, Piscataway 08854, NJ, USA. E-mail: ewsun@263.net

and clearance of apoptotic cells are of paramount importance in the induction of immune tolerance and prevention of autoimmune diseases (Voll et al., 1997) (Fadok VA et al., 1998). The main changes in apoptotic cells are in the cytoplasmic membrane, such as extracellular exposure of phosphatidylserine, and in the nucleus, such as condensation and fragmentation of DNA. The membrane changes serve to detach dying cells from surrounding healthy cells, and to provide an "eat me" signal to phagocytes for identification and clearance (Hengartner, 2001). Receptors on phagocytes recognize molecules specifically expressed on apoptotic cells and rapidly engulf them in hours. Importantly, since phagocytes themselves are antigen presenting cells, it is possible that antigens from apoptotic cells are presented to T cells. Evidence has accumulated, from both clinical practice and experimental studies, to suggest that apoptotic cell mediated cell death (AMIR) plays a crucial role not only in the prevention of autoimmune diseases, but also in transfusion-induced immunosuppression, as well as in the induction of transplantation tolerance. We have shown that apoptotic cells inhibit T cell activation and induce transplantation tolerance (Sun et al., 2000 and unpublished obervations). Therefore, further investigation into the mechanisms of phagocytosis and processing of apoptotic cells and their roles in immune regulation will lead us to a better understanding of immune tolerance and autoimmune diseases and pave the way for new clinical protocols for the treatment of these diseases and for the induction of transplantation tolerance.

Implications from Clinical Transfusion

Early clinical observations revealed that blood transfusion after trauma or operations was often associated with immunosuppression and opportunistic infection. Blood transfusion also reduced the recurrence of autoimmune diseases. In 1960s, blood transfusion was avoided in transplant recipients for fear that recipient lymphocytes might be sensitized by the transfused donor antigens, resulting in accelerated acute rejection. In 1974, however, Opelz reported that blood transfusion did not induce accelerated rejection, but rather prolonged graft survival (Opelz and Terasaki, 1974). The concept that blood transfusion has beneficial effects on transplant recipients was widely accepted and practiced in the late 1970s by most transplant surgeons (Persijn et al., 1979). Therefore, transfusion of whole blood, 3 to 4 units before transplantation, became a preoperative routine in many transplantation units. However, the efficacy of blood transfusion in prolonging graft survival was greatly compromised when a stronger immunosuppressant, cyclosporine (CsA), become clinically available. It was unknown whether the decreased effect of blood transfusion in prolonging graft survival was directly counteracted by CsA or if the effect was only less significant in the shadow of powerful immunosuppression by CsA. In transplant patients receiving CsA treatment, it was found that preoperative transfusion could elicit the production of anti-HLA antibody, called penal reactive antibody (PRA), and increase the risk of accelerated rejection. Why transfusion causes divergent effects on immunity has intrigued transplantation immunologists for years, as yet no satisfactory explanation has been reached. It is possible that immune tolerance induction is compromised by CsA, because the inhibition apoptosis by CsA may prevent AMIR, which may be important in immune tolerance induction (Shi et al., 1989; Well et al., 1999). In addition to these clinical observations, laboratory findings also support the notion that blood transfusion sends inhibitory signals to the immune cells in that decreased proliferation of T cells and NK cells and increased activity of regulatory T cells have been found in transfused patients.

An essential question is, what factor in the transfused blood suppresses the recipient's immune system? Since most of the blood is not autologous, it was postulated that immunosuppression might arise from the foreignness of the transfused blood. This seems implausible because both allogeneic and autologous blood transfusion result in similar infection rates (Nielsen, 1995). Further studies into blood components found that transfusion of red cells alone or white cell-depleted blood caused no immunosuppression, suggesting that the immunosuppressive factors reside in either white blood cells or platelets. Since, in a normal healthy individual, circulating white blood cells or platelets are not immunologically active, why do they become immunosuppressive after being drawn and stored outside the body for a while? A possible explanation is that exposure of white cells and platelets to a non-physiologic environment induces apoptosis and that these apoptotic cells transmit immunosuppressive signals after blood transfusion. This concept is supported by the finding that apoptotic granulocytes and lymphocytes are present in stored blood (Frabetti et al., 2000; Snyder and Kuter, 2000).

The Unique Immune Response in the Liver and Its Relevance to Apoptotic Cells

It has been known for many years that the liver plays a pivotal role in immune tolerance induction. In transplantation models, intraportal vein injection of white blood cells induces tolerance to allogeneic skin, liver, pancreas or small intestine graft, even tolerance to xenogeneic grafts (Fecteau et al., 1994; Gorczynski et al., 1995; Goss et al., 1996; Morita et al., 1998; Nakano et al., 1992). Anastomosis, surgical joining of blood vessels, of the graft vein to the portal vein leads to graft acceptance, while drainage of graft blood directly to the vena cava prevents tolerance induction (Boeckx et al., 1975; Gorczynski et al., 1994; Yang et al., 1994). Bypass of portal blood flow into the inferior vena cava results in loss of tolerance to orally absorbed antigens, suggesting that the liver also plays an important role in oral tolerance (Callery et al., 1989a). In mice and some strains of rats or pigs, liver grafts across the MHC barrier survive permanently without using immunosuppressants, while other solid grafts, such as heart or kidney, are acutely rejected (Kamada et al., 1998). Interestingly, liver transplantation has been shown to reverse other organ rejections (Kamada et al., 1981). Clinically, HLA matching is important in predicting the incidence and severity of rejection for most grafts, such as bone marrow, kidney and heart, but it is not as important in liver transplantation. Simultaneous transplantation of liver and kidney has a lower rejection rate as compared to kidney transplantation alone. Furthermore, some liver transplant recipients can survive permanently, even after withdrawal of immunosuppressants, a phenomenon that seldom occurs in recipients of other organ transplants (Riordan and Williams, 1999).

Lipopolysaccride (LPS) has frequently been used as a stimulator of immune responses in experimental settings (Cella et al., 1997). LPS, as well as some other inflammatory molecules, such as TNF-a , CD40 ligand (CD40L), induces dendritic cells (DC) to express costimulatory molecules, which further activates the adaptive immune response (Inaba et al., 2000; Tough et al., 1997). It is interesting to note that LPS physiologically presents in the portal vein at a concentration as high as 10 pg/ml, due to the large number of microbes in the intestine (Lumsden et al., 1988). However, portal vein LPS does not stimulate inflammation or an immune response. On the contrary, it downregulates immune responses in the liver. The disappearance of LPS in the liver vein indicates that the influence of LPS on immunity is confined to the liver. Portal LPS not only stimulates Kuppfer cells to produce IL-10, but also downregulates antigen presentation by liver sinus endothelial cells (LSEC) (Knolle et al., 1995; Knolle et al., 1999). *In vitro*, LPS similarly downregulates antigen presentation of LSEC by reducing the expression of MHC II, CD80 and CD86 (Knolle et al., 1999). Although TNF-a is produced by LPS-stimulated Kuppfer cells, it has a negative effect on LSEC antigen presentation (Muschen et al., 1998; Muschen et al., 1999).

It was believed that the liver expresses only weak MHC antigens that cannot elicit fierce immune responses, a phenomenon called immune privilege. Purified hepatocytes, however, express sufficient MHC I to stimulate cytotoxic lymphocyte (CTL) *in vitro*. Transplantation of hepatocytes using a sponge matrix also elicits a CTL response (Bumgardner et al., 2000; Bumgardner et al., 1998; Bumgardner et al., 1999; Bumgardner and Orosz, 2000; Muschen et al., 1998; Muschen et al., 1999). Another untenable postulation is that hepatocytes are interrupted by LSEC and Kuppfer cells and cannot access lymphocytes in the blood. It is known that LSEC account for 30 percent of liver cells while Kuppfer cells make up 80 percent of total macrophages in the whole body. Due to their expression of MHC I and MHC II molecules, LSEC and Kuppfer cells should be able to provoke powerful immune responses if in the abscence of other control mechanisms (Bumgardner and Orosz, 2000). The liver is not only prone to tolerance induction, but also gives rise to immune responses. CTL activated in hepatitis patients kill virus-infected hepatocytes, and isolated liver T cells can be induced to differentiate into CTL and to produce cytokines, strongly suggesting that liver is not immune privileged, but rather a specialized immune organ (Cousens and Wing, 2000; Doherty et al., 1999; Seki et al., 2000).

What makes liver so peculiar for immune regulation? Based on the study with liver transplantation models, Starzl proposed a microchimerism hypothesis. He believes that leukocytes migrate from the liver graft into recipient tissues and establish a microchimerism for the induction of graft tolerance. Two major categories of evidence support this hypothesis. First, donor-derived cells have been found in many tissues of some patients with tolerated liver grafts (Starzl et al., 1996; Starzl et al., 1992; Starzl and Zinkernagel, 1998). Second, a specialized subset of DC in liver with weak costimulatory molecules may mediate graft tolerance (Thomson and Lu, 1999). Importantly, in order to induce immune tolerance, donor cells must have the opportunity to access recipient lymphocytes. Few number and random distribution of donor cells in recipients suggest the implausibility of donor cell access to recipient lymphocytes that normally present only in lymphoid tissues. Considering the large lymphocyte repertoire and the continuous new emigrants from thymus, how could so few donor cells induce tolerance in the recipient immune system? If donor cells do really contribute to tolerance induction, liver graft itself can be regarded as a macrochimerism. In liver grafts, donor cells *in situ* are much more likely to contact recipient lymphocytes due to the abundance as well as low speed of blood and lymph flow. This notion of tolerance induction by *in situ* donor liver cells is supported by the finding that liver cells can communicate with naive T cells. Therefore, microchimerism may not be the cause of liver graft tolerance, but the result of tolerance induced by the liver graft itself, making the microchimerism hypothesis an unsatisfactory explanation for the peculiar immune responses occurring in the liver (Wood and Sachs, 1996).

Theoretically, liver may serve as the graveyard of apoptotic cells. Phagocytosis and processing of apoptotic cells by Kuppfer cells and LSEC may transduce unique local immune regulatory signals that contribute to tolerance induction. First, anatomically, a large blood volume, 25 percent of circulating blood, passes through the liver at a very slow flow rate, only one seventh of that in post capillary venules. The special structure of parenchymal cord and sinus promotes contact of liver parenchymal cells with blood cells. Therefore, aged, damaged or apoptotic blood cells may easily contact liver phagocytes. Second, the liver is equipped with plenty of enzymes not only for the metabolism of absorbed carbohydrates, fatty acids and proteins, but also for the synthesis of new molecules for utilization by other parts of the body. Finally, the huge amount of liver lymph fluid, accounting for 20–50 percent of lymph flow, in the thoracic duct, makes it possible for liver cells to have opportunity to contact and induce tolerance of lymphocytes passing through the liver.

New experimental evidence supports the notion that the liver is the accumulation site of apoptotic cells. Injection of specific antigen into TCR transgenic mice deletes specific $CD8^+$ T cells in the periphery as well as in the thymus, but, interestingly, those T cells are found apoptotic and accumulated in the liver (Crispe et al., 2000; Dini and Carla, 1998; Huang et al., 1994). B220 is a marker of impending apoptosis and, in spontaneous diabetic BB rats, B220 positive cells collect in the liver while $CD8^+$ T cells decrease in the periphery. Influenza is confined to the respiratory tract, but following virus infection the apoptotic lymphocytes cannot be cleared in the lungs, but rather in the liver as large numbers of apoptotic $CD8^+$ T cells are found (Belz et al., 1998). In a liver perfusion model, it was discovered that activated T lymphocytes are trapped in the liver and induced to die by apoptosis (Crispe et al., 2000; Mehal et al., 1999). All the evidence strongly suggests that the peculiar liver immune response is related to its ability to intercept and process apoptotic cells.

Autoimmune Diseases and the Phagocytosis of Apoptotic Cells

Autoimmune diseases are etiologically diversified immune disorders that are influenced by genetic and environmental factors. Two important issues in the study of autoimmune diseases are the origin of autoantigens and the activation mechanisms of the immune system. Interestingly, in the past few years, it has been found that apoptotic cells are an important source of autoantigens. When apoptotic cells are not adequately phagocytosed and cleared, secondary necrosis of these cells may activate the immune system to provoke autoimmune diseases (Casciola-Roen et al., 2000; Lorenz et al., 2000; Rodenburg et al., 2000).

Autoantigens in normal cells are not easily accessible to phagocytes due to a lack of membrane "eat me" flags. *In vivo*, apoptotic cell-derived molecules may be an important source of autoantigens. Among the autoimmune diseases, the most intensively studied is systemic lupus erythematosus (SLE) which typically involves many systems and has multi-faceted clinical manifestations. Autoantigens in SLE patients do not come from a special subcellular component, but from almost all substructures, including the nucleus, cytoplasm and cell membrane (Andrade et al., 2000; Casciola-Roen et al., 2000). Accumulating evidence suggests that inadequate phagocytosis of apoptotic cells is one of the potential mechanisms in SLE (Korb and Ahearn, 1997; Navratil et al., 1999). It is known that complement, especially complement 1q (C1q) that specifically binds to blebs of apoptotic cells, participate in the clearance of apoptotic cells. Deficiency of C1q leads to not only human SLE, but also SLE-like immune changes in mice. C1q-deficient mice show increased apoptotic cells in nephrons and display nephritis-like changes (Botto, 1998; Botto et al., 1998; Kishore and Reid, 2000). C1q-deficient patients have decreased phagocytotic capability by monocytes and macrophages that can be reversed by C1q replenishment (Taylor et al., 2000). Likely, C reactive protein (CRP), which binds to C1q and promotes phagocytosis of apoptotic cells, is decreased in SLE patients (Gershov et al., 2000).

Another autoimmune disease that deserves attention is type I diabetes mellitus in which islet inflammation plays an important role in pathogenesis (Rabinovitch et al., 1994). Apoptotic b-cells are increased in diabetic BB rats and NOD mice. On the other hand, macrophages are obviously abnormal in that either reduced number, or deficiency in recognition and phagocytosis of apoptotic cells, is found in the two diabetic models (Luan et al., 1996; Serreze et al., 1993; Trudeau et al., 2000). It is interesting that in chemically-induced diabetic models, anti-CD40 antibody treatment easily induces islet graft tolerance. The same protocol, however, does not work for islet grafts in NOD mice (Markees et al., 1999).

Inhibition on T Cell Function by Apoptotic Cells

The above findings and discussions suggest that apoptotic cells may send immunosuppressive signals to T cells. Therefore, it would be interesting to examine whether apoptotic cells are capable of regulating T cell activation. Ponner et al found that, in contrast to living cells, apoptotic cells have obviously decreased immunogenicity (Ponner et al., 1998). Our results provide direct evidence that apoptotic cells do inhibit T cell activation. When apoptotic human K-562 cells induced by FTY720 were added into a mouse spleen cell culture stimulated by Con A, the expression of CD69, an early T cell activation marker, was inhibited. Further studies revealed that this immunosuppression was not restricted to FTY720 treated K562 cells, since experiments with another apoptotic inducer, cycloheximide and another human cell line, HL-60, produced similar results. Moreover, our results suggest that regulation of immune responses by apoptotic cells is evolutionarily conserved since xenogeneic (human) apoptotic cells can modulate mouse spleen cell activation (Sun et al., 2000). At the same time, we also found that, in the culture of mixed lymphocyte reaction, addition of apoptotic cells can inhibit lymphocyte proliferation. Further studies demonstrated that apoptotic cells prolong allogeneic cardiac graft survival, or even induced long term survival of liver allografts, suggesting that transfusion of apoptotic cells may be an important strategy to induce transplantation tolerance (Sun and Shi, unpublished observation).

Other Evidence of Immune Regulation by Apoptotic Cell

Apart from the evidence mentioned above, other studies also suggest that apoptotic cells are relevant to immunosuppression or immune tolerance: 1) A recent study identified a specialized subset of DC ($CD4^{-}OX41^{-}$) that transports apoptotic intestinal epithelial cells to the mesenteric lymph nodes and are important in oral tolerance (Huang et al., 2000; Steinman et al., 2000); 2) FTY720, a new immunosuppressant that is able to induce lymphocyte apoptosis, induced long-term allogeneic heart graft survival in our studies, suggesting that apoptotic cells can be utilized to induce transplantation tolerance (Sun et al, unpublished results); 3) In the eye anterior chamber, a known immune privileged site, apoptosis induction is critical for the establishment of local immune tolerance, since in Fas or Fas ligand (Fas L)-deficient animals in which the apoptotic machinery is disrupted, local tolerance is hard to induce. These findings suggest that apoptotic cells regulate local immune responses and play an important role in so-called immune privilege (Ferguson and Griffith, 1997; Griffith T.S. et al., 1996).

The Mechanisms of AMIR

As described above, apoptotic cells induce local immunosuppression and immune tolerance. To help elucidate the mechanism, it would be interesting to determine where this regulation takes place, which cells participate, what molecules are involved, and how the signals are transduced. Answering these questions will certainly lead us to a better understanding of the mechanisms.

Macrophages are the major players in the phagocytosis and processing of apoptotic cells. Accumulating evidence shows that macrophages are the major players in the phagocytosis and processing of apoptotic cells. Peritoneal macrophages specifically bind to apoptotic thymic cells when they are co-cultured *in vitro*. This specific recognition is mediated by lectin-like molecules on the cell membrane of apoptotic cells. Several receptors on macrophage membrane are involved in the phagocytosis of apoptotic cells, such as scavenger receptor A (SR-A), CD36, CD14, ABC-1, PStR, MER and ICAM-3 (Devitt et al., 1998; Fadok et al., 2000; Platt et al., 1996; Savill, 1998; Savill and Fadok, 2000; Scitt et al., 2001). These receptors have been found in *C. elegans*, Drosophila, and human, suggesting that they are evolutionarily conserved. In addition to those phagocytic receptors, the complement system is crucial in apoptotic cell phagocytosis. Complement activated by the exposure of PS on apoptotic cells may bind to receptors on phagocytes. Interestingly, complement-deficiency that affects the phagocytic capacity of phagocytes may be one of the mechanisms of autoimmune diseases (Mevorach et al., 1998). Kupffer cells are the main phagocytotic cells in liver. Ten minutes after intraportal injection of activated lymphocytes, approximately 60 percent of the injected cells are bound to Kupffer cells, and after 90 minutes almost all injected cells are bound. Administration of gadolinium chloride, a blocker of phagocytosis, results in a loss of tolerance induction by intraportal injection of donor antigens, further confirming the importance of Kupffer cell phagocytosis in tolerance induction (Callery et al., 1989b; Kamei et al., 1990).

Some evidence suggests that DC also phagocytose apoptotic cells although the outcome, i.e. tolerance induction or promotion of immune response, is controversial (Albert et al., 1998a; Albert et al., 1998b; Shi et al., 2000; Steinman et al., 2000). One explanation of this inconsistency is that the maturation state of DC is crucial, in that immature DC induces tolerance while mature DC elicit an immune response (Steinman et al., 2000). It should be borne in mind that in most studies, DC are isolated and primed with antigen *in vitro* before injection for *in vivo* experiments, so it is likely that DC are activated to mature during the separation and purification process, resulting in immune responses. Therefore, the important point in DC-based immune tolerance protocols is to maintain their *in situ* status while antigens are primed. Recently, a method has been invented to prime DC with antigen *in vivo* without influencing DC maturity. The results showed that antigens presented by DC in this way induced tolerance, not a priming response (Hawiger et al., 2001). This result strongly suggests that presentation of apoptotic cell-derived antigens by *in vivo* DC induces immune tolerance. Another possibility is that excessive apoptotic cells or insufficient phagocytotic capability by macrophages, due to either decrease in number or a functional deficiency, leads to accumulation of un-phagocytosed apoptotic cells that will become necrotic. Some signals released from necrotic cells activate DC to mature and induce immune responses. An example is that while both apoptotic and necrotic cells can be phagocytosed by immature DC, only necrotic cells or their supernatants can activate DC to mature (Rovere et al., 2000; Rovere et al., 1999; Rovere et al., 1998; Sauter et al., 2000). Another report suggests that the status of apoptotic cells determines the consequences of antigen presentation, since heat stress of the cells before they are induced to become apoptotic upregulates heat shock proteins and makes the apoptotic cells immunogenic (Feng et al., 2001). In addition to macrophages and DC, some other cell types, like LSEC in the liver, possess the ability to phagocytose apoptotic cells, although the question of whether phagocytosis of apoptotic cells by tissue cells is coupled with immune regulation awaits further investigation (Dini and Carla, 1998).

Inhibitory cytokines are involved in AMIR. What mediates the immune suppression or tolerance induction after apoptotic cells are phagocytosed? The inhibitory cytokines secreted by apoptotic cells or their phagocytes may play a role. The finding that apoptotic cells release both inactive and activated forms of TGF-b without changes in mRNA transcription indicates that apoptotic cells release TGF-b that is already present in the cells, without *de novo* synthesizes.

TGF-b is a family of three isomers, among them the most extensively studied is TGF-b 1 which has a wide range of activities, such as regulating cell growth and differentiation, modulating immunity and inflammatory response, controlling matrix sedimentation, wound healing, and cell migration and adhesion (Bissell et al., 2001; Chen et al., 2001). Early studies found that mouse B lymphoma cells release TGF-b 1 after being induced into apoptosis. Later, many studies demonstrated that other apoptotic cells also secrete TGF-b 1. *In vitro*, after phagocytosing apoptotic cells, macrophages secrete less inflammatory cytokines, such as growth factors and chemokines, which can be blocked by anti-TGF-b 1 (Chen et al., 2001; Fadok VA et al., 1998; Voll et al., 1997). *In vivo*, diminished secretion of inflammatory cytokines and alleviation of inflammation can be found after the injection of apoptotic cells. Injection of apoptotic cells, but not live cells, enhance the production of TGF-b 1 in macrophages. Most interestingly, although phagocytosis is indispensable to the production of TGF-b 1 by macrophages, opsonized cells that can be processed by macrophages through Fc g -mediated phagocytosis do not induce macrophages to secret TGF-b 1. Therefore, the secretion of TGF-b 1 by macrophages can only be induced by their recognition and engagement of specific molecules expressed on apoptotic cells. Interestingly, some cells, such as Jurkat cells and PBL-985 cells that do not undergo PS exposure after being induced into apoptosis, can hardly induce macrophages to produce TGF-b 1. PS directly transferred into the PBL-985 cell membrane can restore TGF-b 1 production, suggesting a critical role of PS-mediated phagocytosis in the production of TGF-b 1 (Huynh et al., 2002) . In the transfusion-induced immune tolerance model, animals display enhanced expression of intragraft TGF-b 1 and administration of anti-TGF-b 1 elicits rejection, demonstrating that immune tolerance induction in this model is related to enhanced intragraft TGF-b 1 expression (Gagne et al., 2001; Josien et al., 1998). Interestingly, liver Kupffer cells and LSEC contain plenty of TGF-b 1 mRNA while liver parenchymal cells possess hardly any (Bissell et al., 1995). Therefore, secretion of immunosuppressive cytokines, especially TGF-b 1 by macrophages upon phagocytosis of apoptotic cells, plays a critical role in AMIR.

Conclusing Remarks

Apoptosis, a discreet kind of cell death, plays an important role in immune regulation. *In vivo*, appropriate processing of apoptotic cells is critical to maintain immune tolerance to autologous antigens. Inadequate phagocytosis and process of apoptotic cells may drive apoptotic cells into secondary necrotic process, which activate the adaptive immune system and causes autoimmune diseases (Sun and Shi, 2001). The application of apoptotic cells sheds some light on tolerance induction in that phagocytosis and presentation of antigens derived from apoptotic cells are accompanied by secretion of immune inhibitory cytokines. It is necessary to further investigate the mechanisms and key regulators during the phagocytosis of apoptotic cells. In autoimmune diseases, it is important to know if excessive apoptotic cells exist or inadequate phagocytosis occurs and to find out ways for rapid clearance of apoptotic cells. At the same time, it is also interesting to investigate the possibility of utilizing apoptotic cells to induce immune tolerance.

ACKNOWLEDGMENTS: This work was supported by the National Natural Science Foundation of China (39870724, 39970705) and China 973 project (2001CB510009). We thank Arthur I. Roberts for his critical review and helpful discussions with the manuscript.

References

Albert, M. L., Pearce, F. A., Francisco, L. M., Sauter, B., Roy, P., Silverstein, R. L., and Bhardwaj, N. (1998a). Immature dendritic cells phagocytose apoptotic cells via alpha-v-beta-s and CD36, and cross-pressent antigens to cytotoxic T lymphocytes. J Exp Med *188*:1359–1368.

Albert, M. L., Saiter, B., and Bhardwaj, N. (1998b). Dendritic cells acquire antigen from apoptotic cells and induce class I-restricted CTLs. Nature *392*:86–89.

Andrade, F., Casciola-Rosen, L., and Rosen, A. (2000). Apoptosis in systemic lupus erythematosus. Clinical implications. Rheum Dis Clin North Am *26*:215–27, v.

Barrington, R., Zhang, M., Fischer, M., and Carroll, M. C. (2001). The role of complement in inflammation and adaptive immunity. Immunol Rev *180*:5–15.

Belz, G. T., Altman, J. D., and Doherty, P. C. (1998). Characteristics of virus-specific CD8+ T cells in the liver during the control and resolution phases of influenza pneumonia Proc Natl Acad Sci USA *95*:13812–13817.

Bissell, D. M., Roulot, D., and George, J. (2001). Transforming growth factor beta and the liver. Hepatology *34*:859–867.

Bissell, D. M., Wang, S. S., Jarnagin, W. R., and Roll, F. J. (1995). Cell-specific expression of transforming growth factor-beta in rat liver. Evidence for autocrine regulation of hepatocyte proliferation. J Clin Invest *96*:447–455.

Boeckx, W., Sobis, H., Lacquet, A., Gruwez, J., and Vandeputte, M. (1975). Prolongation of allogeneic heart graft survival in the rat after implantation on portal vein. Transplantation *19*:145–149.

Botto, M. (1998). C1q knock-out mice for the study of complement deficiency in autoimmune disease. Exp Clin Immunogenet *15*:231–234.

Botto, M., Dell'Agnola, C., Bygrave, A. E., Thompson, E. M., Cook, H. T., Petry, F., Loos, M., Pandolfi, P. P., and Walport, M. J. (1998). Homozygous C1q deficiency causes glomerulonephritis associated with multiple apoptotic bodies [see comments]. Nat Genet *19*:56–59.

Bumgardner, G. L., Gao, D., Li, J., Baskin, J. H., Heininger, M., and Orosz, C. G. (2000). Rejection responses to allogeneic hepatocytes by reconstituted SCID mice, CD4, KO, and CD8 KO mice. Transplantation *70*:1771–1780.

Bumgardner, G. L., Li, J., Heininger, M., Ferguson, R. M., and Orosz, C. G. (1998). *In vivo* immunogenicity of purified allogeneic hepatocytes in a murine hepatocyte transplant model. Transplantation *65*:47–52.

Bumgardner, G. L., Li, J., Prologo, J. D., Heininger, M., and Orosz, C. G. (1999). Patterns of immune responses evoked by allogeneic hepatocytes: evidence for independent co-dominant roles for CD4+ and CD8+ T-cell responses in acute rejection. Transplantation *68*:555–562.

Bumgardner, G. L., and Orosz, C. G. (2000). Unusual patterns of alloimmunity evoked by allogeneic liver parenchymal cells. Immunol Rev *174*:260–279.

Callery, M. P., Kamei, T., and Flye, M. W. (1989a). The effect of portacaval shunt on delayed-hypersensitivity responses following antigen feeding. J Surg Res *46*:391–394.

Callery, M. P., Kamei, T., and Flye, M. W. (1989b). Kupffer cell blockade inhibits induction of tolerance by the portal venous route. Transplantation *47*:1092–1094.

Casciola-Roen, L. A., Anhalt, G., and Rosen, A. (2000). Autoantigens targeted in systemic lupus erythematosus are clustered in two populations of surface structures on apoptotic keratinocytes. J Exp Med *179*:1317–1330.

Cella, M., Engering, A., Pinet, V., Pieters, J., and Lanzavecchia, A. (1997). Inflammatory stimuli induce accumulation of MHC class II complexes on dendritic cells [see comments]. Nature *388*:782–7.

Chen, W.-J., Frank, M. E., Jin, W., and Wahl, S. M. (2001). TGF-Released by Apoptotic T Cells Contributes to an Immunosuppressive Milieu. Immunity *14*:715–725.

Cousens, L. P., and Wing, E. J. (2000). Innate defenses in the liver during *Listeria* infection. Immunol Rev *174*:150–159.

Crispe, I. N., Dao, T., Klugewitz, K., Mehal, W. Z., and Metz, D. P. (2000). The liver as a site of T-cell apoptosis: graveyard, or killing field? Immunol Rev *174*:47–62.

Devitt, A., Moffatto, O., Raykundalia, C., Capra, J. D., Simmons, D. L., and Gregory, C. D. (1998). Human CD14 mediates recognition and phagocytosis of apoptotic cells. Nature *392*:505–509.

Dini, L., and Carla, E. C. (1998). Hepatic sinusoidal endothelium heterogeneity with respect to the recognition of apoptotic cells. Exp Cell Res *240*:388–393.

Doherty, D. G., Norris, S., Madrigal-Estebas, L., McEntee, G., Traynor, O., Hegarty, J. E., and O'Farrelly, C. (1999). The human liver contains multiple populations of NK cells, T cells, and CD3+CD56+ natural T cells with distinct cytotoxic activities and Th1, Th2, and Th0 cytokine secretion patterns. J Immunol *163*:2314–2321.

Fadok, V. A., Bratton, D. L., Freed, P. W., Westcott, J. Y., and P. M., H. (1998). Macrophages that have ingested apoptotic cells *in vitro* inhibit proinflammatory cytokine production through autocrine/paracrine mechanisms involving TGF-beta, PGE2 AND PAF. J Clin Invest *101*:890–898.

Fadok, V. A., Bratton, D. L., Rose, D. M., Pearson, A., Ezekewitz, R. A., and Henson, P. M. (2000). A receptor for phosphatidylserine-specific clearance of apoptotic cells. Nature *405*:85–90.

Fecteau, A., Tchervenkow, J., Guttman, F., Takara, T., and Rosenmann, E. (1994). Small bowel transplantation: the effects of intraportal donor-specific transfusion 24 hours pretransplant and low-sose cyclosporine. Transplantation *58*:399–402.

Feng, H., Zeng, Y., Whitesell, L., and Katsanis, E. (2001). Stressed apoptotic tumor cells express heat shock proteins and elicit tumor-specific immunity. Blood *97*:3505–3512.

Ferguson, T. A., and Griffith, T. S. (1997). A vision of cell death: insights into immune privilege. Immunol Rev *156*:167–184.

Frabetti, F., Tazzari, P. L., Musiani, D., Bontadini, A., Matteini, C., Roseti, L., Tassi, C., Viggiani, M., Marini, M., and Conte, R. (2000). White cell apoptosis in platelet concentrates. Transfusion *40*:160–168.

Gagne, K., Brouard, S., Guillet, M., Cuturi, M. C., and Souilillou, J. P. (2001). TGF-beta1 and donor dendritic cells are common key components in donor-specific blood transfusion and anti-class II heart graft enhancement, whereas tolerance induction also required inflammatory cytokines down-regulation. Eur J Immunol *31*:3111–3120.

Gershov, D., Kim, S., Brot, N., and Elkon, K. B. (2000). C-Reactive protein binds to apoptotic cells, protects the cells from assembly of the terminal complement components, and sustains an antiinflammatory innate immune response: implications for systemic autoimmunity. J Exp Med *192*:1353–1364.

Gorczynski, R. M., Chan, Z., Chung, S., Cohen, Z., Levy, G., sullivan, B., and Fu, X. M. (1994). Prolongation of rat small bowel or renal allograft survival by pretransplant transfusion and/or by varying the route of allograft venous drainage. Transplantation *58*:816–820.

Gorczynski, R. M., Fu, X. M., Church, D. J., sullivan, B., and Chen, Z. (1995). Manipulation of xenogeneic skin and/or renal graft survival in the rat-mouse concordant combination by portal vein pretransplantation transfusion. Transplant Immunol *3*:321–329.

Goss, J. A., Flye, M. W., and Lacy, P. E. (1996). Induction of allogeneic islet survival by intrahepatic islet preimmunization and transient immunosuppression. Diabetes *45*:144–147.

Griffith, T. S., Yu, X., Herndon, J. M., Green, D. R., and T. A., F. (1996). CD95-induced apoptosis of lymphocytes in an immune privileged site induces immunological tolerance. Immunity *5*:7–16.

Hawiger, D., Inaba, K., Dorsett, Y., Guo, M., Mahnke, K., Rivera, M., Ravetch, J. V., Steinman, R. M., and Nussenzweig, M. C. (2001). Dendritic cells induce peripheral T cell unresponsiveness under steady state conditions *in vivo*. J Exp Med *194*:769–79.

Hengartner, M. O. (2001). Apoptosis: corralling the corpses. Cell *104*:325–328.

Huang, F.-P., Platta, N., Wykesa, M., Majora, J. R., Powella, T. J., Jenkinsa, C. D., and MacPhersona, G. G. (2000). A Discrete Subpopulation of Dendritic Cells Transports Apoptotic Intestinal Epithelial Cells to T Cell Areas of Mesenteric Lymph Nodes. J Exp Med *191*:435–444.

Huang, L., Soldevila, G., Leeker, M., Flavell, R., and Crispe, I. N. (1994). The liver eliminates T cells undergoing antigen-triggered apoptosis *in vivo*. Immunity *1*:741–749.

Huynh, M. L., Fadok, V. A., and Henson, P. M. (2002). Phosphatidylserine-dependent ingestion of apoptotic cells promotes TGF-beta1 secretion and the resolution of inflammation. J Clin Invest *109*:41–50.

Inaba, K., Turley, S., Iyoda, T., Yamaide, F., Shimoyama, S., Reis e Sousa, C., Germain, R. N., Mellman, I., and Steinman, R. M. (2000). The Formation of Immunogenic Major Histocompatibility Complex Class IIC Peptide Ligands in Lysosomal Compartments of Dendritic Cells Is Regulated by Inflammatory Stimuli. J Exp Med *191*:927–936.

Josien, R., Douillard, P., Guillot, C., Muschen, M., Anegon, I., Chetritt, J., Menoret, S., Vignes, C., Soulillou, J. P., and Cuturi, M. C. (1998). A critical role for transforming growth factor-beta in donor transfusion-induced allograft tolerance. J Clin Invest *102*:1920–1926.

Kamada, N., Davies, H. F. S., and Roser, B. (1981). Reversal of transplantation immunity by liver grafting. Nature *292*:840–842.

Kamada, N., Teramoto, K., Baguerizo, A., Ishikawa, M., Suminoto, R., and Ohkouchi, Y. (1998). Cellular basis of transplantation tolerance induced by liver grafting in the rat. Transplantation *46*:165–167.

Kamei, T., Callery, M. P., and Flye, M. W. (1990). Kupffer cell blockade prevents induction of portal venous tolerance n rat cardiac allograft transplantation. J Surg Res *48*:393–396.

Kishore, U., and Reid, K. B. (2000). C1q: structure, function, and receptors. Immunopharmacology *49*:159–170.

Knolle, P., Schlaak, J., Uhrig, A., Kempf, P., Meyer zum Buschenfelde, K. H., and Gerken, G. (1995). Human Kupffer cells secrete IL-10 in response to lipopolysaccharide (LPS) challenge. J Hepatol *22*:226–229.

Knolle, P. A., Germann, T., Treichel, U., Uhrig, A., Schmitt, E., Hegenbarth, S., Lohse, A. W., and Gerken, G. (1999). Endotoxin down-regulates T cell activation by antigen-presenting liver sinusoidal endothelial cells. J Immunol *162*:1401–1407.

Korb, L. C., and Ahearn, J. M. (1997). C1q binds directly and specifically to surface blebs of apoptotic human keratinocytes: complement deficiency and systemic lupus erythematosus revisited. J Immunol *158*:4525–4528.

Lorenz, H. M., Herrmann, M., Winkler, T., Gaipl, U., and Kalden, J. R. (2000). Role of apoptosis in autoimmunity. Apoptosis *5*:443–449.

Luan, J. J., Monteiro, R. C., Sautes, C., Fluteau, G., Eloy, L., Fridman, W. H., Bach, J. F., and Garchon, H. J. (1996). Defective Fc gamma RII gene expression in macrophages of NOD mice: genetic linkage with up-regulation of IgG1 and IgG2b in serum. J Immunol *157*:4707–4716.

Lumsden, A. B., Henderson, J. M., and Kutner, M. H. (1988). Endotoxin levels measured by a chromogenic assay in portal, hepatic and peripheral venous blood in patients with cirrhosis. Hepatology *8*:232–236.

Markees, T. G., Serreze, D. V., Phillips, N. E., Sorli, C. H., Gordon, E. J., Shultz, L. D., Noelle, R. J., Woda, B. A., Greiner, D. L., Mordes, J. P., and Rossini, A. A. (1999). NOD mice have a generalized defect in their response to transplantation tolerance induction. Diabetes *48*:967–74.

Mehal, W. Z., Juedes, A. E., and Crispe, I. N. (1999). Selective retention of activated CD8+ T cells by the normal liver. J Immunol *163*:3202–3210.

Mevorach, D., Mascarenhas, J. O., Gershov, D., and Elkon, K. B. (1998). Complement-dependent clearance of apoptotic cells by human macrophages. J Exp Med *188*:2313–2320.

Morita, H., Sugiura, K., Inaba, M., Jin, T., Ishikawa, J., Lian, Z., Adachi, Y., Sogo, S., Yamanishi, K., Taki, H., *et al.* (1998). A strategy for organ allografts without using immunosuppressants or irrdiation. Proc Natl Acad Sci USA *95*:6947–6952.

Muschen, M., Warskulat, U., Douillard, P., Gilbert, E., and Haussinger, D. (1998). Regulation of CD95 (APO-1/Fas) receptor and ligand expression by lipopolysaccharide and dexamethasone in parenchymal and nonparenchymal rat liver cells. Hepatology *27*:200–208.

Muschen, M., Warskulat, U., Peters, R. T., Bode, J. G., Kubitz, R., and Haussinger, D. (1999). Involvement of CD95 (Apo-1/Fas) ligand expressed by rat Kupffer cells in hepatic immunoregulation. Gastroenterology *116*:666–677.

Nakano, Y., Monden, M., Valdevia, L. A., Gotoh, M., Tono, T., and Mori, T. (1992). Permanent acceptance of liver allografs by intraportal injection of donor spleen cells in rats. Surgery *111*:668–676.

Navratil, J. S., Korb, L. C., and Ahearn, J. M. (1999). Systemic lupus erythematosus and complement deficiency: clues to a novel role for the classical complement pathway in the maintenance of immune tolerance. Immunopharmacology *42*:47–52.

Nielsen, H. J. (1995). Detrimental effects of perioperative blood transfusion. Br J Surg *82*:582–587.

Opelz, G., and Terasaki, P. I. (1974). Poor kidney-transplant survival in recipients with frozen-blood transfusions or no transfusions. Lancet *2*:696–698.

Persijn, G. G., Cohen, B., Lansbergen, Q., and Van Rood, J. J. (1979). Retrospective and prospective studies on the effect of blood transfusions in renal transplantation in The Netherlands. Transplantation *28*:396–401.

Platt, N., Suzuki, H., Kurihara, Y., Kodama, T., and Gordon, S. (1996). Role for the class A macrophage scavenger receptor in the phagocytosis of apoptotic thymocytes *in vitro*. Proc Natl Acad Sci USA *93*:12456–12460.

Ponner, B. B., Stach, C., Zoller, O., Hagenhofer, M., Voll, R. E., Kalden, J. R., and Herrmann, M. (1998). Induction of apoptosis reduces immunogenicity of human T-cell lines in mice. Scand J Immunol *47*:343–347.

Rabinovitch, A., Suarez-Pinzon, W. L., Shi, Y., Morgan, A. R., Bleackley, R. C. DNA fragmentation is an early event in cytokine-induced islet beta-cell destruction. Diabetologia 1994 Aug; *37*:733–738.

Riordan, S. M., and Williams, R. (1999). Tolerance after liver transplantation: does it exist and can immunosuppression be withdrawn? J Hepatol *31*:1106–1119.

Rodenburg, R. J., Raats, J. M., Pruijn, G. J., and van Venrooij, W. J. (2000). Cell death: a trigger of autoimmunity? Bioessays *22*:627–636.

Rovere, P., Sabbadini, M. G., Fazzini, F., Bondanza, A., Zimmermann, V. S., Rugarli, C., and Manfredi, A. A. (2000). Remnants of suicidal cells fostering systemic autoaggression. Apoptosis in the origin and maintenance of autoimmunity. Arthritis Rheum *43*:1663–1672.

Rovere, P., Sabbadini, M. G., Vallinoto, C., Fascio, U., Recigno, M., Crosti, M., Ricciardi-Castagnoli, P., Balestrieri, G., Tincani, A., and Manfredi, A. A. (1999). Dendritic cell presentation of antigens from apoptotic cells in a proinflammatory context: role of opsonizing anti-beta2-glycoprotein I antibodies. Arthritis Rheum *42*:1412–1420.

Rovere, P., Vallinoto, C., Bondanza, A., Crosti, M. C., Rescigno, M., Ricciardi-Castagnoli, P., Rugarli, C., and Manfredi, A. A. (1998). Bystander apoptosis triggers dendritic cell maturation and antigen-presenting function. J Immunol *161*:4467–4471.

Sauter, B., Albert, M. L., Francisco, L., Larsson, M., Somersan, S., and Bhardwaj, N. (2000). Consequences of cell death: exposure to necrotic tumor cells, but not primary tissue cells or apoptotic cells, induces the maturation of immunostimulatory dendritic cells [see comments]. J Exp Med *191*:423–434.

Savill, J. (1998). Apoptosis. Phagocytic docking without shocking [news; comment]. Nature *392*:442–443.

Savill, J., and Fadok, V. (2000). Corpse clearance defines the meaning of cell death Nature *407*:784–788.

Scitt, r. S., McMahon, E. J., Pop, S. M., Reap, E. A., Caricchio, R., Cohen, P. L., Earp, H. S., and Matsisjoma, G. K. (2001). Phagocytosis and clearance of apoptotic cells is mediated by MER. Nature *410*:207.

Seki, S., Habu, Y., Kawamura, T., Takeda, K., Dobashi, H., Ohkawa, T., and Hiraide, H. (2000). The liver as a crucial organ in the first line of host defense:the roles fo Kupffer cells, natural killer (NK) cells and NK1.1Ag+ T cells in T helper 1 immune response. Immunol Rev *174*:35–46.

Serreze, D. V., Gaskins, H. R., and Leiter, E. H. (1993). Defects in the differentiation and function of antigen presenting cells in NOD/Lt mice. J Immunol *150*:2534-2543.

Shi, Y., Zheng, W., and Rock, K. L. (2000). Cell injury releases endogenous adjuvants that stimulate cytotoxic T cell responses. Proc Natl Acad Sci USA *97*:14590–14595.

Shi, Y. F., Sahai, B. M., and Green, D. R. (1989). Cyclosporin A inhibits activation-induced cell death in T-cell hybridomas and thymocytes. Nature *339*:625–626.

Snyder, E. L., and Kuter, D. J. (2000). Apoptosis in transfusion medicine: of death and dying—is that all there is? [editorial; comment]. Transfusion *40*:135–138.

Starzl, T. E., Demetris, A. J., Murase, N., Trucco, M., Thomson, A. W., and Rao, A. S. (1996). The lost chord; microchimerism and allograft survival. Immunol Today *17*:577–584.

Starzl, T. E., Demetris, A. J., Mursase, N., Ildstad, S., Ricordi, C., and Trucco, M. (1992). Cell migration, chimerism and graft tolerance. Lancet *339*:1579.

simmunity and tolerance. N Engl J Med *339*:1905–13.

Steinman, R. M., Turley, S., Mellman, I., and Inaba, K. (2000). The induction of tolerance by dendritic cells that have captured apoptotic cells. J Exp Med *191*:411–416.

Sun, E., Zhang, L., Zeng, Y., Ge, Q., Zhao, M., and Gao, W. (2000). Apoptotic cells actively inhibit the expression of CD69 on con A activated T lymphocytes. Scand J Immunol *51*:231–236.

Sun, E., and Shi, Y. (2001). Apoptosis: the quiet death silences the immune system. Pharmacol Ther *92*:135–145.

Taylor, P. R., Carugati, A., Fadok, V. A., Cook, H. T., Andrew, M., Carroll, M. C., Savill, J. S., Henson, P. M., Botto, M., and Walport, M. J. (2000). A hierarchical role for classical pathway complement proteins in the clearance of apoptotic cells *in vivo*. J Exp Med *192*:359–366.

Thompson, C. B. (1995). Apoptosis in the pathogenesis and treatment of disease. Science *267*:1456–1462.

Thomson, A. W., and Lu, L. (1999). Are dendritic cells the key to liver transplant tolerance? Immunology Today *20*:27–32.

Tough, T. E., Sun, S., and Sprent, J. (1997). T cell stimulation *in vivo* by lipopolysaccharide (LPS). J Exp Med *185*:2089–2094.

Trudeau, J. D., Dutz, J. P., Arany, E., Hill, D. J., Fieldus, W. E., and Finegood, D. T. (2000). Neonatal beta-cell apoptosis: a trigger for autoimmune diabetes? Diabetes *49*:1–7.

Voll, R. E., Herrmann, M., Roth, E. A., Stach, C., and Kalden, J. (1997). Immunosuppressive effects of apoptotic cells. Nature *390*:350–351.

Well, A. D., Li, X. C., Li, Y., Walsh, M. C., Zhang, X. X., Wu, Z., Nunez, G., Tang, A., Sayegh, M., Hancock, W. W., *et al.* (1999). Requirement for T cell apoptosis in the induction of peripheral transplantation tolerance. Nat Med *5*:1303–1307.

Wood, K., and Sachs, D. H. (1996). Chimerism and transplantation tolerance: cause and effect. Immunol Today *17*:584–589.

Yang, R., Liu, Q., Grosfeld, J. L., and Pescovitz, M. D. (1994). Intestinal venous drainage through the liver is a prerequistite for oral tolerance induction. J Pediatr Surg *29*:1145–1148.

Chapter **13**

Neuroprotection against Apoptosis

What Has It Got to Do with the Mood Stabilizer Lithium?

DE-MAW CHUANG[1] AND CHRISTOPHER HOUGH[2]

ABSTRACT: Lithium, a mood-stabilizer, robustly protects cultured rat brain neurons against a variety of apoptotic insults including glutamate-induced excitotoxicity. This neuroprotection involves inactivation of the N-methyl-D-aspartate receptors, presumably through reduced phosphorylation of Tyr1472 in the NR2B subunit. Lithium also up-regulates cytoprotective Bcl-2, but down-regulates pro-apoptotic p53 and Bax. In addition, it activates the cell survival PI-3 kinase/Akt pathway and suppresses glutamate-induced activation of mitogen-activated protein kinase signaling. Release of brain-derived neurotrophic factor and activation of its receptor, TrkB, is a prerequisite for lithium protection against excitotoxicity. In an animal model of acute stroke, post-insult lithium treatment at therapeutic levels reduces ischemia-induced brain infarction and improves behavioral outcome. Lithium also reduces brain damage induced by intra-striatal infusion of an excitotoxin. Collectively, our results suggest that lithium might have clinical utility for the treatment of excitotoxicity-related neurodegeneration.

Key Words: Excitotoxicity; Huntington's Disease; lithium; neuroprotection; stroke

Introduction

What is neuroprotection? A definition might be the prevention or retardation of death or injury to neurons. Fifteen years ago, the notion of neuroprotection rarely passed through our thoughts. Today, this concept is replete with significance attesting to more than a decade of research into the causes of injury and death in neurons and other cells of the brain. Today, one is more likely to ask, "protection from what?" in acknowledgement of the myriad agents and conditions that are now known to lead to neuronal distress and death. Much is now known about the intracellular events that characterize these processes, though this knowledge is still undergoing expansion, distillation and refinement. Agents that cause neuronal distress or death have long been the subject of study, but those which prevent or retard neuronal damage have only recently gained our attention. How these agents prevent or slow distress and death in neurons or, more precisely, how these agents alter those intracellular events

[1]Molecular Neurobiology Section, Mood and Anxiety Disorder Program, National Institute of Mental Health, National Institutes of Health, Bethesda, MD, USA; [2]Department of Psychiatry, Uniformed Services University of Health Sciences, Bethesda, MD, USA. *Correspondence:* De-Maw Chuang, PhD, Molecular Neurobiology Section, Mood and Anxiety Disorder Program, National Institute of Mental Health, 10 Center Dr MSC 1363, Bethesda, MD 20892–1363, USA. Fax: 301-480-9290; e-mail: chuang@helix.nih.gov

characteristic of distress and death have given us an illustrative view of the meaning of those events and the nature of neurological, neurodegenerative and psychiatric illnesses. Lithium is such an agent. Although it has been used extensively as a mood stabilizer against affective mood disorder (also referred to as manic depressive illness), its action as a neuroprotective agent has only recently been brought to light in our laboratory and others.

Neuroprotection against Apoptosis

The study of apoptosis in neurons has been key to our understanding of neuroprotection and its importance in human disease. The distinction between necrosis and apoptosis is now clearly understood to be one of kinetics and degree, with apoptosis being the slower process involving orderly gene expression, protein synthesis and post-translational processing. Depending on the agent or condition, one or more pathways are stimulated that lead to protein degradation, DNA degradation and loss of mitochondrial inner membrane potential. Research has focused on three types of proteins that play key roles in these processes: the transcription factors p53, NF-kB, AP-1 and CREB involved in directing the expression of proteins involved in the prevention or execution of apoptosis, the Bcl-2/Bax family of proteins involved in mitochondrial regulation of apoptosis, and apoptosis-specific proteases, the caspases. The transcription factors are important in the study of apoptosis because they control the response of neurons to cellular stress. In some cases, the response is survival and in others it is apoptosis. The tumor suppressor p53 can be pro-apoptotic, while CREB is recruited as an anti-apoptotic response to cellular distress. Nuclear factor NF-kB and AP-1 can be either pro-apoptotic or cytoprotective, depending on the nature of apoptotic stimuli and gene products induced. The Bcl-2/Bax family contains members that promote or prevent apoptosis by a process that is, as yet, not fully understood. Family members can dimerize with itself or other members within the mitochondrial membrane(s) to promote or inhibit their functions. Bcl-2, a tumor promoter, is anti-apoptotic (Kluck et al., 1997; Susin et al., 1996). Bax is pro-apoptotic and promotes the release of cytochrome c from mitochondria in a signal transduction pathway that involves Apaf-1 and leads to the activation of the caspases (Kuwana et al., 1998; Allen et al., 1998). The caspases are activated in a cascading succession of proteolytic digestions initiated by caspase-9 among others, and subsequent downstream caspases to create a phalanx of proteases that destroy the cell from within. Bcl-2 also regulates other forms of caspase activation independently of the cytochrome c/Apaf-1/caspase-9 apoptosome (Marsden et al., 2002).

It is against the background of the functions of these proteins in apoptosis that neuroprotection is now newly appreciated. Agents or conditions that promote the expression of p53 or Bax, or reduce the expression of Bcl-2, or phosphorylated CREB (the active form of CREB) are neurotoxic because they are pro-apoptotic. Conversely, agents that promote the expression of Bcl-2, or phosphorylated CREB and/or reduce the expression of p53, Bax, or caspases are neuroprotective, because they promote survival and prevent apoptosis.

Neuroprotection against Excitotoxicity

Among neurotoxic agents, there is one that is especially important because it is a neurotransmitter. The agent, glutamate, is responsible for both excitatory neurotransmission and excitotoxicity. The latter is strongly implicated in the pathogenesis of a variety of neurodegenerative diseases including stroke, Huntington's disease, Alzheimer's disease, amyotrophic lateral sclerosis, spinal cord injury, cerebellar degeneration and others. We have extensively studied glutamate excitotoxicity in the primary culture of cerebellar granule cells. These cells synthesize and release glutamate. They also express NMDA receptors, which make them vulnerable to glutamate toxicity and cell death. Exposure of cerebellar granule neurons to 50–100 mM glutamate for a few hours induces apoptosis, as indicated by DNA cleavage and chromatin condensation (Nonaka et al., 1998a). Pretreatment of cerebellar granule cells with LiCl for 7 days results in a dose-dependent protection against excitotoxicity-induced apoptosis with an EC_{50} value of approximately 1.3 mM. The neuroprotective effect of lithium in these cells requires a pretreatment of more than 2 or 3 days and is maximal only after 6 to 7 days of treatment. Excitotoxicity induced by NMDA rather than glutamate is also robustly reduced by lithium pretreatment, but the neurotoxicity of kainic acid is much less reduced by lithium (Nonaka et al., 1998a).

Like cerebellar granule cells, rat embryonic cerebral cortex cells in culture are also subject to glutamate-induced excitotoxicity as a result of NMDA receptor activation. Lithium also suppresses glutamate excitotoxicity nearly completely after 6 days of pretreatment (Hashimoto et al., 2002a). These protective effects with regard to monovalency appeared to be specific for lithium, as neither rubidium nor cesium is effective as a neuroprotective agent. In primary cortical cell cultures, lithium is substantially more potent as a neuroprotective agent than in cerebellar granule cells. Significant protection was observed at 0.2 mM in cortical cultures, and maximal, complete protection was achieved at 1 mM. The neuroprotective action of lithium in neocortical neurons, we noted, occurs within the therapeutic and sub-therapeutic dose ranges for treatments of bipolar affective disorder in humans. We also noted that neuroprotection by lithium depends on the density of the cortical primary cultures with more pronounced effects at higher cell densities. Glutamate excitotoxicity in rat embryonic hippocampal neurons is also prevented by lithium pretreatment (Nonaka et al., 1998a). Our results demonstrate that lithium suppresses excitotoxicity in glutamatergic neurons from several different regions of the brain.

Valproic acid, another mood stabilizer, also has neuroprotective properties. We found it to protect dose-dependently mature cerebellar granule cells and cortical neurons from glutamate excitotoxicity (Kanai et al., 2002; Hashimoto et al., 2002a). Moreover, valproic acid protects cortical neurons from spontaneous cell death induced by aging of the cultures (Jeong et al., 2003). Could neuroprotection be a common action of mood-stabilizers?

Since glutamate excitotoxicity is caused in large part by a prodigious influx of calcium through activated NMDA receptors, we examined the effects of lithium on NMDA receptor-mediated calcium influx. Using fura-2 microphotometry and $^{45}Ca^{2+}$ influx techniques, we found that 7-day lithium treatment causes a dose-dependent reduction in NMDA receptor-mediated Ca^{2+} entry in cerebellar granule cells and cortical neurons (Nonaka et al., 1998a; Hashimoto et al., 2002a). The effect of long-term lithium pretreatment is slow to reverse, because washing out of lithium during the glutamate challenge does not abrogate the neuroprotection. Although the data suggests changes in protein expression, the neuroprotective effect of lithium pretreatment does not appear to be associated with changes in NMDA receptor subunit protein levels (Nonaka et al., 1998a; Hashimoto et al., 2002a). In cortical neurons, however, neuroprotection is correlated with a decrease in the tyrosine phosphorylation of subunit NR2B at residue 1472 (Hashimoto et al., 2002a). Moreover, lithium-induced decrease in NR2B tyrosine phosphorylation is associated with a loss of the activity of Src tyrosine kinase, which is involved in receptor tyrosine phosphorylation (Hashimoto et al., 2003).

Modulation of NMDA receptor function is not the sole mechanism responsible for lithium's neuroprotective effects. Lithium protects neurons from apoptotic agents or conditions that do not rely on NMDA receptor activation in cerebellar granule cells and cortical neurons. These include treatment with anticonvulsants (Nonaka et al., 1998b), b-amyloid peptide (Wei et al., 2000), serum deprivation and low extracellular potassium concentration (Nonaka et al., 1998b; D'Mello et al., 1994). The requirement of chronic treatment in lithium-mediated neuroprotection suggests that alteration of gene expression is involved. We thus focused on the expression of p53, Bax and Bcl-2. The p53 protein promotes the expression of Bax and suppresses the expression of Bcl-2 (Miyashita et al., 1994). We found that long-term, but not acute, lithium treatment dose-dependently decreases p53 and Bax mRNA and protein levels in cerebellar granule cells (Chen and Chuang, 1999). In contrast, Bcl-2 mRNA and protein are time and dose-dependently increased by lithium treatment. Manji and coworkers found that Bcl-2 protein levels were elevated in the brains of rats chronically treated with lithium or valproate (Chen et al., 1999). Glutamate exposure, on the other hand, down-regulates Bcl-2 mRNA and protein levels in cell cultures, but up-regulates p53 and Bax mRNA and protein levels as well as cytochrome c release and caspase activation (Chen and Chuang, 1999). The excitotoxic effects of glutamate are largely suppressed by long-term lithium treatment.

Chronic lithium treatment also alters the intracellular signalling pathways involved in glutamate excitotoxicity. The mitogen-activated protein kinase (MAPK) pathways are protein kinase signalling cascades that are activated in response to extracellular stimuli and cellular distress. Two MAPKs, c-Jun N-terminal kinase (JNK) and p38 kinase, have been implicated in the initiation of apoptosis (for review, Mielke and Herdegen, 2000; McCubrey et al., 2000). Activation of JNK results in phosphorylation of the transcription factor c-Jun (Hibi et al., 1993). This, in turn, increases the capacity of c-Jun to complex with other transcription factors to activate transcription at AP-1 DNA consensus sequence. AP-1 elements reside within the promoters of many genes involved in intracellular stress responses (Hibi et al., 1993). p38 kinase can complex with JNK to alter AP-1 binding activity as well, in a manner that is likely to be related to its pro-apoptotic activity. NMDA receptor activation by glutamate results in a transient activation of JNK and p38, increased phosphorylation of c-Jun and a robust increase in AP-1 binding in cerebellar granule cells (Chen et al., 2003). These effects are largely suppressed by long-term, but not acute, lithium pretreatment or by curcumin, an inhibitor of JNK activity as well as AP-1 and NF-kB binding (Chen and Tan, 2000). Curcumin and SB203580 (an inhibitor of p38 kinase), moreover, dose-dependently protect cerebellar granular neurons from glutamate excitotoxicity (Chen et al., 2003). As discussed above, binding at the AP-1 consensus sequence can lead to either cytotoxicity or cytoprotection [for review, (Karin et al., 1997)]. It is interesting to note that lithium inhibits many forms of AP-1 binding activity induced by stimulation of stress activated signaling pathways(Williams and Jope, 1995), but elevates basal, unstimulated AP-1 binding *in vitro* as well as *in vivo* (Ozaki and Chuang, 1997; Asghari et al., 1998; Yuan et al., 1998). It remains to be established whether the elevation of basal AP-1 binding contributes to the neuroprotective effects of this drug.

The phosphatidylinositol 3-kinase (PI 3-kinase)/Akt signaling pathway is used by growth factors, such as insulin-like growth factor, brain-derived neurotrophic factor (BDNF) and others, to promote cell survival or to stave off apoptosis (Dudek et al., 1997). The products of PI 3-kinase induce the membrane translocation, subsequent phosphorylation (by two other PI-dependent kinases) and activation of Akt. The activation of Akt induces the inactivation of such pro-apoptotic proteins as glycogen synthase kinase-3b (GSK-3b), procaspase 9 and BAD, and the activation of the CREB (the cyclic AMP-response element binding protein), a cytoprotective transcription factor (for review, Dudek et al., 1997). The survival of cerebellar granule cells critically depends on the activity of PI 3-kinase, and selective inhibitors of this kinase rapidly induce neuronal death (Chalecka-Franaszek and Chuang, 1999). Lithium treatment of these neurons results in a rapid increase in PI 3-kinase activity followed by increased phosphorylation and activation of Akt (Chalecka-Franaszek and Chuang, 1999). Akt activation leads to increased phosphorylation and inactivation of GSK-3b. PI 3-kinase inhibitors block these lithium-induced events. This pathway represents an alternative to the well-known, direct inhibitory interaction between GSK-3b and lithium (Klein and Melton, 1996). The inhibition of GSK-3b by either mechanism may contribute to lithium's neuroprotection action, since the activity of GSK-3b is pro-apoptotic by a number of different pathways [for review, (Grimes and Jope, 2001)]. Glutamate, for example, causes rapid but reversible decreases in phosphorylated Akt and Akt activity caused by protein phosphatase 2A activity stimulated by the NMDA receptor (Chalecka-Franaszek and Chuang, 1999). Long-term lithium treatment reduces the loss of Akt activity and facilitates its recovery toward normal levels.

CREB phosphorylation by Akt and other kinases activates this prominent transcription factor. Phospho-CREB (pCREB) is involved in the expression of a variety of proteins, including Bcl-2 and BDNF [for review, see (Finkbeiner, 2000; Tao et al., 1998)]. Glutamate robustly decreases p-CREB levels, largely by the activity of protein phosphatase 1 (Kopnisky et al., 2003). Long-term lithium treatments suppress glutamate-induced loss of p-CREB in part by reducing the activity of this phosphatase and increasing the activity of the MAPK family member, MEK.

The multiple actions of lithium on the activities of PI 3-kinase/Akt, MAPK and CREB may be explained by the induction of BDNF and activation of its receptor TrkB. Indeed, long-term lithium treatment transiently increases intracellular levels of BDNF followed by TrkB activation (Hashimoto et al., 2003b). Moreover, the neuroprotective effects of lithium can be blocked by a Trk receptor blocker K252a or a BDNF-neutralizing antibody. Further, BDNF gene knockout abolishes the neuroprotective action of lithium against excitotoxicity, suggesting an essential role of a critical level of BDNF in lithium neuroprotection. Yamawaki and coworkers have demonstrated that treatment with lithium or valproate for two weeks or longer increases the levels of BDNF in the hippocampus and cerebral cortex (Fukumoto et al., 2001).

Neuroprotection against Ischemia

Glutamate excitotoxicity is thought to be the major source of damage caused by stroke, a major source of morbidity and mortality worldwide. Using animal models of stroke, clinicians have intensively sought potential therapies for the prevention and treatment of this condition. Such a model in rats is the left middle cerebral arterial occlusion (MCAO) model. Significant neurological deficits such as posture abnormality and hemiplegia are induced in the model 24 h after infarct induction (Nonaka and Chuang, 1998). At this time, massive ischemic brain damage occurs in the cerebral cortex of the frontal, sensorimotor and auditory areas and in the lateral segment of the caudate nucleus. We found that a 16-day pretreatment with LiCl significantly reduces the neurological deficits induced by MCAO and markedly decreases infarct volume [a 56% decrease from 215.4 ± 24.7 to 94.2 ± 14.1 mm^3, (Nonaka and Chuang, 1998)]. In a follow-up study, using a transient ischemia model in which rats were subjected to MCAO for 1 h followed by reperfusion for 23 to 72 h, we found that a single dose of lithium administrated after the onset of the ischemic insult still provides a significant reduction in infarct volume (Ren et al., 2003). It is generally believed that the MCAO/ reperfusion animal model is more relevant to the pathophysiology of patients with acute stroke. The effective time window for lithium to be neuroprotective is at least within 3 h of the initiation of MCAO. The range of doses required for neuroprotection are within the concentrations of this drug established for its therapeutic efficacy against bipolar affective illness. More specifically, significant protection occurs at 0.5 mEq/kg of LiCl and its maximal effect at approximately 1.0 mEq/kg.

Heat shock protein 70 (HSP70) is induced during ischemia in brain areas relatively resistant to such insult [for review, (Lipton, 1999)]. We found that HSP 70 immunoreactivity of the brain infarct penumbra of rats subjected to MCAO/reperfusion is increased two-fold over the induction elicited by ischemia alone by lithium treatment (Ren et al., 2003) and is associated with an increase in heat shock factor 1 binding (Ren et al., 2003). The neuroprotective effects of lithium in the MCAO/reperfusion rat model are associated with a marked reduction in the neurological deficits resulting from ischemic injury (Ren et al., 2003).

Neuroprotection against Neurodegenerative Diseases

Glutamate excitotoxicity has also been implicated in Huntington's disease which is an adult-onset, autosomal dominant, fatal neurodegenerative disease. The disease is characterized by hyperkinetic, involuntary movements, cognitive impairment and depression in patients. Although it is a genetic disorder linked to an expansion of the CAG repeat region of the gene huntingtin, there is strong evidence that glutamate excitotoxicity is a contributor to the neurodegeneration observed with the disease. A model of Huntington's disease can be produced in rats by infusing the excitotoxin, quinolinic acid (QA), into the striatum. This induces a loss of medium-sized spiny neurons but leaves large interneurons relatively unaffected (Schwarcz et al., 1983; Beal et al., 1986; Qin et al., 1999). This type of damage to the striatum is very similar to the pathology of Huntington's disease. We found that QA-induced loss of medium-sized spiny neurons, monitored by dopamine D1 receptor autoradiography and glutamic acid decarboxylase (GAD) mRNA *in situ* hybridization, is markedly reduced by lithium pretreatment at therapeutically relevant doses for 1 or 16 days (Wei et al., 2001). Lithium pretreatment reduces QA-induced caspase 3 activation and DNA damage detected by TUNEL staining in brain slices of affected rats (Senatorov et al., 2002). In this model, the neuroprotective effects of lithium are manifested as an increase in the number of Bcl-2 mRNA expressing neurons and Bcl-2 protein immunoreactivity in the striatum and other brain areas (Senatorov et al., 2002). In previous studies, we demonstrated that QA-induced apoptosis of medium-sized spiny striatal neurons is mediated by activation of the transcription factor NF-kB and by the over-expression of such pro-apoptotic proteins as p53, c-Myc and cyclin D1 (Qin et al., 1999). There is evidence that metabotropic glutamate receptor agonists and prostaglandin A1 can protect striatal neurons from QA-induced injury by perturbing the pro-apoptotic pathway of NF-kB (Wang et al., 1999; Qin et al., 2001). Whether a similar mechanism is involved in lithium neuroprotection against excitotoxic neurodegeneration remains to be established.

Neuroprotection against Psychiatric Illness

Bipolar affective disorder is a chronic, devastating and frequently life-threatening disease affecting more than 1% of the world's population. Recent research from several laboratories suggests that neuroanatomical changes within the cerebral cortex accompany the pathogenesis of bipolar mood disorder. Drevets et al. (1997), for example, reported that there is a substantial decrease in the volume of subgenual prefrontal cortex in unipolar and bipolar patients. Interestingly, the loss of subgenual prefrontal cortex volume was much less in patients who received lithium or valproate treatment (Drevets, 2001). Subsequent studies showed that in bipolar subjects, this volume decrease is associated with a reduction in the number of glial cells (Öngür et al., 1998). Using computer-assisted cell counting, a significant atrophy and loss of neuronal and glial density were also found in the prefrontal cortex of postmortem samples of unipolar and bipolar patients (Rajkowska et al., 1999).

Lithium has been one of the major drugs used over the past half century for the treatment of this illness. Lithium is effective as an anti-manic drug, a prophylactic agent for mania and depression and as an augmentation to classical antidepressants. The molecular and cellular mechanisms underlying lithium's therapeutic actions remain poorly understood despite many years of intensive research. It has long been hoped that an elucidation of the mechanism of action of lithium's therapeutic efficacy against bipolar illness would shed light on the pathogenesis of the illness. In light of the neuroprotective actions of lithium, it now seems possible that bipolar affective disorder involves some form of neurodegeneration. A decade ago, such a notion would have been hard to accept. Our challenge is to determine whether this occurs and how.

Conclusions

Lithium is neuroprotective against excitotoxicity mediated by NMDA receptors in rat cerebellar granule cell and cerebral cortical neuron cultures. These effects are long-lasting, reach a maximum after long-term treatment and occur in the range of concentrations that this drug is used to treat bipolar affect disorder in humans. Lithium-mediated neuroprotection against excitotoxicity involves multiple mechanisms. These include the inhibition of NMDA receptors, the up-regulation of Bcl-2 and down-regulation of p53 and Bax gene expression, the activation of anti-apoptotic Akt and CREB, and the modulation of the MAPKs JNK and p38. A growing list of studies by us and by others indicates that lithium inhibits cell death induced by a wide spectrum of insults to neurons in culture and *in vivo* (for review, Chuang et al., 2002). Lithium reduces neurological deficits and decreases the brain infarct volume in the rat MCAO model of stroke, even when the drug was administrated after the onset of ischemia. Lithium neuroprotection against ischemia-induced brain injury is concurrent with up-regulation of the heat shock responses. In the rat QA model of Huntington's disease, lithium reduces excitotoxin-induced apoptosis of medium-sized spiny striatal neurons presumably by up-regulating Bcl-2 and down-regulating caspase 3 expression.

Our notion of neuroprotection has arisen out of a better understanding of exactly what happens during cell death gained over the last fifteen years of research on apoptosis. This notion of neuroprotection and how neuroprotective agents such as lithium prevent cell death have illuminated issues that have given us a new perspective on intractable diseases such as Huntington's disease and bipolar affective disorder.

ACKNOWLEDGMENTS: We thank Shigeyuki Nonaka, Ryota Hashimoto, Ming Ren, Vladimir Senatorov, Lori Christ, Ren-Wu Chen, Hua-Feng Wei, Elziebeta Chalecka-Franaszek, Kathy Kopnisky, Hirohiko Kanai, Koichiro Fujimaki, MiRa Jeong, and Peter Leeds of the Molecular Neurobiology Section, NIMH, NIH for their contributions to the neuroprotective studies discussed in this report.

References

Allen, R.T., Cluck, M.W., and Agrawal, D.K. (1998). Mechanisms controlling cellular suicide: role of Bcl-2 and caspases. Cell. Mol. Life Sci. 54, 427–445.

Asghari, V., Wang, J.F., Reiach, J.S., and Young, L.T. (1998). Differential effects of mood stabilizers on Fos/Jun proteins and AP-1 DNA binding activity in human neuroblastoma SH-SY 5Y cells. Mol. Brain Res. 58, 95–102.

Beal, M.F., Kowall, N.W., Ellison, D.W., Mazurek, M.F., Swartz, K.J., and Martin, J.B. (1986). Replication of the neurochemical characteristics of Huntington's disease by quinolinic acid. Nature 321, 168–171.

Chalecka-Franaszek, E., and Chuang, D.-M. (1999). Lithium activates the serine/threonine kinase Akt-1 and suppresses glutamate-induced inhibition of Akt-1 activity in neurons. Proc. Natl. Acad. Sci. USA 96, 8745–8750.

Chuang, D.-M., Chen, R.-W., Chalecka-Franaszek, E., Ren, M., Hashimoto, R., Senatorov, V., Kanai, H., Hough, C., Hiroi, T., and Leeds, P. (2002). Neuroprotective effects of lithium in cultured cells and animal models of diseases. Bipolar Disorder. 4, 129–136.

Chen, R.-W., Qin, Z.-H., Ren, M., Kanai, H., Chalecka-Franaszek, E., Leeds, P., and Chuang, D.-M. (2003). Regulation of c-Jun N-Terminal kinase, p38 kinase and AP-1 DNA Binding in cultured brain neurons: Roles in glutamate Excitotoxicity and lithium Neuroprotection. J. Neurochem. 84, 566–575.

Chen, R.-W., and Chuang, D.-M. (1999). Long-term lithium treatment suppresses p53 and Bax expression but increases Bcl-2 expression. J. Biol. Chem. 274, 6039–6042.

Chen, G., Zeng, W.Z., Jiang, L., Yuan, P.X., Zhao, J., and Manji, H.K. (1999). The mood-stabilizing agents lithium and valproate robustly increase the levels of the neuroprotective protein bcl-2 in the CNS. J Neurochem. 72, 879–882.

Chen, Y.R., and Tan, T.H. (2000). The c-Jun N-terminal kinase pathway and apoptosis signaling. Int. J. Oncol. 16, 651–662.

D'Mello, S.R., Anelli, R., and Calissano, P. (1994). Lithium induces apoptosis in immature cerebellar granule cells but promotes survival of mature neurons. Exp Cell Res. 211, 332–338.

Drevets, W.C., Price, J.L., Simpson, J.R., Todd, R.D., Reich, T., Vannier, M., and Raichle, M.E. (1997). Subgenual prefrontal cortex abnormalities in mood disorders. Nature 386, 824–827.

Drevets, W.C. (2001). Neuroimaging Studies of Major Depression. In Advances in Neuroimaging of the American Psychiatric Press Annual Review of Psychiatry (Oldham, J.M., Series ed). J. Morihisa ed. (Washington, D.C.: American Psychiatric Press Inc.), pp. 123–170.

Dudek, H., Datta, S.R., Franke, T.F., Birnbaum, M.J., Yao, R., Cooper, G.M., Segal, R.A., Kaplan, D.R., and Greenberg, M.E. (1997). Regulation of neuronal survival by the serine-threonine protein kinase Akt. Science 275, 661–665.

Finkbeiner, S. (2000). CREB couples neurotrophin signals to survival messages. Neuron 25, 11–14.

Fukumoto, T., Morinobu, S., Okamoto, Y., Kagaya, A., and Yamawaki, S. (2001). Chronic lithium treatment increases the expression of brain-derived neurotrophic factor in the rat brain. Psychopharmacology (Berl) 158, 100–106.

Grimes, C.A., and Jope, R.S. (2001). The multifaceted roles of glycogen synthase kinase 3b in cellular signaling. Prog. Neurobiol. 65, 391–426.

Hashimoto, R., Fujimaki, K., Jeong, M.R., Christ, L., and Chuang, D.-M. (2003). Lithium-induced inhibition of SRC tyrosine kinase in rat cerebral cortical neurons: a role in neuroprotection against N-methyl-D-aspartate mediated excitotoxicity. FEBS Lett. 538, 145–148.

Hashimoto, R., Hough, C., Nakazawa, T., Yamamoto, T., and Chuang, D.-M. (2002a). Lithium protection against glutamate excitotoxicity in rat cerebral cortical neurons: Involvement of NMDA receptor inhibition possibly by decreasing NR2B tyrosine phosphorylation. J. Neurochem. 80, 589–597.

Hashimonto, R., Takei, N., Shimazu, K., Christ, L., Lu, B., and Chuang, D.-M. (2002b). Lithium induces brain-derived neurotrophic factor and activates TrkB in rodent cortical neurons: an essential step for neuroprotection against glutamate excitotoxicity. Neuropharmacology. 43, 1173–1179.

Hibi, M., Lin, A., Smeal, T., Minden, A., and Karin, M. (1993). Identification of an oncoprotein- and UV-responsive protein kinase that binds and potentiates the c-Jun activation domain. Genes Dev. 7, 2135–2148.

Jeong, M.-R., Hashimoto, R., Senaortov, V., Fujimaki, K., and Chuang, D.-M. (2003). Valproic acid, a mood stabilizer and anticonvulsant, protects rat cerebral cortical neurons from spontaneous cell death: a role of histone deacetylase inhibition. FEBS Lett. 542, 74–78.

Kanai, H., Sawa, A., and Chuang, D.-M. (2002). Possible involvement of histone deacetylase inhibition in the neuroprotective and neurotoxic effects of valproate in cultured cerebellar granule cells. Society for Neuroncience Annual Meeting Abstract 701.2.

Karin, M., Liu, Z.-G., and Zandi, E. (1997). AP-1 function and regulation. Curr. Opin. Cell Biol. 9, 240–246.

Klein, P.S., and Melton, D.A. (1996). A molecular mechanism for the effect of lithium on development. P. Natl. Acad. Sci. USA 93, 8455–8459.

Kluck, R.M., Bossy-Wetzel, E., Green, D.R., and Newmeyer, D.D. (1997). The release of cytochrome c from mitochondria: A primary site for Bcl-2 regulation of apoptosis. Science 275, 1132–1136.

Kopnisky, K.L., Chalecka-Franaszek, E., Gonzalez-Zulueta, M., and Chuang, D.-M. (2003). Chronic lithium treatment antagonizes glutamate-induced decrease of phosphorylated CREB in neurons via reducing protein phosphatase 1 and increasing MEk activities. Neuroscience. 116, 425–435.

Kuwana, T., Smith, J.J., Muzio, M., Dixit, V., Newmeyer, D.D., and Kornbluth, S. (1998). Apoptosis induction by caspase-8 is amplified through the mitochondrial release of cytochrome c. J. Biol. Chem. 273, 16589–16594.

Lipton, P. Ischemic cell death in brain neurons. (1999). Physiol. Reviews 79, 1431–1568.

Marsden, V.S., O'Connor, L., O'Reilly, L.A., Silke, J., Metcalf, D., Ekert, P.G., Huang, D.C.S., Cecconi, F., Kuida, K., Tomaselli, K.J., Roy, S., Nicholson, D.W., Vaux, D.L., Bouillet, P., Adams, J.M., and Strasser, A. (2002). Apoptosis initiated by Bcl-2-regulated caspase activation independently of the cytochrome c/apaf-1/ caspase-9 apoptosome. Nature 419, 634–637.

McCubrey, J.A., May, W.S., Duronio, V., and Mufson, A. (2000). Serine/threonine phosphorylation in cytokine signal transduction. Leukemia 14, 9–21.

Mielke, K., and Herdegen, T. (2000). JNK and p38 stress kinases-degenerative effectors of signal transduction-cascades in the nervous system. Prog. Neurobiol. 61, 45–60.

Miyashita, T., Krajewski, S., Krajewski, M., Wang, H.G., Lin, H.K., Liberman, D.A., Hoffman, B., and Reed, J.C. (1994). Tumor-suppressor p53 is a regulator of Bcl-2 and Bax gene-expression in-vitro and in-vivo. Oncogen 9, 1799–1805.

Nonaka, S., Hough, C.J., and Chuang, D.-M. (1998a). Chronic lithium treatment robustly protects neurons in the central nervous system against excitotoxicity by inhibiting N-methyl-D-aspartate receptor-mediated calcium influx. Proc. Natl. Acad. Sci. USA 95, 2642–2647.

Nonaka, S., Katsube, N., and Chuang, D.-M. (1998b). Lithium protects rat cerebellar granule cells against apoptosis incuced by anticonvulsants, Phenytoin and carbamazepine. J. Pharmacol. Exp. Thera. 286, 539–547.

Nonaka, S., and Chuang, D.-M. (1998). Neuroprotective effects of chronic lithium on focal cerebral ischemia in rats. NeuroReport 9, 2081–2084.

Öngür, D., Drevets, W.C., and Price, J.L. (1998). Glial reduction in the subgenual prefrontal cortex in mood disorders. Proc. Natl. Acad. Sci. USA 95, 13290–13295.

Ozaki, N., and Chuang, D.-M. (1997). Lithium increases transcription factor binding to AP-1 and cyclic AMP-responsive element in cultured neurons and rat brain. J. Neurochem. 69, 2336–2344.

Qin, Z.-H., Chen, R.-W., Wang, Y., Nakai, M., Chuang, D.-M., and Chase, T.N. (1999). Nuclear Factor kappa B nuclear translocation upregulates c-Myc and p53 expression during NMDA receptor-mediated apoptosis in rat striatum. J. Neurosci. 19, 4023–4033.

Qin, Z.-H., Wang, Y., Chen, R.-W., Wang, X., Ren, M., Chuang, D.-M., and Chase, T.N. (2001). Prostaglandin A_1 protects striatal neurons against excitotoxic injury in rat striatum. J. Pharmacol. Exp. Thera. 297, 78–87.

Rajkowska, G., Miguel-Hidalgo, J.J., Wei, J., Dilley, G., Pittman, S.D., Meltzer, H.Y., Overholser, J.C., Roth, B.L., and Stockmeier, C.A. (1999). Morphometric evidence for neuronal and glial prefrontal cell pathology in major depression. Biol. Psychiatry 45, 1085–1098.

Ren, M., Senatorov, V.V., Chen, R.-W., and Chuang, D.-M. (2003). Post-insult treatment with lithium reduces brain damage and facilitates neurological recovery in a rat ischemia/reperfusion model. Proc. Natl. Acad. Sci. USA 100, 6210–6215.

Schwarcz, R., Whetsell, W.O., and Mangano, R.M. (1983). Quinolinic acid: an endogenous metabolite that produces axon-sparing lesions in rat brain. Science 219, 316–318.

Senatorov, V.V., Ren, M., Kanai, H., and Chuang, D.-M. (2002). Short-term lithium treatment promotes neuronal survival and proliferation in rat striatum infused with quinolinic acid. Society for Neuroscience Annual Meeting Abstract 417.3.

Susin, S.A., Zamzami, N., Castedo, M., Hirsh, T., Marchetti, P., Macho, A., Daugas, E., Geusken, M., and Kroemer, G. (1996). Bcl-2 inhibits the mitochondrial release of an apoptogenic protease. J. Exp. Med. 184, 1331–1341.

Tao, X., Finkbeiner, S., Arnold, D.B., Shaywitz, A.J., and Greenberg, M. (1998). Ca^{2+} influx regulates BDNF transcription by a CREB family transcription factor-dependent mechanism. Neuron 20, 709–726.

Wang, Y., Qin, Z.-H., Nakai, M., Chen, R.-W., Chuang, D.-M., and Chase, T.N. (1999). Co-stimulation of cyclic AMP-linked metabotropic glutamate receptors in rat striataum attenuates excitotoxin-induced NF-kB activation and apoptosis. Neuroscience 94, 1153–1162.

Wei, H., Leeds, P.R., Qian, Y., Wei, W., Chen, R.-W., and Chuang, D.-M. (2000). b-Amyloid peptide-induced death of PC 12 cells and cerebellar neurons is inhibited by long-term lithium treatment. Eur. J. Pharmacol. 392, 117–123.

Wei, H., Qin, Z.-H., Senatorov, V.V., Wei, W., Wang, Y., Qin, Y., and Chuang, D.-M. (2001). Lithium suppresses excitotoxicity-induced striatal lesions in rats, an animal model of Huntington's disease. Neuroscience 106, 603–612.

Williams, M.B., and Jope, R.S. (1995). Circadian variation in rat brain AP-1 DNA binding activity after cholinergic stimulation: modulation by lithium. Psychopharmacology 122, 363–368.

Yuan, P.-X., Chen, G., Huang, L.-D., and Manji, H.K. (1998). Lithium stimulates gene expression through the AP-1 transcription factor pathway. Mol. Brain. Res. 58, 225–230.

Chapter **14**

Apoptosis, Cancer, and Cancer Therapy

XIAO QIANG FAN, HAO WANG, WEIZHU QIAN, AND YAJUN GUO*

ABSTRACT: Apoptosis is a complex process involving a large array of genes and mutation of any of these genes may lead to malignancy formation. Tumor cells may actively evade the immune surveillance by actively inducing the apoptosis of effector lymphocytes, facilitating mass formation and metastasis. Apart from traditional therapies inducing tumor cell apoptosis, new strategies employing death receptor ligands and making use of tumor counterattack are also being developed.

Apoptosis is a kind of cell death that is extensively implicated in the life of multiple-cellular organisms (Abigail Hunt and Gerard Evan, 2001). Its importance lies not only in the hosts' development but also in their homostasis maintenance. The malfunctioning of apoptosis during development will lead to abortion or abnormalities, while failure of DNA-damaged cells to kill themselves via apoptosis may result in malignancy formation. Apoptosis is also one of the weapons immune system employs to eliminate virus-infected and transformed cells, but unfortunately this weapon is also taken by tumor cells to counter-attack and sometimes the latter even gains a super arm in the combat. It is by inducing apoptosis of target cells that most anti-cancer drugs function. However, mutations of apoptosis-related genes may nullify these efforts. Further understanding the molecular mechanisms of apoptosis may help to expand our knowledge about tumor biology and put forward new effective ways to treat cancer. Herein, we will firstly review the recent progresses made in the apoptotic signaling transduction pathways and then their application in cancer therapy.

Cell Death Signaling Transduction

Recently, it has been commonly recognized that apoptotic signals have two entry sites, namely, the mitochondria and death receptor pathway. Mitochondria pathway is often activated by DNA damages, while death receptor pathways are activated by the engagement of the death receptors. However, the two pathways are not distinctly separated. Studies showed

*International Joint Cancer Institute, The Second Military Medical University, Shanghai 200433, People's Republic of China, and Epply Cancer Institute, University of Nebraska Medical Center, Omaha, NE. 68198-6805. *Correspondence:* Dr. Yajun Guo, International Joint Cancer Institute, 800 Xiang Yin Road, Shanghai 200433, People's Republic of China; Tel: 0086-21-25070241; Fax: 0086-21-25074349; E-mail: yguo@unmc.edu there exists crosstalk between these two pathways, bridged mainly by a member of Bcl-2 family, which is called Bid (Eskes, R. Desagher, 2000).

Mitochondria Pathway

Mitochondria are crucial organelles in that they not only provide energies for cells to live on, but also produce essential molecular mediators for cell apoptosis, among which Bcl-2 family members are the main components. In humans, more than 20 members of this family have been identified, including proteins that suppress (Bcl-2, Bcl-XL, Mcl-1, Bfl-1/A1, Bcl-W, Bcl-G) and proteins (Bax, Bak, Bok, Bad, Bid, Bik, Bim, Bcl-Xs, Krk, Mtd, Nip3, Nix, Noxa, Bcl-B) that promote apoptosis (Antonsson, B., 2000; Guo, B., 2001; Ke, N., 2001; Oda, E., 2000.). These apoptotic modulatory proteins are generally located in the inner membrane of mitochondria and function largely by keeping and altering the concentration ratio of its pro- and anti-apoptotic members, especially that of Bcl-2 and Bax. Bcl-2 protein resides on the membranes of mitochondria, smooth endoplasmic reticulum and perinuclear membrane with its signal anchoring sequence (Zamzami, N., et al., 1996). When co-expressed with Apaf-1, it will arrest Apaf-1 in the intracellular membranes, and when co-expressed with Bax, it will dimerize with Bax and leave Apaf-1 free. The released Apaf-1 has the propensity to oligomerize and bind to an initiator caspase, procaspase-8 (Muzio, M., et al., 1998). The aligned procaspases will then be activated by self-cleavage and apoptosis occurs. In addition, Bcl-2 can also inhibit apoptosis by blocking the opening of PT hole. The opening of PT hole may lead to two important sequences. It may result in the decrease of Δc m and the damage of the respiratory chain, which increases ROX. On the other hand, it results in the ascending of matrix osmotic pressure, which enlarges the matrix space. While crests on the inner membrane allows its spreading, the outer membrane remains rigid and unchanged, resulting in breaking of the outer membrane and further release of inter membrane proteins including AIF (apoptosis inducing factor), which in turn activate apoptosis via caspase-3, and cyto c et al. The sequence of the cyto c release is variable in different cell types. In cells lacking cyto c, it will result in the breaking down of respiratory chain, overproduction of ROX and hamper of ATP production, which lead to non-apoptotic cell death. However, in cells with large quantity of cyto c, the release of inter-membrane space cyto c is not enough to deprive the cell of the ability to respire and to produce enough ATP, so the released cyto c gets complexed with Apaf-1 and procaspase-9 to form an apoptotsome, which activates the caspase and induces apoptosis (Srinivasula, S.M., et al., 1998). Activated caspases can also feedback on the PT hole, accelerating the process of apoptosis.

Death Receptor Pathway

Apart from the mitochondria pathway activated by DNA damages, apoptosis may also be induced by the ligation of death receptors (Fan, X., et al., 2001). Death receptors generally belong to the tumor necrosis factor (TNF) receptor gene superfamily and their ligands are all members of TNF superfamily, except NGF. Most TNF family cytokines are expressed as type II transmembrane proteins, whose C-terminal extracellular domain is processed proteolytically to form a soluble homotrimeric molecule. In contrast, the members of the

TNF receptor (TNFR) family are type I transmembrane proteins. In both families, sequence homology is found mainly in the extracellular regions, which mediate ligand-receptor binding. Although death receptors are characteristic of the intracellular death domain, which is critical in death signal transduction, members of the TNF family have diverse biological activities. In addition to inducing the apoptosis of target cells, they also take part in T cell co-stimulation, B cell proliferation and differentiation, and macrophage activation.

Among these death receptors and their ligands, Fas/FasL, TNFR/TNF and TRAIL and its receptors are best-characterized couples. All ligands are homotrimeric molecules and bind to three counterpart receptor molecules as suggested by the crystal structure of lymphotoxin a in complex with TNFR1. These receptors then recruit adapter proteins with their cytosolic domains that bind specific initiator caspases. The bound caspases are activated by an "induced proximity" mechanism involving clustering of procaspase zymogens around aggregated receptor complexes, which is also called death signal inducing complex (DISC) (Salvesen, G.S. & Dixit, V.M. 1999). Here we will take the activation of Fas as an example. Fas ligation leads to clustering of the receptors' death domains (DD), to which an adapter protein called FADD (Fas-associated protein with death domain; also called MORT1) then binds through its own death domain. FADD also contains a 'death effector domain' (DED) that binds to an analogous domain repeated in tandem within the zymogen form of caspase-8, which is also called FLICE, or MACH (Boldin, M.P., et al., 1996), a typical example of initiator caspases. Upon recruitment by FADD, caspase-8 oligomerization drives its activation through self-cleavage. Caspase-8 then activates downstream effector caspases such as caspase-9, committing the cell to death. Signal transduction of ligated TNFR1 is similar to that of ligated Fas, only different in that the adapter protein binding to clustered death receptor death domains is TRADD (TNFR-associated death domain) (Hsu, H., et al., 1995). TRADD functions as a platform adaptor that recruits several signaling molecules to the activated receptor including FADD, TNFR-associated factor-2 (TRAF2) and receptor-interacting protein (RIP), starting apoptosis machinery as well as NF-kappa B and JNK/AP-1 activation, which leads to induction of proinflammatory and immunomodulatory genes, whereas FADD only mediates activation of apoptosis (Guoqing Chen and David V. Goeddel, 2002).

It is understandable that the engagement of TNF family receptors does not always lead to cell death. The dichotomy in death signal transduction may be explained by the regulated expression of inhibitors of death receptor pathway. An important regulator of Fas and TNFR1 signaling is the cellular homolog of viral FLICE-inhibitory protein (c-FLIP, also called I-FLICE, FLAME, CASH, CLARP, MRIT, and Usurpin). Cellular FLIP is a homolog of caspase-8 that consists of two DED domains and a caspase proteolytic domain. Unlike caspase 8, however, c-FLIP is short of a functional caspase enzymatic active site and therefore unable to cleave cellular substrates. Two alternatively spliced forms of c-FLIP have been described, a short form and a long one. The short c-FLIP (c-FLIPs) includes only the two DED domains while the long one (c-FLIP$_L$) also includes a caspase-like domain. Upon receptor engagement and DISC formation, both forms can be recruited to FADD by DED interaction (Krammer, 2000).

In the DISC, c-FLIP has been described seeming paradoxically both as an inhibitor and an activator of caspase-8 processing and apoptosis. As an inhibitor of cell death, c-FLIP may act by blocking caspase 8 recruitment to the receptor complex or may inhibit full proteolytic processing of caspase 8 to its active form, while as an activator of cell death, it is proposed to act as a scaffold, enhancing recruitment of caspase 8 to the DISC and promoting caspase 8 autoprocessing. The distinction between these two distinct effects may be determined in part by the splice variant of c-FLIP recruited to the DISC. c-FLIPs appears antiapoptotic because it effectively blocks caspase-8 recruitment to the DISC. c-FLIP$_L$, in contrast, has been found in some cases to increase caspase-8 aggregation and autoproteolytic activation, although this seems to depend on the cell type in study and c-FLIP$_L$ expression levels. However, in vivo experiments have shown c-FLIP$_L$ inhibites the activation of caspase 8 by blocking recruitment of caspase 8 to DISC and may instead promote survival by activating the NF-kB and Erk signaling pathways (Kataoka et al., 2000).

Crosstalk between the Two Pathways

Recent studies have revealed that the two pathways are not distinctly separated but related to each other. Crosstalk between the death receptor and mitochondria pathways is mediated mainly by caspase-8 cleavage of the protein Bid, which contains Bcl-2 homoloy domain 3(BH3). Trancated Bid can trigger mitochodrial cytochrome c release through the proapoptotic Bcl-2 homologs Bax and Bak (Eskes R. Desagher, et al., 2000; Wei, M., et al., 2000; Zamzami, N. 2000), thereby initiate a mitochondrial amplification loop. In addition, cleaved Bid forms ion channels in liposomes, suggesting that it can disrupt the mitochondria independently and induce cytochrome c-release (Schendel, S.L., et al., 1999; Tang, D., et al., 2000). Scaffidi et al. classified Fas+ cells into two categories. Type 1 cell shows optimal formation of Fas DISC, initiating a direct caspase cascade independently of mitochondria, while type 2 cell shows weak DISC formation and requires the crosstalk amplification of death signals (Scaffidi, C., et al., 1999). Conversely, both pathways can be inhibited by BAR. BAR was identified by screening for cDNAs capable of inhibiting Bax-induced apoptosis. It contains a protein-interacting domain that binds and inhibits procaspase-8 and thereby blocks receptor-mediated pathway. BAR also inhibits the mitochondrial-mediated pathway through an unknown mechanism involving interactions with Bcl-2 and Bax. Physically, BAR bridges the 2 pathways by forming a complex with procaspase-8 and Bcl-2, however, the significance of this complex is yet to be known (Zhang, H., et al., 2000).

Caspase and Execution of Cells

Despite the diverse apoptosis stimuli and different death signal pathways, cells may commit suicide mainly by the same group of proteins, as suggested by the similar morphological and biochemical changes, such as DNA fragmentation, chromatin condensation, membrane blebbing, cell shrinkage, and membrane-enclosed vesicles (apoptotic bodies) (Sharad Kumar and David L. Vaux. 2002). It is now clear these proteins defined now as effector caspases (cysteinyl aspartate-specific protease) in deed have much similarity in structure and specificity for substrates.

There are at least 14 caspases that have been identified in mammalian cells, which are generally classified into three groups: cytokine activator, apoptosis effector, and apoptosis initiator (Chang, H.Y., et al., 2000). Effector caspases function by several means. One role of such caspases is to inactivate proteins that protect living cells from cell death. For example, there exists in nonapoptotic cells a nuclease responsible for DNA fragmentation called CAD (caspase-activated deoxyribonuclease) but in an inactive form complexed with its inhibitor ICAD/DFF45 (Enari, M., et al., 1998). During apoptosis, ICAD is inactivated by caspases, leaving CAD free to function as a nuclease. Surprisingly, CAD synthesized in the absence of ICAD is not active, implying that the CAD-ICAD complex is formed co-translationally, and that ICAD is required by both the activity and inhibition of this nuclease. Other negative regulators of apoptosis cleaved by caspases are Bcl-2 proteins, and cleavage of these proteins not only inactivates their activity, but also produces a fragment that promotes apoptosis.

Caspases also contribute to apoptosis through directly disassembling and indirectly reorganizing cell structures, as illustrated by the destruction of nuclear lamina, a rigid structure that underlies the nuclear membrane and is involved in chromatin organization. Lamina is formed by head-to-tail polymers of intermediate filament proteins called lamins. During apoptosis, lamins are cleaved at a single site by caspases, causing lamina to collapse and contributing to chromatin condensation. Cleavage of several proteins involved in cystoskeleton regulation, including gelsolin, focal adhesion kinase (FAK), and p21-activated kinase 2(PAK2) may also lead to reorganization of cell structures (Rudel, T., et al., 1997).

Dissociation of regulatory and effector domains is hallmark of caspase function. For example, they inactivate or deregulate proteins involved in DNA repair (such as DNA-PKCS), mRNA splicing (such as U1-70K), and DNA replication (such as replication factor C), facilitating the disassembly of cellular organization (Cryns, Yuan, J. 1998).

Caspase-Independent Apoptosis

Athought caspases are the main executors of cell death, there also exist other executors as suggested by cells dying with cytoplasmic and membrane changes seen in apoptosis while not exhibiting DNA and/or nuclear fragmentation. (Borner, C., Monney, L. 1999; Chautan, M., et al., 1999; Kitanaka, C., et al., 1999). Also, z-VAD-fmk, which is an effective blocker for caspases, could not rescue cells from apoptosis induced by the overexpression of Bax despite that caspase-3 activation and nuclear fragmentation were clearly blocked (Miller, T.M., et al., 1997; Xiang, J., et al., 1996), suggesting that activation of caspases seems to be dispensable in certain types of apoptosis (Vercammen, D., et al., 1998). The cellular components of caspase-independent apoptosis are not identified as yet. In most apoptotic systems, z-VAD-fmk does not block mitochondrial changes, such as loss of the membrane potential, production of ROS, or the release of apoptogenic factors such as cytochrome c and AIF (Schulze-Osthoff, K., et al., 1994). Therefore, in spite of the caspase inhibition, apoptotic morphology may still be induced by AIF and Bax or Bax-like proteins. The mechanism of how AIF and other released proteins trigger apoptosis in the absence of caspases is still unknown. Some experiments suggest these molecules may activate proteases such as the calcium-dependent calpain proteinases or the lysosomal cathepsins, which can partially substitute caspases though in a less efficient way (Borner, C., Monney, L. 1999). Cathepsins released from lysosomes cleave Bid, which activates the mitochondrial apoptosis pathway (Stoka, V., et al., 2001), while calpains cleave Bax, promoting cytochrome c release (Gao, G., Dou, P. 2000; Wolf, B.B., et al., 1999 Wood, D.E., 1998). And calpains are also known to cleave and thereby inactivate the cytoprotective endoplasmic reticulum (ER) chaperone glycoprotein GRP94, which contributes to apoptosis (Reddy, R.K., Jun, L., Lee, A.S. 1999).

A brief survey of apoptotic process suggests that this process occurs in a manner reminiscent of a well-orchestrated military operation. Modulatory molecules act as commanders at different levels and coordinate each other and caspases are fighting soldiers. They cut off contacts with surrounding cells, reorganize the cytoskeleton, shut down DNA replication and repair, interrupt splicing, destroy DNA, disrupt the nuclear structure, induce the cell to display signals that mark it for phagocytosis, and disintegrate the cell into apoptotic bodies. As systematic approaches for further research are developed, better understanding of the mechanisms will emerge.

Aberrant Apoptotic Molecules Contribute to Tumorigenesis

Since apoptosis is a gene-involved programmed cell death, any mutation of the involved genes may interfere in the process and lead to the survival of transformed cells and malignancy formation. The most common mutated genes found in cancer cells include p53, c-myc, Bcl-2 family members and ndm23. Fas-related genes mutation such as deletion or down-regulation of Fas, truncation of Fas in its cytoplastic domain and up-regulation of negative regulatory proteins also abrogate the Fas-mediated cell death. These genes interact each other to decide the fate of normal and transformed cells.

P53 is a very important regulatory protein targeting many other apoptosis regulatory genes such as Bcl-2 and Bax. Wild-type p53 can down-regulate the expression of Bcl-2 and up-regulate that of Bax, altering the balance of the couple genes in favor of apoptosis. Many kinds of human tumors such as colorectal carcinoma, brain and lung cancer, mammary carcinoma, skin and bladder carcinomas are found to be p53 mutated. The mutated p53 loses the function of binding DNA and exerting control over the expression of target genes. What's more, wide-type p53 will be deprived of the ability to bind DNA and activate target gene expression when complexed with mutated p53, which is termed dominant negative. Thus cells with DNA damages, which should have been checked by p53 on stage G1 and compulsory to be repaired, will progress into S stage and in some circumstances further deteriorate into malignancies. Other factors besides mutation may also lead to the loss of function of p53. Large T antigen, Ltag of SV40, E1B of adenovirus and E6 of human herpse viruses et al. also encode proteins that directly or indirectly inhibit the function of p53, making the normal p53 unable to transactivate the target gene specifically. For example, the product of E6 can facilitate its degradation when binding to p53. Other effective means blocking p53 include gene product of mdm2, which is commonly seen in human osteosarcoma.

Bcl-2 family members are expressed in most normal human tissues, although in low quantity. Bcl-2 is significantly overexpressed in acute myoleukia and its expression is related to the cell differentiation, i.e., the expression of Bcl-2 is in a positive proportion to its degree of malignancy and negative proportion to its clinical responses (Porwit MacDonald, A., et al. 1995). In acute B lymphocyte leukemia, the life of lymphocytes with high expression of Bcl-2 can be postponed over one week, even after the born marrow culture is removed, the cells still remain alive. And it is also found that bcl-2 over-expression and downexpression of Bax exist in 1/3 mammary carcinomas.

Apoptosis and Immune Surveillance Evasion

Apoptosis is a powerful weapon immune system employs to kill cancer. However, this weapon may also be used by the treacherous enemies, who not only escape the immune surveillance by down-regulating their antigen presenting and T cell stimulating ability but also actively attack the immune system by expressing apoptosis-inducing molecules, among which Fas ligand is a mostly studied cripple, and other death factors such as TNF and TRAIL may also contribute to this model (Reviewed by Frederik H. Igney, et al., 2000).

FasL is firstly found to be a determinant of immune-privileged status in the eyes and later revealed to be expressed constitutively in many types of tumors (Joe O' Connell, et al., 1999). Katsuya, et al., made investigations into the expression of FasL in human colorectal carcinoma and liver metastasis foci. The result showed that among the 7 cases of the primary foci, 2 express FasL and in all of the 4 patients with liver metastasis, the FasL expression of metastasis are FasL positive, suggesting that the FasL expressing tumor cells may be tougher. The explanation is that FasL-negative tumor cells are more likely to be attacked by the activated T lymphocytes while FasL expressing tumor cells may escape the killing of lymphocytes by fighting back. And also they may induce the apoptosis of neighboring liver cells to facilitate the formation of neoplasm. Melanoma, pulmonary carcinoma, colorectal carcinoma, glioma and hepatoma are five typical examples. Noticeably, chemotherapy (e.g. bleomycin, cisplatin, etoposide) strongly induces the expression of Fas lignand on tumor cells, perhaps contributing to tumor resistance to drugs or implying chemotherapy as a 'double edged sword'.

Someone argues that since both tumor cells and activated CTLs express FasL, why are CTLs always killed or prohibited but not vise versa? In deed, when tumor cells and antigen-specific CTL express both Fas and FasL, bi-directions interactions can lead to killing in both cell types (Zeytun, A., et al., 1997). There is, however, an important difference between the conditions used for measuring cytotoxicity in vitro and those present at the tumor site. The number of effector cells (E) used in the standard cytotoxicity assay always exceeds that of tumor targets (T), while the E: T ratios in the established tumor, almost invariably are in favor of tumor cells, even when considerable TIL infiltrates are present. Therefore, it is likely that, in vivo, FasL+ tumor will always have an advantage in perpetrating the death of Fas+ immune cells. Besides, FasL expressing tumor cells are often Fas down regulated or nullified because of mutation of genes involved in Fas-mediated death signal transduction.

Nullified immune effector cells by FasL expressed on the surface of tumor cells include not only activated T lymphocytes that express Fas but also NK cells. It was demonstrated that murine AK-5 tumor cells transiently upregulated FasL when grown in peritoneal cavity of syngeneic mice, and that this FasL upregulation coincided with depletion of the intraperitoneal NK cell population. Depletion of NK cells was local to the tumor microenvironment, as splenic and peripheral NK cells were unaffected. The same AK-5 tumor cells did not express FasL when injected subcutaneously. FasL-negative, subcutaneous AK-5 tumors showed about 70% regression, mediated largely by NK cells, whereas FasL-expressing intraperitoneal tumors grew successfully, always resulting in death of host. These findings suggest that FasL-mediated counterattack against anti-tumor NK cells may contribute to the successful immune evasion of FasL-expressing tumors in vivo (O'Conell, et al., 1999).

The high prevalence of expression of FasL in several diverse tumor types, including colon cancer, esophageal cancer, melanoma, liver cancer, astrocytoma and lung cancer, suggests that FasL is a general, perhaps essential factor in the inhibition of anti-tumor responses. As such, the Fas counterattack might be a rational target for therapeutic intervention.

Despite the wealth of data accumulated in the past years, there are also some experiments inconsistent with the Fas counter-attack model. (Nicholas P. Restifo, 2000) Hiroshi

Arai, et al. found that FasL expression on the surface of tumor cells promotes tumor regression through apoptosis or inflammation rather than enhanced tumor growth through its effects on immune suppression. The mechanism of antitumor effect was dependent on Fas expression of tumor cell. In the case of Fas+ tumors, there appeared to be direct induction of apoptosis. In cells resistant to lysis by FasL, tumor regression was induced through an independent mechanism involving FasL-induced inflammation and a potential "by-stander effect". The mechanism of the proinflammatory effect of FasL remains to be elucidated. Neutrophils, monocytes, or both play a role in this process. The histologic findings suggested that the neutrophil was involved as an effector cell in this process. Monocytes produce a variety of chemokines and cytokines, including interleukin 8, a potent chemoattractant for neutrophil, and they have been reported susceptible to FasL stimulation. It is likely that FasL expression on tumor cells induced apoptosis of infiltrating monocytes, causing release of these inflammatory mediators that promote further inflammation, neutrophil migration, and tumor destruction. However, neutrophil recruitment in these instances may be an unwelcome consequence of an excessive level of FasL. A modest level of recombinant FasL expression was shown to protect liver allografts, whereas higher levels led to graft destruction (Li, X.K., et al., 1998).

However, a recent work by Dong et al. adds new evidences into the tumor counterattack concept. B7-H1, a recently described member of the B7 family of costimulatory molecules, is found abundantly expressed in human carcinomas of lung, ovary and colon and in melanomas while not in normal human tissues except cells of the macrophage lineage. Cancer cell–associated B7-H1 increases apoptosis of antigen-specific human T-cell clones in vitro, and the apoptotic effect of B7-H1 is mediated largely by one or more receptors other than PD-1. In addition, expression of B7-H1 on mouse P815 tumor increases apoptosis of activated tumor-reactive T cells and promotes the growth of highly immunogenic B7-1 + tumors in vivo (Haidong Dong, et al. 2002). These findings have implications for the design of T cell–based cancer immunotherapy.

Apoptosis in Cancer Treatment

Since some malignancy formation is caused by the hamper of apoptosis, apoptotic machinery becomes a cancer therapeutic target (G. Wu, Z. Ding, 2002). Pro-apoptotic genes and antisense of anti-apoptotic genes have been extensively transferred into tumor cells to examine their therapeutic effects. The treatment of Experimental Prostate Cancers by Use of Apoptotic Genes bax and bad Driven by the Prostate-Specific Promoter ARR(2)PB suggests that polygene therapy with apoptotic molecules is more effective in experimental models of androgen-dependent or -independent prostate cancer than monogene therapy (Zhang, Y., et al., 2002). In fact, it is one of the mechanisms by which traditional radiotherapy and chemotherapy take effects. Sakakua-C et al. examined the histological response, rate of apoptosis, DNA fragmentation and p53 status in tumors from 28 patients with rectal cancer undergoing HCR therapy (combined hyperthermia, chemotherapy and radiation) before surgery and from 22 patients who did not have preoperative treatment. The therapeutic effect of HCR therapy was closely correlated with the rate of apoptosis and related to the p53 status of the tumors to some extent. Photodynamic therapy is also demonstrated to be able to induce apoptosis in lymphoma cells mainly due to the leakage of cytochrome c into the cytosol (Varnes. M.E., et al., 1999). Certain chemotherapeutic drugs have been noted to induce apoptosis by altering surface expression of death receptors. For example, doxorbicin can up-regulates the Fas expression on the surface of tumor cells, suggesting that chemotherapy may have a role in regulating responsiveness of tumors to Fas/FasL-mediated apoptosis and incubation of cell lines with doxorubicin or 5-fluorouracil significantly augments TRAIL-induced apoptosis in most breast cell lines (Keane, M.M., et al., 1999). Aaron C Spalding's work showed that paracrine TRAIL-induced apoptosis is important for genotoxins like etoposide and doxorubicin but less important for the apoptotic response to micro-tubule toxins as evidenced by DcR1, DR4:Fc and anti-TRAIL antibody suppression of genotoxin-induced apoptosis in epithelial-derived carcinomas (Aaron C. Spalding, et al. 2002). Besides, irradiation and some chemotherapeutic drugs may also down-regulate the expression of Bcl-2, allowing cells to progress into apoptosis.

However, mutations of apoptosis-related genes or any factors influencing their activity may endow tumor cells the ability to decline being executed. For example, the overexpression of Bcl-2 results in many tumor cells' resistance against therapeutic drugs and further experiments confirmed this phenomenon has no direct relation with the expression of multiple drug resistance gene resulting from drugs accumulation. However, leukemia cells transduced with antisense oligonucleotide of Bcl-2 will regain the sensitivity towards chemotherapeutic drugs. Estrogen can up-regulate the expression of Bcl-2 in mammary carcinoma cells, so the use of Tamoxifen, which can act as an antagonist against estrogen, helps the cells reacquire the sensitivity to amycin by down-regulating the expression of Bcl-2. Apart from that, mutations that inhibit TRAIL receptor signaling or diminish TRAIL expression should promote resistance to chemotherapy.

Cell death induced by some certain chemotherapeutic drugs are p53 dependent. Studies show that in response to stimuli such as chemotherapy and radiation, p53 activates the mitochondrial-mediated pathway of apoptosis (Norbury, C., Hickson, I. 2001; Daniel, P. 2000). p53 induces mitochondrial-mediated apoptosis through multiple mechanisms including the transcription of the proapoptotic protein Bax (Zhan, Q., et al., 1994), enhancement of reactive oxygen species (ROS) (Asher, G., et al., 2000) and another mechanism that remains ill-defined but is independent of the nucleus (Hermeking, H., Eick, D., 1994). Therefore, p53 mutation will lead to drug resistance in some circumstances. However, as there also exists a non-p53-dependent pathway towards DNA-damage-induced cell death, it doesn't mean all cells with p53 loss of function acquire the ability to refuse death. For those cells whose apoptosis may be p53 dependent and at the same time p53 deficiency does exist, it may be advisable to transduce wide type p53 gene into these cells to re-endow them with sensitivity to chemotherapy. Experiments showed that p53 transduced vectors can induce the apoptosis of p53 mutated non-small-cell carcinoma of lung, squamous epithelial carcinoma and mammary planted carcinoma, and some time even inhibit their metastasis (Liu, T.J., et al., 1995). AdCMp53 can inhibit the growth of melanoma cells both in vivo and in vitro.

Death Receptor Ligand in Cancer Treatment

The idea of targeting specific death receptors to induce apoptosis in tumors is attractive as death receptors have direct access to the caspase machinery. Moreover, unlike many chemotherapeutic agents or radiation therapy, death receptors initiate apoptosis independently of p53 tumor suppressor gene, which is inactivated by mutation in more than half of human cancers. TNF, Fas Ligand and TRAIL are three death receptor ligands that are studied most extensively.

TNF was initially identified in 1975 on the basis of its tumor-killing activity (Carswell, E.A., et al., 1975) but later revealed to be identical with a molecular called cachectin, which is a key mediator in shock, wasting and inflammation (Beutler, B. & Cerami, A. 1986). Thereafter, the deleterious side effects of systemic administration of TNF dampen the hopes for a general application of TNF in tumor therapy and the only clinical success has achieved with isolated limb perfusion for a limited subset of susceptible tumors, such as melanoma and sarcoma (Fraker, D.L. & Alexander, H.R. 1994). It will be ideal to block the systemic toxicity of TNF while allow the anti-neoplastic effects to act, which Van Molle et al. successfully achieved by treating tumor-bearing mice with a humid 42°C oven for a short period of time repeatedly (Wim Van Molle, 2002). It is deduced that repeated HS (heat shock) treatment induced elevated HSP70, which enabled normal cells more resistant for TNF, while tumors have already played the joker and used it for securing their basic survival and another stress-induced upregulation of this protein became futile. Therefore, HS preconditioning might ultimately turn out to be a valuable addition to the oncologists' choices for therapy refinement. But this may be difficult to achieve with heat. A reverse strategy lowering the level of HSP70 in tumor cells may work better.

During the past several years, TRAIL has offered the most promising hope for tumor therapy using death receptor ligand. The ligand was firstly isolated by Robert M. Pitti et al. in 1996 as a new member of TNF family via an expressed sequence tag. Several differences between TRAIL and TNF or Fas suggest that TRAIL may be a safer agent for cancer therapy. First, although its death receptors DR4 and DR5 can activate NF-kB upon overexpression, TRAIL itself induces this response only weakly, and activation requires doses that are considerably higher than those of TNF that activate a strong NF-kB response. Second, TRAIL mRNA is expressed in many tissues, suggesting that the ligand may be nontoxic to normal cells. Repeated intravenous injections of TRAIL in nonhuman primates did not cause detectable toxicity to tissues and organs examined. Third, DR4 and DR5 are expressed in normal tissues and in many types of tumor cells, whereas DcR1 and DcR2, which are called decoy receptors because the former lacks the cytoplasmic region and the latter has a truncated death domain, are expressed frequently in normal tissues but infrequently in tumor cells (Reviewed by Shkenazi, A. & Dixit, V. M. 1998). This differential expression of death and decoy receptors might enable TRAIL to induce apoptosis in tumors while sparing normal cells. Promising preclinical study has achieved in multiple tumors such as melanoma, human mammary adenocarcinoma cell line MDA-231, human brain tumors and malignant gliomas. However, many recent reports also provide alarming evidences that there may be a long way to go before TRAIL can be successful applied for cancer therapy. Firstly, TRAIL may be not as safe as had been expected. TRAIL has been found deleterious to some normal human cells, including primary human keratinocytes (Kothakota, et al., 1997), hepatocytes (Jo, et al., 2000) and prostate epithelial cells (A. Nesterov, et al., 2002). Different preparation of TRAIL may explain part but not all of the reasons. Secondly, tumors with MMR deficiency and a crosstalk-dependent apoptosis-signaling character may develop resistance to death receptor-targeted therapy through Bax mutation. Sequential combination of chemotherapy and death receptor-directed agents may help circumvent the resistance associated with loss of Bax and enhance tumor responsiveness to both types of apoptotic stimuli (Heidi Leblance, et al., 2002). However, Kelly M.M. and his collegues demonstrated that pretreatment with doxorubicin is sufficient to sensitize cells to TRAIL, but doxorubicin did not affect steady state levels of Bax, Bcl-2 and Bcl-X(L) in the majority of the prostate cancer cell lines but down-regulated an anti-apoptotic protein c-FLIP(S), which predisposes cells to TRAIL-induced apoptosis. (Kelly, M.M., et al., 2002).

Smac and Cancer Therapy

As induction of apoptosis in target cells is a key mechanism for most anti-tumor therapies, including chemotherapy, g -irradiation, immunotherapy and cytokines, defects in apoptosis programs may cause resistance. IAPs (Inhibitors of apoptosis proteins) contribute to the resistance of cancers, as they are highly expressed in many tumors. Inhibition of caspases by IAPs targets the core of the apoptotic machinery and therefore therapeutic modulation of IAPs could perturb a key control point in deciding cell fate. Recently, Smac (second mitochondria-derived activator of caspase) was identified as a mitochondrial protein that is released into the cytosol in response to apoptotic stimuli and promotes caspase activation by antagonizing IAPs. FULDA et al. (2002) reported that transfer of the gene encoding Smac or Smac peptides sensitized various tumor cells in vitro and malignant glioma cells in vivo for apoptosis induced by death-receptor ligation or cytotoxic drugs. Expression of a cytosolic active form of Smac or cell-permeable Smac peptides bypassed the Bcl-2 block, which prevented the release of Smac from mitochondria, and also sensitized resistant neuroblastoma or melanoma cells and patient-derived primary neuroblastoma cells ex vivo. Most importantly, Smac peptides strongly enhanced the antitumor activity of Apo-2L/tumor necrosis factor–related apoptosis-inducing ligand (TRAIL) in an intracranial malignant glioma xenograft model in vivo. Complete eradication of established tumors and survival of mice was only achieved upon combined treatment with Smac peptides and Apo2L/TRAIL without de-tectable toxicity to normal brain tissue. Thus, Smac agonists are promising candidates for cancer therapy by potentiating cytotoxic therapies.

Therapeutic Implications from Tumor-Immune Counterattack

Many current attempts at harnessing the immune system to kill tumors focus on generating tumor vaccines, which are supposed to produce sufficient tumor-specific CTLs. Other manipulation includes adoptively transfer effectoer T cells into cancer patients to promote tumor regression. However, the tumor counterattack concept implies that these two approaches may both become futile without further modulation on CTLs or tumor cells (Gordon J. Freeman, et al., 2002). Depriving the tumor cells of their newly acquired FasL, grinding the edge of weapon of FasL on the surface of CTL, re-endowing tumor cells with sensitivity to Fas-mediated cell death, and strengthening CTL to be tolerant of the Fas-mediated apoptosis may all become plausible strategies.

In more details, the use of neutralizing antibodies specific to FasL or their fragments on the surface of tumor cells might be helpful in removing their fighting back ability. The use of the soluble form of Fas, such as Fas-Fc, targeted to the tumor cells, could result in blocking the ligand and thus preventing its binding to receptors on lymphocytes. In tumors that are Fas-sensitive, agonistic anti-Fas antibodies, able to induce apoptosis, could be of use. Immune cells may also be protected from Fas-mediated apoptosis by administration of antagonistic anti-Fas antibodies directed to the Fas antigen on the surface of immune cells, antisense oligonucleotides specific for Fas, transducing lymphocytes with an "apoptosis-resistance" gene and treating patient with cytokines that selectively protect immune cells from apoptosis, typical of which are IL-2, IL-12, and IL-15. To modify Fas expression or enhance sensitivity to Fas-mediated signals in tumor cells, administration to cancer patients with cytokines such as IFNg (Simone Fulda and Klaus-Michael Debatin, 2002) or genetic engineering through transduction of tumor cells with genes such as Fas or Bax can be applied to up-regulate pro-apoptotic molecule expression.

Conclusion

In summary, substantial progress has been made in understanding the mechanism of apoptosis and much effort has been done to treat cancer via apoptosis. Despite enthusiasm for apoptosis-based cancer therapies, possible limitations of such approaches can be anticipated, such as selection of apoptosis-resistant tumor cells and systemic toxicity of death receptor ligands. Other problem includes the smooth transfer of the cancer therapy protocol for xerograft models to human patients in clinic. The further investigation with mechamisms by which cell apoptosis can be manipulated should be useful for overcoming the cloud in its clinical applications.

ACKNOWLEDGMENT: This work is partly supported by the grants from National Science Foundation of China, Dept of Health of PLA, and Shanghai Science & Technology Commission.

References

Aaron C. Spalding, Robert M. Jotte, Robert I. Scheinman, Mark W. Geraci, Penny Clarke, Kenneth L. Tyler and Gary L. Johnson. (2002). TRAIL and inhibitors of apoptosis are opposing determinants for NF-kB-dependent, genotoxin-induced apoptosis of cancer cells. Oncogene. *21*, 260–271.

Abigail Hunt and Gerard Evan. (2001). Apoptosis: Till Death Us Do Part, Science. *293*, 1784–5.

Alexandre Nesterov, Yuri Ivashchenko, and Andrew S. Kraft (2002). Tumor necrosis factor-related apoptosis-inducing ligand (TRAIL) triggers apoptosis in normal prostate epithelial cells. Oncogene. *21*, 1135–1140.

Antonsson, B., and Martinou, J.C. (2000). The Bcl-2 protein family. Exp Cell Res. *256*, 50–57.

Asher, G., Lotem, J., Cohen, B., and Shaul, Y. (2000). Regulation of p53 stability and p53-dependent apoptosis by NADH quinone oxidoreductase 1. Proc Natl Acad Sci USA. *98*, 1188–1193.

Beutler, B. and Cerami, A. (1986). Cachetin and tumor necrosis factor as two sides of the same biological coin. Nature. *320*, 584–588.

Boldin, M.P., Goncharov, T.M., Goltsev, Y.V., and Wallach, D. (1996). Involvement of MACH, a novel MORT1/FADD-interacting protease, in Fas/APO-1- and TNF receptor-induced cell death. Cell. *85*, 803–15.

Borner, C., and Monney, L. (1999). Apoptosis without cas-pases: an inefficient molecular guillotine? Cell Death Diff. *6*, 497–507.

Carswell, E.A., Old, L.J., Kassel, R.L., Green, S., Fiore, N., and Williamson, B. (1975). An endotoxin-induced serum factor that causes necrosis of tumors. Proc. Natl. Acad. Sci. USA. *72*, 3666–3670.

Chautan, M., Chazal, G., Cecconi, F., Gruss, P., and Golstein, P. (1999). Interdigital cell death can occur through a necrotic and caspase-independent pathway. Curr Biol. *9*, 967–970.

Cryns, V., and Yuan, J. (1998). Proteases to die for. Genes Dev. *12*, 1551–70.

Daniel, P. (2000). Dissecting the pathways to death. Leukemia. *14*, 2035–2044.

Enari, M., Sakahira, H., Yokoyama, H., Okawa, K., Iwamatsu, A., and Nagata, S. (1998). A caspase-activated DNase that degrades DNA during apoptosis, and its inhibitor ICAD. Nature. *391*, 43–50.

Eskes, R. Desagher, S. Antonsson, B. and Martinou, J. (2000). Bid induces the oligomerization and insertion of Bax into the outer mitochondrial mebrane. Mol. Cell. Biol. *20*, 929–935.

Fan Xiaoqiang, and Yajun Guo. (2001). Apoptosis in oncology, Cell Research. *11*, 1–7.

Fraker, D.L. and Alexander, H.R. (1994). Isolated limb perfusion with high-dose tumor necrosis factor for extremity melanoma and sarcoma., Important adv. Oncol. 179–192.

Frederik H. Igney, Christian K. Behrens and Peter H. Krammer. (2000). Tumor counterattack-concept and reality. Eur. J. Immunol. 30, 725–731.

Gao, G., and Dou, P. (2000). N-terminal cleavage of bax by cal-pain generates a potent proapoptotic 18-kDa fragment that promotes Bcl-2-independent cyto-chrome c release and apoptotic cell death. J Cell Biochem. *80*, 53–72.

Gen Sheng Wu and Zhenhua Ding. (2002). Caspase 9 is required for p53-dependent apoptosis and chemosensitivity in a human ovarian cancer cell line. Oncogene. *21*, 1–8.

Gordon J. Freeman, Arlene H. Sharpe and Vijay K. Kuchroo. (2002). Protect the killer: CTLs need defences against the tumor. Nat. Med. *8*, 787–789.

Guo, B., Godzik, A., and Reed, J.C. (2001). Bcl-G, a novel pro-apoptotic member of the Bcl-2 family. J Biol Chem. *276*, 2780–2785.

Guoqing Chen and David V. Goeddel (2002). TNFR1 signalling: a beautiful pathway, Science, *296*, 1634–1635.

HaidongDong, Scoot E. Strome, Diva R. Salomao, Hideto Tamura, Fumiya Hirano, Dallas B. Flies, Patrick C. Roche, Jun Lu, Gefeng Zhu, Koji Tamada, Vanda A. Lennon, Esteban Celis, and Lieping Chen. (2002). Tumor-associated B7-H1 promotes T-cell apoptosis: A potential mechanism of immune evasion, Nature Medicine. *8*, 793–800.

Heidi Leblance, David Lawrence, Eugene Varfolomeev, Klara Totpal, John Morlan, Peter Schow, Sharon Fong, Ralph Schwall, Dominick Sinicropi and Avi Ashkenzi. (2002). Tumor-cell resistance to death receptor-induced apoptosis through mutational inactivation of the proapoptotic Bcl-2 homolog Bax. Nat Med. *8*, 274–281.

Hermeking, H., and Eick, D. (1994). Mediation of c-Myc induced apoptosis by p53. Science. *265*, 2091–2093.

Hiroshi Arai, David Gordon, Elizabeth G. Nabel, and Gary J. Nabel. (1997). Gene transfer of Fas ligand induces tumor regression in vivo, Proc. Natl. Acad. Sci. USA. *94*, 13862–13867.

Howard Y. Chang and Xiaolu Yang (2000). Proteases for Cell Suicide: Functions and Regulation of Caspases. Micro Mol Biol Rev. *64*, 821–846.

Hsu, H., Xiong, J., and Goeddel, D.V. (1995). The TNF receptor 1-associated protein TRADD signals cell death and NF-kappa B activation. Cell. *81*, 495–504.

Jo, M., Kim, T.H., Seol, D.W., Esplen, J.E., Dorko, K., Billiar, T.R., and Strom, S.C. (2000). Apoptosis induced in normal human hepatocytes by tumor necrosis factor-related apoptosis-inducing ligand. Nat. Med. *6*, 564–567.

Joe O'Connell, Michael W. Bennett, Gerald C., O'Sullivan, J. Kevin Collins and Fergus Shanahan. (1999). Fas counterattack-the best form of tumor defense? Nat. Med. *5*, 267–268.

Kataoka, T., Budd, R.C., Holler, N., Thome, M., Martinon, F., Irmler, M., Burns, K., Hahne, M., Kennedy, N., Kovacsovics, M., and Tschopp, J. (2000). The caspase inhibitor FLIP promotes activation of NF-KB and Erk signaling pathways. Curr. Biol, *10*, 640–648.

Ke, N., Godzik, A., and Reed, J.C. (2001). Bcl-B, a novel Bcl-2 family member that differentially binds and regu-lates Bax and Bak. J Biol Chem. *276*, 12481–12484.

Keane, M.M., Ettenberg, S.A., Nau, M.M., Russell, E.K., and Lipkowitz, S. (1999). Chemotherapy augments TRAIL-induced apoptosis in breast cell lines. Cancer Res. *59*, 734–41.

Kelly, M.M., Hoel, B.D., and Voelkel Johnson C. (2002). Doxorubicin Pretreatment Sensitizes Prostate Cancer Cell Lines to TRAIL Induced Apoptosis Which Correlates with the Loss of c-FLIP Expression. Cancer Biol. Ther. *1*, 520–7.

Kitanaka, C., and Kuchino, Y. (1999). Caspase-independent programmed cell death with necrotic morphology. Cell Death Differ. *6*, 508–515.

Krammer, P.H. (2000). CD95's deadly mission in the immune system. Nature. *407*, 789–795.

Li, X.K., Okuyama, T., Tamura, A., Enosawa, S., Kaneda, Y., Takahara, S., Funashima, N., Yamada, M., Amemiya, H., and Suzuki, S. (1998). Prolonged survival of rat liver allografts transfected with Fas ligand-expressing plasmid. Transplantation. *66*, 1416–1423.

Liu, T.J., el-Naggar, A.K., McDonnell, T.J., Steck, K.D., Wang, M., Taylor, D.L., and Clayman, G.L. (1995). Apoptosis induction mediated by wild type p53 adenoviral gene transfer in squamous cell carcinoma of the head and neck. Cancer Res. *55*, 3117–22.

Miller, T.M., Moulder, K.L., Knudson, C.M., Creedon, D.J., Deshmukh, M., Korsmeyer, S.J., and Johnson, E.M. Jr. (1997). Bax deletion further orders the cell death pathway in cerebellar granule cells and suggests a caspase-independent pathway to cell death. J Cell Biol. *139*, 205–217.

Xiang, J., Chao, D.T., and Korsmeyer, S.J. (1996). Bax-induced cell death may not require interleukin 1-convert-ing enzyme-like proteases. Proc Natl Acad Sci USA. *93*, 14559–14563.

Muzio, M., Stockwell, B.R., Stennicke, H.R., Salvesen, G.S., and Dixit, V.M. (1998). An induced proximity model for caspase-8 activation. J Biol Chem. *273*, 2926–30.

Nicholas P. Restifo. (2000). Not so Fas: Re-evaluating the mechanisms of immune privilege and tumor escape. Nat. Med. *6*, 493–5.

Norbury, C., and Hickson, I. (2001). Cellular responses to DNA damage. Annu Rev Pharmacol Toxicol. *41*, 367–401.

O'Conell, J., Bennett. M.W., O'Sullivan G.C., Collin. J.K., and Shanahan F. (1999). The Fas counterattack: cancer as a site of immune privilege. Immunol. Today. *20*, 46–52.

Oda, E., Ohki, R., Murasawa, H., Nemoto, J., Shibue, T., Yamashita, T., Tokino, T., Taniguchi, T., and Tanaka, N. (2000). Noxa, a BH3-only member of the Bcl-2 family and candidate mediator of p53-induced apoptosis. Science. *288*, 1053–1058.

Porwit MacDonald, A., Ivory, K., Wilkinson, S., Wheatley, K., Wong, L., and Janossy, G. (1995). Bcl-2 protein expression in normal human bone marrow precursors and in acute myelogenous leukemia. Leukemia. *9*, 1191–8.

Reddy, R.K., Jun, L., and Lee, A.S. (1999). The endoplasmic reticulum chaperone glycoprotein GRP94 with Ca21-binding and antiapoptotic properties is a novel proteolytic target of calpain during etopo-side induced apoptosis. J Biol Chem. *274*, 28476–28483.

Pitti, R.M., Marsters, S.A., Ruppert, S., Donahue, C.J., Moore, A., and Ashkenazi, A. (1996). Induction of Apoptosis by Apo-2 Ligand, a New Member of the Tumor Necrosis Factor Cytokine Family. J. Biol. Chem, *271*, 12687–12690.

Rudel, T., and Bokoch, G.M. (1997). Membrane and morphological changes in apoptotic cells regulated by caspase-mediated activation of PAK2. Science. *276*, 1571–4.

Salvesen, G.S., and Dixit, V.M. (1999). Caspase activation: the induced proximity model. Proc. Natl. Acad. Sci. USA. *96*, 10964–10967.

Scaffidi, C. et al. (1999). Differential modulation of apoptosis sensitivity in CD95 type 1 and type 2 cells. J. Biol. Chem. *274*, 22532–22538.

Schendel, S.L., Azimov, R., Pawlowski, K., Godzik, A., Kagan, B., and Reed, J.C. (1999). Ion channel activity of the BH3 only Bcl-2 family member Bid. J Biol Chem. *274*, 21932–21936.

Sharad Kumar and David L. Vaux. (2002). A Cinderella Caspase Takes Center Stage. Science. *297*, 1290–1.

Shkenazi, A., and Dixit, V. M. (1998). Death receptors: signaling and modulation. Science. *281*, 1305–1308.

Simone Fulda and Klaus-Michael Debatin. (2002). IFN g sensitizes for apoptosis by upregulating caspase-8 expression through the Stat1 pathway Oncogene. *21*, 2295–2308.

Schulze Osthoff, K., Krammer, P.H., and Droege, W. (1994). Divergent signaling via APO-1/Fas and the TNF re-ceptor, two homologous molecules involved in physiological cell death. EMBO J. *13*, 4587–4596.

Simone Fulda, Wolfgang Wick, Michael Weller and Klaus. Michael Debatin. (2002). Smac agonists sensitize for Apo2L/TRAIL- or anticancer drug-induced apoptosis and induce regression of malignant glioma in vivo Nat. Med. *8*, 808–815.

Srinivasula, S.M., Ahmad, M., Fernandes-Alnemri, T., and Alnemri, E.S. (1998). Autoactivation of procaspase-9 by Apaf-1-mediated oligomerization. Mol Cell. *1*, 949–57.

Stoka, V., Turk, B., Schendel, S.L., et al. (2001). Lysosomal protease pathways to apoptosis. J Biol Chem. *276*, 3149–3157.

Tang, D., Lahti, J.M., and Kidd, V. (2000). Caspase-8 activation and bid cleavage contributes to MCF7 cellular execution in a casapse-3-dependent manner dur-ing staurosporine-mediated apoptosis. J Biol Chem. *275*, 9303–9307.

Wim Van Molle, Ben Wielockx, Tina Mahieu, Masuhiro Takada, Takahide Taniguchi, Kenji Sekikawa, and Claude Liber (2002). Immunity, *16*, 685–695.

Varnes, M.E., Chiu, S.M., Xue, L.Y., and Oleinick, N.L. (1999). Photodymic therapy-induced apoptosis in lymphoma cells: translocation of cytochrome c causes inhibition of respiration as well as caspase activation. Biochem. Biophys. Res. Column. *255*, 673–9.

Vercammen, D., Beyaert, R., Denecker, G., Goossens, V., Van Loo, G., Declercq, W., Grooten, J., Fiers, W., and Vandenabeele, P. (1998). Inhibition of caspases increases the sensitivity of L929 cells to necrosis mediated by tumor necro-sis factor. J Exp Med. *187*, 1477–1485.

Wei, M.C., Lindsten, T., Mootha, V.K., Weiler, S., Gross, A., Ashiya, M., Thompson, C.B., and Korsmeyer, S.J. (2000). tBid, a membrane targeted cell death ligand, oligoerizes Bak to release cytochrome c. Genes Dev. *14*, 2060–2071.

Wolf, B.B., Goldstein, J.C., Stennicke, H.R., Beere, H., Amarante-Mendes, G.P., Salvesen, G.S., and Green, D.R. (1999). Calpain functions in a caspase-independent manner to promote apoptosis-like events during platelet activation. Blood. *94*, 1683–1692.

Wood, D.E., Thomas, A., Devi, L.A., Berman, Y., Beavis, R.C., Reed, J.C., and Newcomb, E.W. (1998). Bax cleavage is mediated by calpain during drug-induced apoptosis. Oncogene. *17*, 1096–1078.

Zamzami, N., El Hamel, C., Maisse, C., Brenner, C., Munoz-Pinedo, C., Belzacq, A.S., Costantini, P., Vieira, H., Loeffler, M., Molle, G., and Kroemer, G. (2000). Bid acts on the permeability transition pore complex to induce apoptosis. Oncogene. *19*, 6342–635.

Zamzami, N., Susin, S.A., Marchetti, P., Hirsch, T., Gemez-Monterrey, I., Castedo, M., and Kroemer, G. (1996). Mitochondrial control of nuclear apoptosis. J Exp Med. *183*, 1533–44.

Zeytun, A., Hassuneh, M., Nagarkatti, M., and Nagarkatti, P. (1997). Fas-Fas ligand-based interactions between tumor cells and tumor-specific cytotoxic T lymphocytes: a lethal two-way street. Blood *90*, 1952–1959.

Zhan, Q., Fan, S., Bae, I., Guillouf, C., Liebermann, D.A., O'Connor, P.M., and Fornace, A.J. Jr. (1994). Induction of bax by genotoxic stress in human cells correlates with normal p53 status and apoptosis. Oncogene. *9*, 3743–3751.

Zhang, H., Xu, Q., Krajewski, S., Krajewska, M., Xie, Z., Fuess, S., Kitada, S., Godzik, A., and Reed, J.C. (2000). BAR: an apoptosis regulator at the intersection of caspases and Bcl-2 family proteins. Proc Natl Acad Sci USA. *97*, 2597–2602.

Zhang, Y., Yu, J., Unni, E., Shao, T.C., Nan, B., Snabboon, T., Kasper, S., Andriani, F., Denner, L., and Marcelli, M. (2002). Monogene and Polygene Therapy for the Treatment of Experimental Prostate Cancers by Use of Apoptotic Genes bax and bad Driven by the Prostate-Specific Promoter ARR(2)PB. Hum Gene Ther. *13*, 2051–64.

Chapter 15

DNA Fragmentation in Mammalian Apoptosis and Tissue Homeostasis

MING XU AND JIANHUA ZHANG*

ABSTRACT: Apoptosis is a highly regulated physiological process critical in development and tissue homeostasis. Abnormal apoptosis can lead to disease conditions including neurodegeneration, autoimmunity and cancer. DNA fragmentation is an integral part of apoptosis and has long been suspected to be of critical importance in cleaning up potentially antigenic DNA and genetic materials capable of inducing neoplasmic transformation in neighboring cells. Direct evidence for this role of DNA fragmentation in apoptosis however, is still lacking. The identification of a heterodimeric DNA fragmentation factor composed of a 45 and 40 kDa subunit (termed DFF45 and DFF40, or ICAD for Inhibitor of Caspase Activated DNase and CAD for Caspase Activated DNase, respectively) as well as endonuclease G (EndoG) provides a timely opportunity for addressing the physiological significance of DNA fragmentation in apoptosis and tissue homeostasis. We previously generated a DFF45 mutant mouse in which the DFF activity is abolished. We found that DFF45-deficient thymocytes are resistant to DNA fragmentation both *in vivo* and in cultured primary cells exposed to various apoptotic stimuli. Interestingly, DFF45-deficient thymocytes and mouse embryonic fibroblasts (MEFs) are partially resistant to apoptosis in response to several apoptotic-inducing agents. There are more granule cells in the dentate gyrus of the hippocampal formation in DFF45 mutant mice than in normal control mice. This increased neuronal cell number correlates with enhanced spatial and non-spatial learning and memory retention in DFF45 mutant mice compared with control mice. These results suggest that DFF45 is critical for DNA fragmentation and a deficiency in DFF45 can affect timely completion of apoptosis and consequently tissue homeostasis and proper cellular function. Likely due to the unaffected EndoG activity however, residual DNA fragmentation can be found in DFF45-deficient splenocytes and MEFs. In a collaborative effort, we are generating EndoG mutant mice and mice with combined deficiencies of DFF45 and EndoG to investigate how DFF and EndoG jointly function to insure proper apoptosis and tissue homeostasis.

Introduction

Apoptosis plays an essential role in development and tissue homeostasis. A cellular suicide program that is highly conserved from *C. elegans* to humans exists for the

*Department of Cell Biology, Neurobiology and Anatomy, University of Cincinnati Medical Center, Cincinnati, OH 45267-0521. *Correspondence:* Ming Xu at ming.xu@uc.edu or Jianhua Zhang at jianhua.zhang@uc.edu regulation and execution of apoptosis (Horvitz, 1999; Abrams, 1999; Budihardjo et al., 1999). In mammals, at least two general apoptosis pathways exist. One pathway involves activation of death receptors, such as binding of the Fas ligand to its receptor or tumor necrosis factor (TNF) to the TNF receptor (Nagata, 1997; Budihardjo et al., 1999; Strasser et al., 2000). The other pathway involves mitochondrial membrane permeabilization in response to growth factor withdrawal or exposure to DNA damaging chemicals and other stress signals (Adams and Cory 1998; Gross et al., 1999; Wang, 2001). The mitochondron-mediated pathway also amplifies death signals from activation of death receptors through a Bid-mediated mechanism (Wang, 2001). Upon activation of apoptosis, cytochrome c, apoptosis-inducing factor (AIF) and other mitochondrial factors are released (Susin et al., 1999; Wang, 2001), caspases are activated (Budihardjo et al., 1999), chromosomal DNA is cleaved into nucleosomal sized fragments by endonucleases (Peitsch et al., 1994; Counis and Torriglia, 2000; Nagata, 2000; Zhang and Xu, 2000; Zhang and Xu, 2002), and apoptotic cells are phagocytosed by scavenger cells (Fadok and Henson, 1998; Hengartner, 2001). Fig. 1 highlights some of the main molecular events involved in apoptosis.

Cleavage of chromosomal DNA into nucleosomal sized fragments is a hallmark of apoptosis. Failure of clearance of DNA during apoptosis has been suspected to result in development of anti-DNA antibodies and to play a role in tumorigenesis. Multiple endonucleases with different enzymatic properties and residing in different cellular compartments have been suggested to play a role in DNA degradation during apoptosis (Peitsch et al., 1994; Counis and Torriglia, 2000; Nagata, 2000; Zhang and Xu, 2000; Zhang and Xu, 2002). One major endonuclease, DFF, was identified as a heterodimeric protein consisting of the DFF45 and DFF40 subunits (Liu et al., 1997; Enari et al., 1998). DFF40 contains an intrinsic endonuclease activity while its synthesis and folding critically depends on the chaperone and inhibitor subunit, DFF45 (Liu et al., 1997; Enari et al., 1998; Halenbeck et al., 1998; Sakahira et al., 1998; Liu et al., 1998). DFF is activated exclusively upon induction of apoptosis when DFF45 is cleaved by caspase-3 and dissociates from DFF40. DFF40 is then capable of inducing chromatin condensation and cleaving DNA into nucleosomal sized fragments (Figure 1, Liu et al., 1997; Enari et al., 1998; Halenbeck et al., 1998; Sakahira et al., 1998; Liu et al., 1998).

EndoG is an evolutionarily conserved nuclear-encoded endonuclease that was previously suspected to be involved in mitochondrial DNA replication (Cote and Ruiz-Carrilo, 1993). Upon activation of apoptosis, EndoG is released from the mitochondrial intermembrane space with kinetics similar to that of cytochrome c release and travels to the nucleus to elicit DNA fragmentation (Li et al., 2001; Parrish et al., 2001). In DFF45-deficient MEFs, in response to TNF-caspase 8-Bid signaling, UV exposure or Bim stimulation, EndoG cleaves chromosomal DNA in a caspase-independent manner upon release from mitochondria (Li et al., 2001). DFF40 and EndoG have different substrate preferences and enzymatic activities, suggesting that they may perform complementary functions (Ruiz-Carrillo and Renaud, 1987; Ikeda and Ozaki, 1997; Widlak et al., 2000; Widlak et al., 2001). An EndoG mutation in *C.elegans* results in ineffective generation of TUNEL-negative DNA from TUNEL-positive DNA, as well as delayed apoptosis (Parrish et al., 2001). Whether EndoG and DFF play complementary roles in DNA fragmentation and apoptosis in different mammalian tissues in response to diverse apoptotic signals, and whether a deficient EndoG activity leads to insufficient DNA degradation and contributes to the development of autoimmunity and tumorigenesis are unknown.

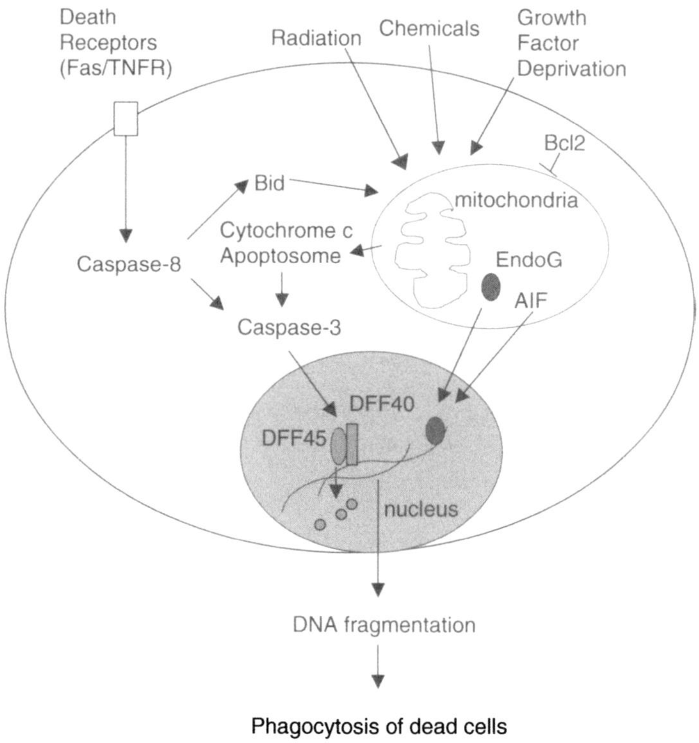

Figure 1. Major molecular events involved in apoptosis. There are two general apoptotic pathways in mammals. One pathway involves activation of death receptors, such as stimulation of the Fas receptor or the TNF receptor (TNFR). Death signals are communicated and regulated by adaptor proteins that in turn activate certain caspases. The other pathway involves mitochondrial membrane permeabilization in response to radiation, UV exposure, growth factor deprivation and other stress signals. Bcl-2 family proteins regulate mitochondrial membrane permeabilization. Among others, cytochrome c, AIF, and EndoG are released from the mitochondria upon activation of apoptosis. The released cytochrome c, AIF, and EndoG activates caspases, unknown targets, and cleaves DNA, respectively. Chromosomal DNA is cleaved into nucleosomal sized fragments by activated DFF40 and EndoG. Dead cells are phagocytosed by scavenger cells.

Genetically engineered mouse models have been very useful in studying molecular mechanisms of apoptosis. With the help of the gene targeting approach, various apoptotic molecules have been shown to play significantly different or redundant roles in different cell types in response to different apoptotic stimuli *in vivo* (Lindsten et al., 2000; Zheng et al., 2000; Ranger et al., 2001). Moreover, dysregulation or mutations of apoptotic genes can result in abnormal development or diseased conditions (Strasser et al., 1991; Adachi et al., 1995; Ranger et al., 2001). To determine the molecular and cellular mechanisms of apoptotic DNA fragmentation *in vivo* and to understand the role of deficient DNA fragmentation and dysregulated apoptosis in the predisposition to and progression of diseased conditions, we previously generated DFF45 mutant mice. Our collective analyses of the mutant mice indicate that DFF45 is important in DNA fragmentation, apoptosis and tissue homeostasis. Moreover, additional endonucleases exist to degrade DNA during apoptosis. We have started to generate an EndoG mutant mouse to further investigate its role in DNA fragmentation and the physiological importance of complete DNA fragmentation in tissue homeostasis.

Results

Generation of DFF45 Mutant Mice

We designed a DFF45 gene targeting construct to delete the first three exons of the DFF45 gene that encode the first 142 amino acids, including the first caspase-3 cleavage site. DFF45 mutant mice were generated through standard gene targeting. The DFF45 gene mutation was confirmed by genomic Southern blotting and was further verified by western blot analysis (Zhang et al., 1998).

DFF45-Deficient Tissue Extracts Lack DFF40 Endonuclease Activity

To investigate the effects of the DFF45 deficiency on DFF activity, we assayed for DFF40 endonuclease activity *in vitro* using extracts from spleen, thymus and testis from DFF45 mutant and wild-type control mice. We found that there was abundant DFF40 nuclease activity in those tissues from control mice in the presence of activated caspase-3, while such activity was completely absent in tissues from DFF45-deficient mice (Zhang et al., 1998). We chose to target the DFF45 gene at a time when which of the DFF subunit encodes an intrinsic endonuclease activity was still unclear. Nonetheless, mutating the DFF45 gene led to a loss of the DFF40 endonuclease activity, indicating that DFF45 is essential for the normal expression of this activity *in vivo* (Zhang et al., 1998).

DFF45 Is Important for DNA Fragmentation upon Activation of Apoptosis

To determine whether the DFF40 endonuclease activity is required for DNA fragmentation *in vivo*, we treated both DFF45 mutant and wild-type control mice with dexamethasone for different periods of time. Genomic DNA was isolated from the thymi of treated mice and resolved by gel electrophoresis. We found that DNA samples from thymi of DFF45 mutant mice exhibit a resistance to DNA fragmentation compared to those from wild-type mice in response to dexamethasone treatment, indicating that DFF45 is important for DNA fragmentation *in vivo* (Zhang et al., 1999). Nagata and colleagues generated a mouse carrying a dominant-negative mutant form of DFF45/ICAD (McIlroy et al., 2000). They found that thymocytes from this line of mouse also lack DNA fragmentation activity when treated with apoptotic inducing agents (McIlroy et al., 2000). This is consistent with our findings. We also exposed primary cells from the spleen, thymus, and bone marrow granulocytes from DFF45 mutant and wild-type mice to actinomycin D, etoposide, staurosporine, or cyclohexamide respectively. We found that DNA samples of these cells from DFF45 mutant mice were devoid of DNA fragmentation (Zhang et al., 1998). In contrast, DNA samples from wild-type control cells exhibited standard DNA laddering under identical experimental conditions (Zhang et al., 1998). Together, these results suggest that DFF plays a major role in DNA fragmentation upon induction of apoptosis.

Degradation of chromosomal DNA into 50 kilobase pair (kb) fragments has been suggested to occur prior to nucleosomal DNA fragmentation during apoptosis (Oberhammer et al., 1993). To determine whether such degradation requires a functional DFF45, we treated thymocytes from wild-type and DFF45 mutant mice with various apoptosis-inducing agents and performed gel electrophoresis. We found that, in contrast to that found in the wild-type control samples, chromosomal DNA in DFF45-deficient thymocytes was resistant to cleavage to 50 kb fragments (Zhang et al., 2000). Consistent with our finding, cells overexpressing a dominant negative mutant form of DFF45/ICAD and DFF45-deficient MEFs were also defective in their ability to cleave DNA into 50 kb fragments (Sakahira et al., 1999; Boulares et al., 2001). Together, these results demonstrate that DFF45 is required for DNA degradation into large chromosomal fragments upon activation of apoptosis.

Dying DFF45-Deficient Thymocytes Exhibit Different Morphology Compared to Wild-Type Control Thymocytes

To detect possible cellular morphological consequences of the DFF45 mutation in apoptosis, we stained DFF45-deficient and wild-type splenocytes and thymocytes with 4,6-diamidino-2-phenylindole (DAPI) after exposure to various apoptotic agents. We found that DFF45-deficient cells exhibit reduced chromatin condensation compared to the control cells (Zhang et al., 1998). Electron microscopic studies showed that, whereas wild-type thymocytes exhibit highly condensed chromatin and cytoplasmic disintegration upon exposure to dexamethasone, DFF45-deficient thymocytes exhibit disaggregated and poorly condensed chromatin and prominent intranuclear cytoplasmic invagination (Zhang et al., 1999). Moreover, while wild-type thymocytes exhibit nuclear fragmentation or apoptotic body formation, DFF45-deficient thymocytes do not (Zhang et al., 1999). Whereas the significance of these observations is still unclear, they suggest the importance of DFF45 in normal apoptosis.

DFF45-Deficient Cells Are More Resistant to Cell Death Than Wild-Type Cells upon Apoptotic Stimulation

Although DNA fragmentation is a downstream event during apoptosis, double stranded breaks of DNA generated by the DNA fragmentation factor are themselves apoptotic signals and can feed forward to ensure speedy apoptosis. Therefore, the lack of DNA fragmentation can result in a deficiency of the amplification of death signaling by DNA fragmentation, leading to a slower death process. To test this possibility, we compared the extent of cell death in DFF45-deficient and wild-type control cells in response to several apoptotic stimuli. We found that the viability of thymocytes from DFF45 mutant mice, as measured by both the trypan blue exclusion and the FITC-Annexin V/propidium iodide staining methods, was higher than that of thymocytes from wild-type mice after dexamethasone, etoposide or staurosporine treatments (Zhang et al., 1999). These results demonstrate that DFF45-deficient cells are more resistant to cell death than normal control cells upon activation of apoptosis.

Additional studies also demonstrated that DFF45-deficient MEFs are more resistant to granzyme B (Thomas et al., 2000) and TNF-a -induced cell death (Boulares et al., 2001) than wild-type MEFs, respectively. Moreover, the loss of liver natural killer T cells after a lymphocytic choriomeningitis virus infection was reduced in DFF45 mutant mice compared to wild-type mice, suggesting a reduced activation-induced cell death (Hobbs et al., 2001). DFF45 mutant mice are also more resistant to kainic acid-induced neuronal cell death in the hippocampus than in wild-type control mice (Zhang et al., 2001). Together, these results suggest that DFF contributes to the timely completion of apoptosis.

DFF45 May Be Involved in Tissue Homeostasis and Function

Timely apoptosis is critical for development and function particularly in the central nervous system. The hippocampal formation has been shown to be critical for spatial and nonspatial learning behaviors (Milner et al., 1998). It is also a region where neurogenesis and apoptosis occur in both developing and adult brains (Kempermann et al., 1997). To understand the potential neurobiological consequences of the resistance to DNA fragmentation and apoptosis due to the DFF45 mutation, we compared neuronal cell numbers in the hippocampal formation of DFF45 mutant and wild-type mice. We found that both the granule cell density and the total granule cell number in the dentate gyrus are higher in DFF45 mutant brains than in wild-type brains (Slane et al., 2000). The cell density and number in other hippocampal regions including the CA1, CA2 and CA3 regions are similar in DFF45 mutant mice compared to wild-type control mice (Slane et al., 2000). Interestingly, the increased neuronal cell number correlates with enhanced spatial learning, as judged by the Morris water maze performance, and longer memory retention, as demonstrated by the novel object recognition performance, in DFF45 mutant mice compared to wild-type control mice (Slane et al., 2000; McQuade et al., 2002). Thus, a lack of proper DNA fragmentation and apoptosis can also affect tissue homeostasis and function.

Apparent Normal General and Immune System Development in DFF45 Mutant Mice

We previously found that DFF45 is ubiquitously expressed throughout mouse development, suggesting that it may also function during the entire time window (Zhang et al., 1999). To test this possibility, we performed a general survey of all major organ systems using DFF45 mutant mice either two months or fifteen months of age. Essentially all DFF45-deficient organs, including the brain, muscle, eye, thymus, spleen, liver, heart, lung, kidney and colon, appear to be normal when compared with those from the wild-type littermates (Zhang et al., 1998; Zhang et al., 2000). Peripheral blood cell counts from DFF45 mutant mice at 2 months of age, including total white blood cells, red blood cells and platelets, as well as percentages of neutrophils, lymphocytes, monocytes, eosinophils and basophils, are also comparable to those from the wild-type mice. These results indicate that the DFF45 mutation apparently did not affect the normal development of mice despite defects in proper DNA fragmentation and cell death in response to apoptotic stimuli.

Apoptosis plays a crucial role in immune system development and function. As our DFF45-deficient thymocytes exhibit more resistance to cell death than wild-type cells, we analyzed the effects of the DFF45 deficiency on development of various cells in lymphoid tissues. We found that thymus, spleen, lymph nodes and Peyer's patches from DFF45 mutant mice were similar in size to those from wild-type mice at 2 months of age (Zhang et al., 1998). The distribution of T, B and myeloid cells in all lymphoid compartments tested appeared to be similar in DFF45 mutant and wild-type mice at this age. Whether DFF45 mutant mice develop autoimmunity at an older age is under investigation.

DNase I heterozygous and homozygous mutant mice develop autoimmunity around 6–8 months of age due to an insufficient DNA clearance (Napirei et al., 2000). However, the symptoms in DNase I mutant mice are heterogeneous and show incomplete penetrance. To understand whether a DFF45 deficiency contributes to systemic lupus erythematosus (SLE)-like symptoms, we obtained the DNase I mutant mice from Dr. Moroy (Ruhr-Universitat Bochum, Bochum, Germany) and crossed this line of mouse with the DFF45 mutant mouse. We are in the process of investigating whether the DFF45 mutation contributes to the development of autoimmunity and exacerbates the development of lupus-like symptoms in DNase I mutant mice.

DNA Fragmentation Can Occur upon Prolonged Cultures of DFF45 Mutant Splenocytes

We previously found that DNA fragmentation can occur upon prolonged cultures of DFF45 mutant splenocytes (Figure 2, Zhang et al., 1998). This result suggests that additional endonucleases participate in DNA fragmentation under such conditions. In another study, Faden and colleagues found that traumatic brain injury (TBI)-induced DNA fragmentation is independent of DFF, although the identity of the endonuclease is unknown (Yakovlev et al., 2001). Wang and colleagues showed that, although at a reduced level compared to the wild-type cells, DFF45-deficient MEFs also exhibit DNA fragmentation upon UV exposure or TNF and cycloheximide treatment (Li et al., 2001). The groups of Wang and Xue used biochemical and genetic methods, respectively, and identified a mitochondrial apoptotic endonuclease, EndoG (Li et al., 2001; Parrish et al., 2001). Wang and colleagues showed that upon activation of apoptosis, EndoG is released from the mitochondrial intermembrane space with kinetics similar to that of cytochrome c release and travels to the nucleus to elicit nucleosomal DNA fragmentation (Li et al., 2001). Wang's group further demonstrated that DNA fragmentation induced by tBID or BIM can be blocked by an anti-EndoG antibody in DFF45-deficient MEFs, suggesting that the residual DNA fragmentation activity can be largely attributed to EndoG at least in those cells (Li et al., 2001). The presence of EndoG may provide sufficient DNA fragmentation to sustain apparent normal developmental phenotypes of the DFF45 mutant mice.

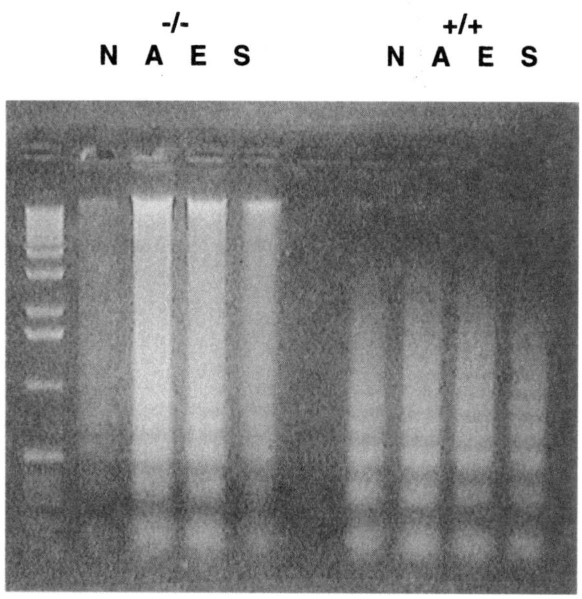

Figure 2. DNA fragmentation can occur in DFF45-deficient splenocytes after prolonged culture. Splenocytes from DFF45 mutant and wild-type mice were cultured for 6 days before being treated with actinomycin (A, 500 nM), etoposide (E, 50 mM), or staurosporine (S, 2 mM) for 10 hours. DNA was extracted, resolved on an agarose gel and visualized by ethidium bromide staining.

DFF40 and EndoG Co-express in Most Mouse Tissues

To understand how DFF and EndoG may function in different tissues at different developmental stages and in response to different environmental signals, we examined the expression profiles of DFF40 and EndoG and 107 other apoptosis-related genes in 82 mouse tissues and experimental conditions from C57BL/6 mouse, including several in which there is active apoptosis, using the Incyte Mouse GemI microarray chips (Zhang et al., 2002). We found that both DFF40 and EndoG are relatively ubiquitously expressed, suggesting parallel functions of these endonucleases. We noted that the expression ratios of the two endonucleases are not identical in all tissues. This result suggests that these two endonucleases may make differential contributions to DNA fragmentation in different tissues upon activation of apoptosis (Zhang et al., 2002).

Generation of EndoG Mutant Mice

To determine the functions of EndoG in DNA fragmentation, apoptosis, and tissue homeostasis, and to determine how EndoG and DFF function together in these processes, we have started to generate EndoG mutant mice by standard gene targeting. In collaboration with Wang and colleagues at U. T. Southerwestern, we transfected embryonic stem cells with an EndoG gene targeting construct. Genomic Southern analyses identified homologous recombinants. We then obtained chimeric male mice and viable germline EndoG heterozygous mutant mice, and mice heterozygous for both EndoG and DFF45 mutations. We are in the process of making EndoG mutant and EndoG/DFF45 double mutant mice and MEFs. Analyses of these mice together with wild-type and DFF45 mutant mice will allow us to further determine the importance of a complete DNA fragmentation in apoptosis and tissue homeostasis.

Discussion

DNA fragmentation is a biochemical hallmark of apoptosis and incomplete DNA fragmentation has long been suspected to be involved in development of disease conditions. The identification of DFF and Endo G provides us an opportunity to investigate the importance of these endonucleases in DNA fragmentation, apoptosis and tissue homeostasis.

DFF Plays A Critical Role in DNA Fragmentation In Vivo

To specifically investigate how various endonucleases function *in vivo* and how deficiencies of endonucleases contribute to abnormal development and diseases, we previously generated DFF45 mutant mice. We found that DFF45 is important for DNA fragmentation in response to apoptotic stimuli *in vivo*. Moreover, DFF45 is required for DNA degradation into large chromosomal fragments upon activation of apoptosis. Nagata and colleagues also reported that overexpression of a caspase-resistant mutant form of DFF45/ICAD inhibits both chromosomal sized and nucleosomal sized DNA fragmentation in staurosporine-treated Jurkat cells (Sakahira et al., 1999). Transgenic mice overexpressing a dominant-negative mutant form of DFF45/ICAD still exhibit DNA fragmentation during normal tissue turnover, likely due to the actions of DNase II in lysosomes of the phagocytes after the apoptotic cells are phagocytosed (McIlroy et al., 2000; Kawane et al., 2001). However, a clear deficit in DNA fragmentation becomes apparent in these transgenic mice when massive apoptosis occurs after g-irradiation (McIlroy et al., 2000). Together, these results suggest that DFF plays a critical role in DNA fragmentation *in vivo*.

Additional Endonucleases Exist to Degrade Chromosomal DNA upon Activation of Apoptosis

DFF45-deficient splenocytes and MEFs exhibit residual DNA fragmentation. Partially stimulated by these findings, Wang and colleagues identified EndoG that travels from the mitochondrial intermembrane space to the cell nucleus to elicit DNA fragmentation upon activation of apoptosis. They further demonstrated that, using extracts from DFF45-deficient MEFs, an anti-EndoG antibody can block the DNA fragmentation activity. Xue and associates found that EndoG can further digest degraded DNA that has been cleaved by an unknown endonuclease(s) upon activation of apoptosis (Parrish et al., 2001). Moreover, EndoG also resides in mitochondria in *C. elegans*, indicating that the involvement of mitochondria in apoptosis is conserved from *C. elegans* to mammals. Future *in vivo* studies will help to understand the relative roles of DFF and EndoG in mammalian DNA fragmentation and apoptosis.

Horvitz and colleagues demonstrated that, in *C. elegans*, a complete DNA degradation requires activities both from engulfing cells and from NUC-1 in dying cells (Wu et al., 2000). NUC-1, which is a DNase II homolog, plays an important role in an intermediate step of DNA degradation (Wu et al., 2000). Nagata and colleagues showed that DNase II resides in the lysosomes of phagocytes and degrades DNA after the engulfment of apoptotic cells (McIlroy et al., 2000; Kawane et al., 2001). In the absence of DFF activity, such as in mice carrying a dominant-negative form of DFF45/ICAD, DNase II provides a backup mechanism for degrading chromosomal DNA after the apoptotic cells are phagocytosed (McIlroy et al., 2000). DNase II-deficient embryos develop severe anemia and die on embryonic day 17.5 due to defects in definitive erythropoiesis (Kawane et al., 2001). The number of F4/80-positive macrophages is reduced in DNase II-deficient embryonic liver, and the nuclear DNA expelled from erythroid precursor cells is not cleared (Kawane et al., 2001). Because of the embryonic lethality, whether down-regulation or mutations of DNase II can contribute to SLE development in adult mice is not known.

Additional endonucleases have been proposed to function in apoptosis (Peitsch et al., 1994; Counis and Torriglia, 2000; Zhang and Xu, 2000; Zhang and Xu, 2002). DNase I degrades extracellular DNA and a failure to clean up nuclear debris in DNase I mutant mice rendered them with SLE-like autoimmunity (Napirei et al., 2000). Although DNase I stimulates EndoG endonuclease activity *in vitro* (Widlak et al., 2001) and it is required for apoptotic DNA fragmentation in certain cell lines (Oliveri et al., 2001), whether DNase I plays a role in regulating EndoG activity *in vivo* is unknown. Cyclophilins and L-DNase II can degrade DNA (Counis and Torriglia, 2000; Zhang and Xu, 2000; Zhang and Xu, 2002). However, how these various endonucleases functionally interact *in vivo* in DNA fragmentation, apoptosis, and homeostasis is still unknown.

A Deficiency in DNA Fragmentation Can Have Physiological Consequences

We observed that DFF45-deficient cells are partially resistant to apoptosis (Zhang et al., 1999), providing the first evidence that apoptotic endonucleases may be involved in regulation of apoptosis process. An EndoG mutation in *C. elegans* also results in delayed apoptosis (Parrish et al., 2001). One possible mechanism for the resistance to apoptosis is that cleaved DFF45 or EndoG released from the mitochondria may exert feedback regulation of activities of upstream apoptotic regulators and executors. The DFF45- and EndoG-deficient MEFs will provide useful cellular models to investigate such potential regulatory mechanisms.

Earnshaw and colleagues generated a DFF40-deficient chicken cell line (Samejima et al., 2001). They showed that, whereas resistant to DNA degradation into nucleosomal sized fragments, the mutant cells exhibit normal DNA fragmentation into chromosomal sized fragments and cell death compared to control cells (Samejima et al., 2001). These results suggest that DFF45 and DFF40 may not have identical influences on DNA fragmentation and apoptosis.

Mammalian development depends on proper apoptosis (Jacobson et al., 1997). DFF45 mutant mice exhibit apparent normal development. The existence of other endonucleases may explain the apparent normal development in DFF45 mutant mice. We found an increased neuronal cell number in the dentate gyrus region of the hippocampal formation correlates with enhanced spatial learning and longer memory retention in DFF45 mutant mice compared to wild-type control mice (Slane et al., 2000; McQuade et al., 2002). This result suggests that defective DNA fragmentation can affect timely apoptosis, tissue homeostasis and function. Analysis of the EndoG mutant mice will provide additional insights into how an endonuclease normally localized in the mitochondria may contribute to apoptosis and development.

Human DFF45 and DFF40 genes are both mapped to 1p36 (Leek et al., 1997; Judson et al., 2000), a chromosomal region deleted in many forms of cancer and implicated in SLE development in human patients (Caron et al., 1995; Imyanitov et al., 1999; Ohira et al., 2000; Judson et al., 2000; Chen et al., 2001). Similarly, mouse DFF45 and DFF40 genes have been mapped to the distal region of chromosome 4 (Kawane et al., 1999), also a region suggested to contribute to lupus development (Wakeland et al., 2001). Despite this knowledge, a direct link between abnormal functions of DFF and SLE and tumorigenesis is still missing. Mice deficient in DFF45 and EndoG will provide useful *in vivo* model systems for investigating the importance of proper DNA fragmentation and apoptosis in the development of diseased conditions including cancer, autoimmune diseases and neurodegeneration.

Potential Influence of the Mouse Genetic Background on Our Studies

For all our studies, we used mice with a 129Sv/C57BL/6 mixed genetic background. Different genetic background can influence the outcome of experiments. To minimize such an influence, we made an effort to use littermates from multiple litters of the same generation of breeding to perform all the studies. In the mean time, we are breeding the mutant mice into the C57BL/6 genetic background for future experiments.

Experimental Procedures

All experimental procedures have been described (Zhang et al., 1998; Zhang et al., 1999; Slane et al., 2000; Thomas et al., 2000; Zhang et al., 2000; Yakovlev et al., 2001; Zhang et al., 2001; Boulares et al., 2001; Hobbs et al., 2001; McQuade et al., 2002; Zhang et al., 2002).

ACKNOWLEDGMENTS: We thank Xiaodong Wang for advice, stimulating discussions, and sharing preliminary results. We also thank the people working in our labs, and all of our collaborators. This study was supported by grants from the NIH (M.X. and J.Z) and DOD (J.Z.).

References

Abrams, J.M. (1999). An emerging blueprint for apoptosis in *Drosophila*. Trends Cell Biol. *9*, 435–440.

Adachi, M., Suematsu, S., Kondo, T., Ogasawara, J., Tanaka, T., Yoshida, N., and Nagata, S. (1995). Targeted mutation in the Fas gene causes hyperplasia in peripheral lymphoid organs and liver. Nat. Genet. *11*, 294–300.

Adams, J.M., and Cory, S. (1998). The Bcl-2 protein family: arbiters of cell survival. Science *281*, 1322–1326.

Boulares, A.H., Zoltoski, A.J., Yakovlev, A., Xu, M., and Smulson, M.E. (2001). Roles of DNA fragmentation factor and poly(ADP-ribose) polymerase in an amplification phase of TNF-induced apoptosis. J. Biol. Chem. *276*, 38185–38192.

Budihardjo, I., Oliver, H., Lutter, M., Luo, X., and Wang, X. (1999). Biochemical pathways of caspase activation during apoptosis. Annu. Rev. Cell Dev. Biol. *15*, 269–290.

Caron, H., Peter, M., van Sluis, P., Speleman, F., de Kraker, J. et al. (1995). Evidence for two tumour suppressor loci on chromosomal bands 1p35–36 involved in neuroblastoma: one probably imprinted, another associated with N-myc amplification. Hum. Mol. Genet. *4*, 535–539.

Chen, Y.Z., Soeda, E., Yang, H.W., Takita, J., Chai, L., Horii, A., Inazawa, J., Ohki, M., and Hayashi, Y. (2001). Homozygous deletion in a neuroblastoma cell line defined by a high-density STS map spanning human chromosome band 1p36. Genes Chromosomes Cancer *31*, 326–332.

Cote, J., and Ruiz-Carrilo, A. (1993). Primers for mitochondrial DNA replication generated by endonuclease G. Science *261*, 765–769.

Counis, M.F., and Torriglia, A. (2000). Dnases and apoptosis. Biochem. Cell Biol. *78*, 405–414.

Enari, M., Sakahira, H., Yokoyama, H., Okawa, K., Iwamatsu, A., and Nagata, S. (1998). A caspase-activated DNase that degrades DNA during apoptosis, and its inhibitor ICAD. Nature *391*, 43–50.

Fadok, V.A., and Henson, P.M. (1998). Apoptosis: getting rid of the bodies. Curr. Biol. *8*, R693–695.

Gross, A., McDonnell, J.M., and Korsmeyer, S.J. (1999). Bcl-2 family members and the mitochondria in apoptosis. Genes Dev. *13*, 1899–1911.

Halenbeck, R., MacDonald, H., Roulston, A., Chen, T.T., Conroy, L., and Williams, L.T. (1998). CPAN, a human nuclease regulated by the caspase-sensitive inhibitor DFF45. Curr. Biol. *8*, 537–540.

Hengartner, M.O. (2001). Apoptosis: corralling the corpses. Cell *104*, 325–328.

Hobbs, J.A., Cho, S.Y., Roberts, T.J., Sriram, V., Zhang, J., Xu, M., and Brutkiewicz, R.R. (2001). Selective loss of NKT cells by apoptosis following infection with lymphocytic choriomeningitis virus. J. Virol. *75*, 10746–10754.

Horvitz, H.R. (1999). Genetic control of programmed cell death in the nematode *Caenorhabditis elegans*. Cancer Res. *59*, 1701s–1706s.

Ikeda, S., and Ozaki, K. (1997). Action of mitochondrial endonuclease G on DNA damaged by L-ascorbic acid, peplomycin, and cis-diamminedichloroplatinum (II). Biochem. Biophys. Res. Comm. *235*, 291–294.

Imyanitov, E.N., Birrell, G.W., Filippovich, I., Sorokina, N., Arnold, J., Mould, M.A., Wright, K., Walsh, M., Mok, S.C., Lavin, M.F. et al. (1999). Frequent loss of heterozygosity at 1p36 in ovarian adenocarcinomas but the gene encoding p73 is unlikely to be the target. Oncogene *18*, 4640–4642.

Jacobson, M.D., Weil, M., and Raff, M.C. (1997). Programmed cell death in animal development. Cell *88*, 347–354.

Judson, H., van Roy, N., Strain, L., Vandesompele, J., Van Gele, M., Speleman, F., and Bonthron, D.T. (2000). Structure and mutation analysis of the gene encoding DNA fragmentation factor 40 (caspase-activated nuclease), a candidate neuroblastoma tumour suppressor gene. Hum. Genet. *106*, 406–413.

Kawane, K., Fukuyama, H., Adachi, M., Sakahira, H., Copeland, N.G., Gilbert, D.J., Jenkin, N.A., and Nagata, S. (1999). Structure and promoter analysis of murine CAD and ICAD genes. Cell Death Differ. *6*, 745–752.

Kawane, K., Fukuyama, H., Kondoh, G., Takeda, J., Ohsawa, Y., Uchiyama, Y., and Nagata, S. (2001). Requirement of DNase II for definitive erythropoiesis in the mouse fetal liver. Science *292*, 1546–1549.

Kempermann, G., Kuhn, H.G., and Gage, F.H. (1997). More hippocampal neurons in adult mice living in an enriched environment. Nature *386*, 493–495.

Leek, J.P., Carr, I.M., Bell, S.M., Markham, A.F., and Lench, N.J. (1997). Assignment of the DNA fragmentation factor gene (DFFA) to human chromosome bands 1p36.3—>p36.2 by in situ hybridization. Cytogenet. Cell Genet. *79*, 212–213.

Li, L.Y., Luo, X., and Wang, X. (2001). Endonuclease G (EndoG) is an apoptotic DNase when released from Mitochondria. Nature *412*, 95–99.

Lindsten, T., Ross, A.J., King, A., Zong, W.X., Rathmell, J.C., Shiels, H.A., Ulrich, E., Waymire, K.G., Mahar, P., Frauwirth, K., et al. (2000). The combined functions of proapoptotic Bcl-2 family members bak and bax are essential for normal development of multiple tissues. Mol. Cell *6*, 1389–1399.

Liu, X., Zou, H., Slaughter, C., and Wang, X. (1997). DFF, a heterodimeric protein that functions downstream of caspase-3 to trigger DNA fragmentation during apoptosis. Cell *89*, 175–184.

Liu, X., Li, P., Widlak, P., Zou, H., Luo, X., Garrard, W.T., and Wang, X. (1998). DFF40 induces DNA fragmentation and chromatin condensation during apoptosis. Proc. Natl. Acad. Sci. USA *95*, 8461–8466.

McIlroy, D., Tanaka, M., Sakahira, H., Fukuyama, H., Suzuki, M., Yamamura, K., Ohsawa, Y., Uchiyama, Y., and Nagata, S. (2000). An auxiliary mode of apoptotic DNA fragmentation provided by phagocytes. Genes Dev. *14*, 549–558.

McQuade, J.S., Vorhees, C., Xu, M., and Zhang, J. (2002). Enhanced non-spatial learning and memory in DFF45 knockout mice compared to wild-type mice. Physiol. Beh. *76*, 315–320.

Milner, B., Squire, L.R., and Kandel, E.R. (1998). Cognitive neuroscience and the study of memory. Neuron *20*, 445–468.

Nagata, S. (1997). Apoptosis by death factor. Cell *88*, 355–365.

Nagata, S. (2000). Apoptotic DNA fragmentation. Exp. Cell Res. *256*, 12–18.

Napirei, M., Karsunky, H., Zevnik, B, Stephan, H., Mannherz, H.G., and Moroy, T. (2000). Features of systemic lupus erythematosus in Dnase1-deficient mice. Nat. Genet. *25*, 177–181.

Oberhammer, F., Wilson, J.W., Dive, C., Morris, I.D., Hickman, J.A., Wakeling, A.E., Walker, P.R., and Sikorska, M. (1993). Apoptotic death in epithelial cells: cleavage of DNA to 300 and/or 50 kb fragments prior to or in the absence of internucleosomal fragmentation. EMBO J. *12*, 3679–3684.

Ohira, M., Kageyama, H., Mihara, M., Furuta, S., Machida, T., Shishikura, T., Takayasu, H., Islam, A., Nakamura, Y., Takahashi, M. et al. (2000). Identification and characterization of a 500-kb homozygously deleted region at 1p36.2–p36.3 in a neuroblastoma cell line. Oncogene *19*, 4302–4307.

Oliveri, M., Daga, A., Cantoni, C., Lunardi, C., Millo, R., and Puccetti, A. (2001). DNase I mediates internucleosomal DNA degradation in human cells undergoing drug-induced apoptosis. Eur. J. Immunol. *31*, 743–751.

Parrish, J., Li, L., Klotz, K., Ledwich, D., Wang, X., and Xue, D. (2001). C. *elegans* mitochondrial endonuclease G is important for DNA fragmentation and progression of apoptosis. Nature *412*, 90–94.

Peitsch, M.C., Mannherz, H.G., and Tschopp, J. (1994). The apoptosis endonucleases: cleaning up after cell death? Trends Cell Biol. *4*, 37–41.

Ranger, A.M., Malynn, B.A., and Korsmeyer, S.J. (2001). Mouse models of cell death. Nat. Genet. *28*, 113–118.

Ruiz-Carrillo, A., and Renaud, J. (1987). Endonuclease G: a (dG)n X (dC)n-specific DNase from higher eukaryotes. EMBO J. *6*, 401–407.

Sakahira, H., Enari, M., and Nagata, S. (1998). Cleavage of CAD inhibitor in CAD activation and DNA degradation during apoptosis. Nature *391*, 96–99.

Sakahira, H., Enari, M., Ohsawa, Y., Uchiyama, Y., and Nagata, S. (1999). Apoptotic nuclear morphological change without DNA fragmentation. Curr. Biol. *9*, 543–546.

Samejima, K., Tone, S., and Earnshaw, W.C. (2001). CAD/DFF40 nuclease is dispensable for high molecular weight DNA cleavage and stage I chromatin condensation in apoptosis. J. Biol. Chem. *276*, 45427–45432.

Slane, J., Lee, H., Vorhees, C., Zhang, J., and Xu, M. (2000). DNA fragmentation factor 45 deficient mice exhibit enhanced spatial learning and memory compared to wild-type control mice. Brain Res, *867*, 70–79.

Strasser, A., Whittingham, S., Vaux, D.L., Bath, M.L., Adams, J.M., Cory, S., and Harris, A.W. (1991). Enforced BCL2 expression in B-lymphoid cells prolongs antibody responses and elicits autoimmune disease. Proc. Natl. Acad. Sci. USA *88*, 8661–8665.

Strasser, A., O'Connor, L., and Dixit, V.M. (2000). Apoptosis signaling. Annu. Rev. Biochem. *69*, 217–245.

Susin, S.A., Lorenzo, H.K., Zamzami, N., Marzo, I., Snow, B.E., Brothers, G.M., Mangion, J., Jacotot, E., Costantini, P., Loeffler, M., et al. (1999). Molecular characterization of mitochondrial apoptosis-inducing factor. Nature *397*, 441–446.

Thomas, D., Du, C., Xu, M., Wang, X., and Ley, T.J. (2000). DFF45/ICAD can be directly process by granzyme B during the induction of apoptosis. Immunity *12*, 621–632.

Wakeland, E.K., Liu, K., Graham, R.R., and Behrens, T.W. (2001). Delineating the genetic basis of systemic lupus erythematosis. Immunity *15*, 397–408.

Wang, X. (2001). The expanding role of mitochondria in apoptosis. Genes Dev. *15*, 2922–2933.

Widlak, P., Li, P., Wang, X., and Garrard, W.T. (2000). Cleavage preferences of the apoptotic endonuclease DFF40 (caspase-activated DNase or nuclease) on naked DNA and chromatin substrates. J. Biol. Chem. *275*, 8226–8232.

Widlak, P., Li, L.Y., Wang, X., and Garrard, W.T. (2001). Action of recombinant human apoptotic endonuclease G on naked DNA and chromatin substrates: cooperation with exonuclease and DNase I. J. Biol. Chem. *276*, 48404–48409.

Wu, Y.-C., Stanfield, G.M., and Horvitz, H.R. (2000). NUC-1, a Caenorhabditis elegans Dnase II homolog, functions in an intermediate step of DNA degradation during apoptosis. Genes Dev. *14*, 536–548.

Yakovlev, A.G., Di, X., Movsesyan, V., Mullins, P.G.M., Wang, G., Boulares, H., Zhang, J., Xu, M., and Faden, A.I. (2001). Presence of DNA fragmentation and lack of neuroprotective effect in DFF45 knockout mice subjected to traumatic brain injury. Mol. Medicine *7*, 205–216.

Zhang, J., Liu, X., Scherer, D.C., Van Kaer, L., Wang, X., and Xu, M. (1998). Resistance to DNA fragmentation and chromatin condensation in mice lacking the DNA fragmentation factor 45. Proc. Natl. Acad. Sci. USA *95*, 12480–12485.

Zhang, J., Wang, X., Bove, K.E., and Xu, M. (1999). DNA fragmentation factor 45-deficient cells are more resistant to apoptosis and exhibit different dying morphology than wild-type control cells. J. Biol. Chem. *274*, 37450–37454.

Zhang, J., Lee, H., Lou, D.W., Boivin, G., and Xu, M. (2000). Lack of obvious 50 kilobase pair DNA fragments in DNA fragmentation factor 45-deficient thymocytes upon activation of apoptosis. Biochem. Biophys. Res. Comm. *274*, 225–229.

Zhang, J., and Xu, M. (2000). DNA fragmentation in apoptosis. Cell Res. *10*, 205–211.

Zhang, J., Lee, H., Agarwala, A., Lou, D.W., and Xu, M. (2001). DNA fragmentation factor 45 mutant mice exhibit resistance to kainic acid-induced neuronal cell death. Biochem. Biophy. Res. Comm. *285*, 1143–1149.

Zhang, J., and Xu, M. (2002). Apoptotic DNA degradation and tissue homeostasis. Trends Cell Biol. *12*, 84–89.

Zhang, J., Xu, M., and Aronow, B. (2002). Expression profiles of 109 apoptosis pathway-related genes in 82 mouse tissues and experimental conditions. Biochem. Biophy. Res. Comm. *297*, 537–544.

Zheng, T.S., Hunot, S., Kuida, K., Momoi, T., Srinivasan, A., Nicholson, D.W., Lazebnik, Y., and Flavell, R.A. (2000). Deficiency in caspase-9 or caspase-3 induces compensatory caspase activation. Nat. Med. *6*, 1241–1247.

Chapter 16

Ubiquitin and Intracellular Aggregation

A Common Pathway of Neurodegeneration in Chronic Dementia?

SUNGMIN SONG AND YONG-KEUN JUNG*

ABSTRACT: The ubiquitin-proteasome system (UPS) is involved in many biological pathways via the degradation of short-lived and regulatory proteins important in cellular processes. Moreover, in the most of the chronic human neurodegenerative diseases, including Alzheimer's disease (AD) and Parkinson's disease (PD), protein aggregation co-localized with ubiquitin is a common feature and may be caused by the abrogation of UPS. *Parkin* gene isolated from autosomal recessive-juvenile parkinsonism (AR-JP) is ubiquitin-protein ligase (E3) and many mutations of *parkin* in familial PD disrupted the ubiquitin-protein ligase activity to eventually accumulate its substrates in intracellular aggresome (Lewy body). Pathogenic proteins of AD, including presenilin and Amyloid-b Precursor Protein (APP), are the substrates of UPS and the mutations related with AD pathogenesis give resistance to their degradation by proteasome. Therefore, UPS is the important target of therapeutic trial for neurodegenerative disorders.

The ubiquitinylation, the conjugation of ubiquitin to other target proteins produced by the successive actions of ubiquitin-activating (E1), ubiquitin-conjugating (E2), and ubiquitin-ligase (E3) enzymes, is the central signal for ATP-dependent non-lysosomal proteolysis catalyzed by 26S proteasome (reviewed by Hershko and Ciechanover, 1998). The ubiquitin-proteasome system (UPS) is involved in many biological pathways including regulation of the cell cycle, modulation of cell surface receptors and ion channels, cell death, and antigen presentation via the degradation of short-lived and regulatory proteins important in cellular processes (reviewed by Schwartz and Ciechanover, 1999; Jesenberger and Jentsch, 2002). Moreover, an un-ignorable function of UPS is the degradation of unwanted proteins, such as proteins affected by abnormal folding, unfolding, or damages by reactive oxygen species, which may eventually form intracellular aggregation, aggresomes and russel bodies (reviewed by Kopito and Sitia, 2000).

Protein aggregation is a common feature of most of the chronic human neurodegenerative diseases, including Alzheimer's disease (AD), Parkinson's disease (PD), prion disease, polyglutamine disease, tauopathy, and familial amyotrophic lateral sclerosis (reviewed by

*Department of Life Science, Kwangju Institute of Science and Technology, 1 Oryong-dong, Puk-gu, Gwangju 500-712, Korea. *Correspondence:* Department of Life Science, Kwangju Institute of Science and Technology, 1 Oryong-dong, Puk-gu, Gwangju 500-712, Korea Tel: 82-62-970-2492; Fax: 82-62-970-2484; E-mail: ykjung@kjist.ac.kr

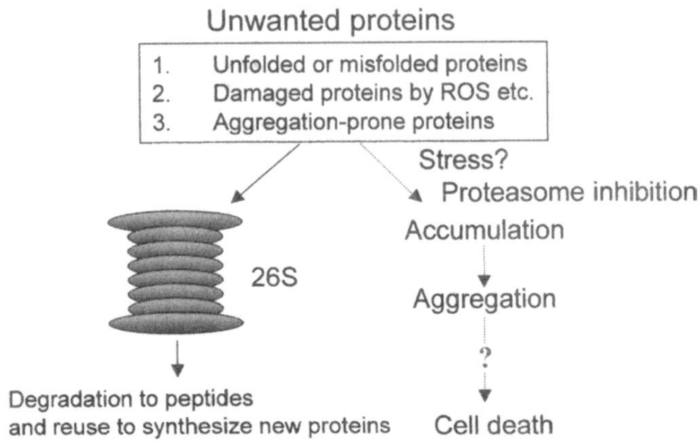

Figure 1. Schematic diagram of cell death induced by cellular indigestions. In living cells, proteins are constantly synthesized and also degraded for clearance of cellular garbage or unwanted proteins. Signals abrogating proteasome-dependent degradation (PDD) can cause accumulation of those proteins and they eventually aggregates to be able to induce cell death whose mechanism is not fully understood.

Taylor et al., 2002). Significantly, the intraneuronal inclusions in many of these diseases contain deposits of ubiquitinylated proteins, indicating that perturbations of ubiquitin-dependent proteolysis may be occurring (reviewed by Layfield et al., 2001). The UPS is so efficient to degrade ubiquitinylated proteins. The suppression of proteasome-dependent degradation (PDD) using proteasome inhibitors was reported to be sufficient to induce neuronal death and cellular aggresomes composed of aggregation-prone proteins (Wyttenbach et al., 2000; McNaught et al., 2002). From these evidences, we can expect that the degradation of some proteins may be important for the maintenance of cellular system as much as the synthesizing new proteins. Thus, the cellular accumulation of ubiquitinylated proteins by the abrogation of UPS function is often prone to cause diseases. Also, it is speculated that the toxicity by intracellular indigestion, generated by the inhibition of PDD or by pathogenic processes in chronic neurodegenerative diseases, to cells may be triggered by the accumulations of abnormal proteins, such as damaged and mis-folded proteins, or signaling regulatory proteins used already. The accumulation of those proteins may then interfere their essential roles or give stresses to cell by the excessive signaling or garbage to be unable to be contained (Figure 1).

Here we review the pathogenic mechanisms of neurodegenerative diseases in the view of the accumulation and aggregation of toxic proteins, UPS, and the toxic effects of the aggregation, and speculate whether the abrogation of UPS and PDD may be a cause of neurodegenerative diseases and their neuronal loss.

Intraneuronal Inclusion and Neuronal Loss

What is the meaning of the toxicity of PDD inhibition and aggregations in *in vivo* pathology of diseases? Interestingly, the histological characteristics of a broad range of neurodegenerative disorders are dead neurons or damaged neurons that contain aggregations, which are intracellular accumulation of toxic and aggregation-prone proteins (reviewed by Taylor et al., 2002). PDD activity seems to be closely correlated with the stabilities of the neurodegenerative disease-associated proteins and their fragments, including presenilin,

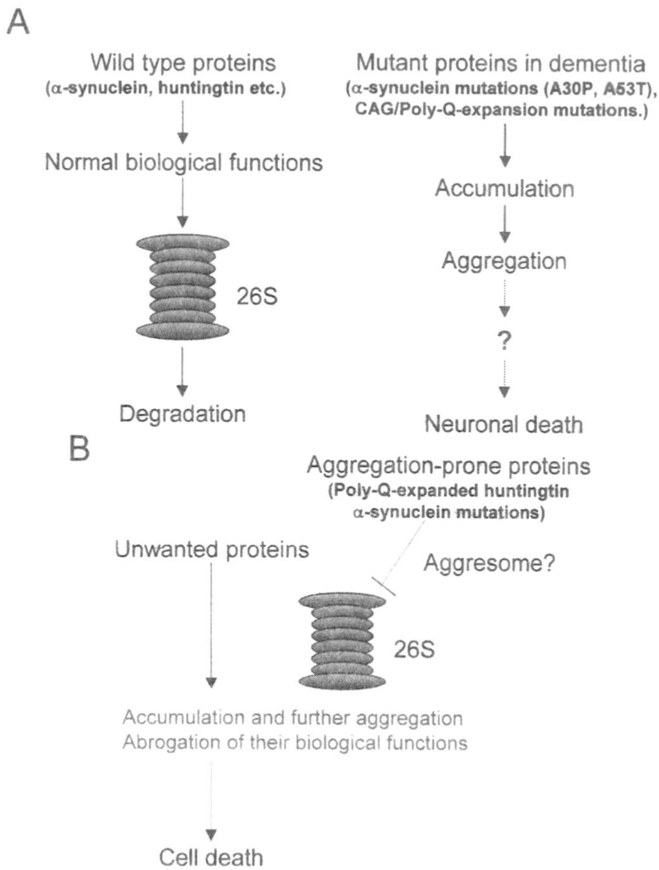

Figure 2. Neuronal loss in dementia caused by mutations of aggregation-prone proteins. (A) Neurodegenerative disease-associated proteins, a -synuclein, SCA1, SCA3, and huntingtin etc., are constantly synthesized and degraded for their normal functions. However, mutations give them resistance to PDD and cause them accumulated in cells (Bennett et al., 1999). (B) Aggregation-prone proteins transformed by mutations are readily accumulated to inhibit 26S proteasome and induce further cellular indigestions (Lee et al., 2002).

amyloid-b precursor protein (APP), Parkin, and prion protein which are responsible to generate related disease pathologies (Fraser et al., 1998; Jin et al., 2000; Cupers et al., 2001). Also, mutations in proteins pathologically associated with neurodegenerative disease, such as a -synuclein, Huntingtin and so on, are reported to give resistance against PDD and readily aggregate in cells (Bennett et al., 1999;Tanaka et al. 2001; Lee et al., 2002) (Figure 2A). The toxicity of aggregation-prone proteins stabilized by their resistance to PDD may be caused indirectly by their inhibition of general PDD. For instances, expression of huntingtin containing a pathologic polyglutamine or a -synuclein caused inhibition of UPS, which may contribute to further aggregation of unwanted proteins involved in fundamental cellular events, causing cellular deregulation and pathological processes (Bence et al., 2001; Tanaka et al., 2001) (Figure 2B).

Based on these observations, researches for the prevention of dementia have been focused on the mechanisms of the aggregation and neuronal death induced by aggregation-prone proteins and proteasome inhibition in neuronal system (Zhou et al., 2001; Nishitoh et al., 2002; Ding et al., 2002). A recent report showed apoptosis singal-regulatory kinase 1 (ASK1) (−/−) primary neurons have resistance to toxicity of aggregation-prone protein and the proteasome inhibitor as a downstream molecule of ER stress signal (Nishitoh et al., 2002). Unfortunately, the understanding of mediator or regulator proteins, which are responsible for aggregations and neurotoxicity, still is insufficient to cure the patients and remains to be progressed. Identification of such modulators in neurotoxicity deserves for further investigation.

Role of Ubiquitin-Proteasome System in Parkinson's Disease

PD is a common neurodegenerative disorder with a lifetime incidence of approximately 2 percents. PD is characterized by loss of dopaminergic neurons in the substantia nigra, formation of filamentous intraneuronal inclusions (Lewy bodies) and movement disorder. Mutation analysis showed that mutations in the *parkin* and a *-synuclein* genes are implicated in the pathogenesis of early- and adult-onset familial Parkinson's disease, respectively (Bennett et al., 1999; Lucking et al., 2000; Polymeropoulos et al., 2000). While PD has more than 1 genetic and/or environmental cause, Parkin and a -synuclein are proposed to be functionally associated with pathogenesis of PD via abrogation of UPS and intracellular aggregation.

Parkin as E3 Ligase Functioning in UPS: Parkin's Substrates and Aggregation

Parkin is a RING-finger-containing protein whose function was proposed to play a role in E2-dependent ubiquitinylation (Lorick et al., 1999). Parkin was shown to bind to the E2 ubiquitin-conjugating human enzyme(UbcH8) through its C-terminal RING-finger and shows ubiquitin-protein ligase (E3) activity. Familial mutation was found in RING-finger of Parkin isolated from autosomal recessive-juvenile parkinsonism (AR-JP) and disrupted Parkin's ubiquitin-protein ligase activity, indicating that E3 ubiquitin ligase activity of Parkin is associated with PD.

Parkin as an E3 ubiquitin-protein ligase may control protein levels via ubiquitinylation and may have specific substrate that is critical to exert its pathogenic activity. Recently, a putative G protein-coupled transmembrane polypeptide, Pael receptor, was identified as an interacting protein with and ubiquitinylated by Parkin (Imai et al., 2001). Imai et al. (2001) proposed that the insoluble Pael receptor accumulated in the brains of AR-JP patients and induced endoplasmic reticulum stress, leading to unfolded protein-induced selective neuronal death (Figure 3). Thus, Parkin seems to promote the degradation of insoluble Pael receptor, resulting in suppression of the cell death induced by Pael receptor. While the direct contribution of Pael-R to pathogenesis of PD waits to be clarified, Parkin's ability to function as E3 ligase in UPS and its role in coordinating aggregation of the putative pathogens may be crucial in the development of PD. Similarly, synphilin-1 was suggested to be ubiquitinylated by Parkin and together with a -synuclein and parkin, forms Lewy body-like ubiquitin-positive cytosolic inclusion, providing a molecular basis for the ubiquitinylation of intracytoplasmic inclusion and the absence of Lewy-body detected by ubiquitin antibody in patients with Parkin mutation (Chung et al., 2001). In addition, the synaptic vesicle-associated protein, CDCrel-1 was identified as another substrate of Parkin but the

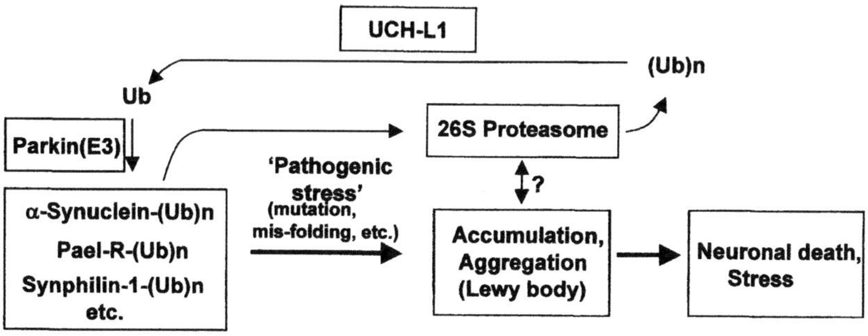

Figure 3. Functional contribution of Parkin and a -synuclein to aggregation/UPS-mediated neuronal cell death. Parkin as an E3 ubiquitin ligase functions to ubiquitinylate its substrates, including a -synuclein, Pael-R, and synphilin-1, for their degradations by proteasome. Under pathogenic condition such as stress and mutation, the ubiquitinylated adducts accumulate and aggregate, leading to neuronal death. Also, modulation of PDD activity may result from or in aggregation. Inactivation of ubiquitin carboxy terminal hydrolase (UCH-L1) inhibits UPS.

function is not clear yet. Also, Parkin ubiquitinylates itself and promotes its own degradation (Zhang et al., 2000). Whether there are more target proteins of E3 ligase Parkin, critical for the aggregation-associated pathogenesis of PD, remains to be further clarified.

a -Synuclein and UPS: Aggregation-Associated Pathogenesis

a*–Synuclein* gene encodes a pre-synaptic protein and is believed to be involved in neuronal plasticity. In the *Drosophila* and mice model of PD generated by wild type and mutant forms of a -synuclein, loss of dopaminergic neurons and formation of inclusions containing a -synuclein (Feany et al., 2000; Masliah et al., 2000) were observed. In these animal models, neuronal inclusions were associated with loss of dopaminergic terminals and contained a -synuclein and ubiquitin in the neocortex, hippocampus, and substantia nigra. In the sporadic disease, a -synuclein is also found with ubiquitin in Lewy bodies, suggesting that UPS is functionally associated with the pathogenesis of PD.

Then, a pathway in which a -synuclein induces formation of neuronal inclusion and loss of dopaminergic neurons needs to be uncovered. There are several possibilities. First, under stress condition, pathogenic protein itself becomes resistant to degradation mediated by UPS. There seems to be subtle difference in the UPS-mediated-degradation between wild type and mutant forms of a -synuclein: A53T a -synuclein mutant was degraded with slow kinetics, affecting its intracellular accumulation and favoring its aggregation (Bennett et al., 1999). Such accumulation of a -synuclein aggregates could also be triggered by extracellular stress (Giasson et al., 2000). Using antibodies specific to nitrated tyrosine residues in a -synuclein, accumulations of nitrated a -synuclein was detected in the signature inclusions of PD as well as in the insoluble fractions of the affected brain. Nitration may render a -synuclein more resistant to proteolysis or alter other properties of a -synuclein, thereby playing a role in the formation and/or stability of a -synuclein and the onset of a -synuclein pathogenesis. Second, pathogenic protein may directly inhibit PDD activity, leading to accumulations of intracellular neurotoxic proteins. In this view, Tanaka et al. (2002) observed that the forced expression of a -synuclein inhibited PDD. In addition, PDD activity may be indirectly modulated by ill-regulation of components of UPS. As a component of UPS, the reduced enzyme activity of ubiquitin carboxyl terminal hydrolase

(UCH-L1) cleaving ubiquitin-ubiquitin to produce monoubiquitin was found in UCH-L1 mutant obtained from PD patients (Wilkinson, 1997), presumably leading to decrease of ubiquitin-dependent proteolysis and accumulations of aggregations associated with PD. Thus, alterations in the degradation of a -synuclein may contribute to its aggregation or inhibition of UPS and the pathogenesis of PD, linking aggregation of a -synuclein to the onset or progression of PD (Figure 3). Mutation in *parkin* and a *-synuclein* gene, which are implicated in the pathogenesis of PD, apparently causes to favor increase of intracellular inclusions.

Regulated ubiquitinylation of critical proteins as a targeting signal for degradation by UPS is important and its ir-regulation contributes to pathogenesis of neurodegenerative diseases. Auluck et al. (2002) recently showed that chaperone molecules including HSP70 mitigated mutant a -synuclein toxicity in *Drosophilla* model, indicating proper control of folding and aggregation modulates associated pathogenesis. Thus, genetic disorder or extracellular stress affecting the normal UPS induces accumulation of mis-folded protein or toxic protein, leading to cellular toxicity for the development of associated disease.

Aggregation and Ubiquitin in Alzheimer's Disease Pathogenesis

AD is the most common dementia in neurodegenerative diseases. Progressive loss of cognitive function, the main characteristic in AD patients, occurs with pathological hallmarks including widespread neuronal degeneration, neurofibrillary tangle (NFT) and neuritic plaque containing ubiquitinylated and hyper-phosphorylated microtubule binding protein tau (Morishima-Kawashima et al., 1993) and amyloid-b (Ab) generated from proteolytic processing of APP by b -secretase (BACE) and g -secretase (reviewed by Yankner et al., 1996). The intra- and extra-cellularly accumulated deposits were estimated as causes to generate pathological processes in AD. For a long time, many researchers have focused on the role of their components in AD deposits and the roles were suspected to induce neuronal loss to promote cognitive impairment (reviewed by David et al., 2001).

One of the striking characteristics of the senile deposits is co-aggregation with ubiquitin and ubiquitin conjugates (Perry et al., 1987) but the pathogenic involvement of ubiquitin was unknown until 1998. What is the function of ubiquitin and ubiquitin conjugates accumulated in the deposits of AD brains, just byproducts of Ab and tau aggregations or neuronal death? If ubiquitin or ubiquitin conjugate is involved in neurodegeneration of AD, the therapeutic approaches have to be focused onto ubiquitin pathology as much as Ab peptide and tau, major components in senile plaque and NFT. The evidence of ubiquitin in AD pathogenesis, reported in 1998, was UBB+1, mutant form generated by molecular misreading, dinucleotide deletion of ubiquitin mRNA, found in early-onset AD patients (van Leewen et al., 1998). Recently we could meet some molecular mechanistic explanations about the neurotoxicity of UBB+1, suggesting the abrogation of UPS may be one of the causes of AD pathology (Lindsten et al., 2002). They showed that the UBB+1 mutant was resistant to PDD and its ubiquitin conjugated form inhibited PDD, which might promote neuronal death (Lindsten et al., 2002) (Figure 4). These evidences are strongly suggesting that the abrogation of PDD can be a cause of AD pathogenesis.

In speculation of the relationship between AD pathogenesis and PDD, we can meet this question: How can the ubiquitin conjugates be accumulated in AD? The amount of ubiquitin conjugates is regulated by PDD activity and the activity of ubiquitin ligase (UL) complex, E1, E2, and E3. If the activity of PDD is higher than UL activity, ubiquitin conjugates

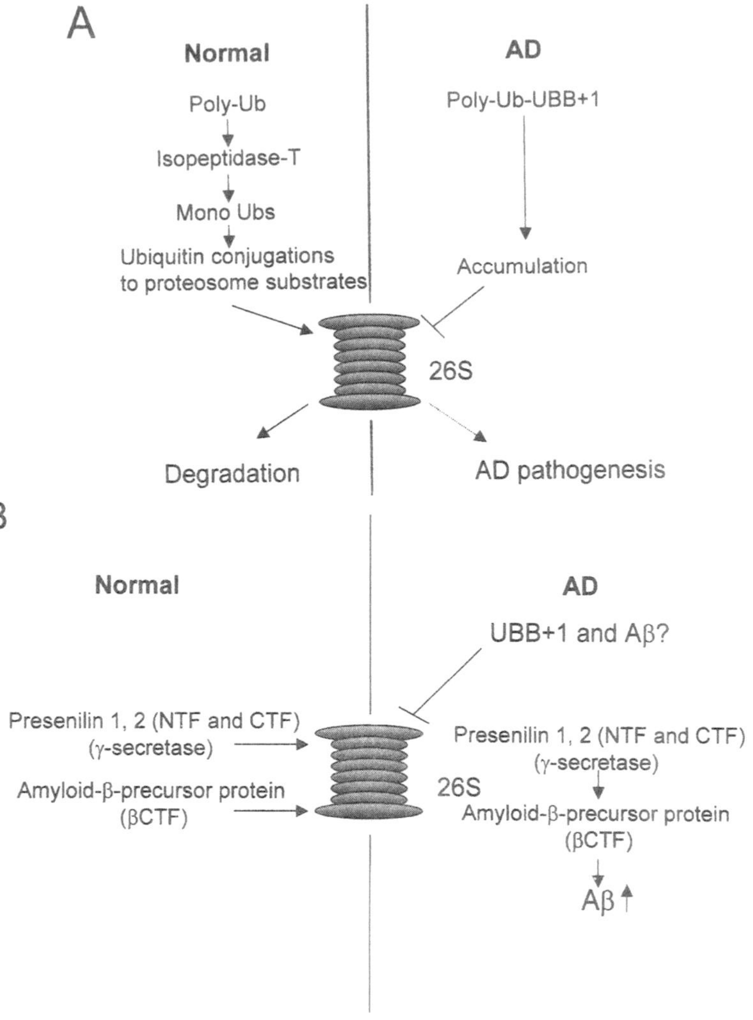

Figure 4. Role of ubiquitin mutant in AD pathogenesis. (A) Poly-ubiquitin is readily degraded by isopeptidase-T to mono-ubiquitin. UBB+1, mutant form of ubiquitin, is ubiquitinylated and the resulting poly-ubiquitin-UBB+1 becomes resistant to PDD. The ubiquitin conjugates of UBB+1 are able to bind to and inhibit 26S PDD activity *in vitro* (Lindsten et al., 2002). (B) Possible effects of PDD inhibition on the Ab generation in AD pathogenesis. PDD inhibition associated by poly-Ubiquitin-UBB+1 or low concentration of intracellular Ab may be a cause of AD pathogenesis by inhibiting degradation of AD gene products, presenilins and APP, and facilitating proteolytic production of Ab.

will be decreased and at inverse situations ubiquitin conjugates will be increased as you can expect. Interestingly, there were reports that PDD and UL activities were decreased in AD patients (Keller et al., 2000; Lopez et al., 2000). So, it can be expected that the reduction of PDD activity may precede that of UL activity in AD to generate accumulation of ubiquitin and ubiquitin conjugates, which was consistent with the expected mechanisms of UBB+1-mediated AD pathogenesis (Lindsten et al., 2002).

The most famous AD genes are presenilin (PS) and APP, which are actively degraded by proteasome (Fraser et al., 1998; Da Costa et al., 1999; Chen et al., 2002). Identification of UBB+1 in the AD brain and its prospective role in PDD inhibition may link UPS to the known pathogenic process of AD via the stabilization of APP, its b-cleavage product (Chen et al., 2002), and PS fragments, strongly correlated with g-secretase activity, which may increase Ab production and thus indirectly contributes to Ab neurotoxicity (Fraser et al., 1998; da Costa et al., 1999) (Figure 4B). Also, the major neurotoxic protein Ab, a model of neurodegeneration in AD, was known to suppress PDD activity *in vitro* (Gregori et al., 1997), suggesting that abrogation of PDD activity may be crucial in Ab neurotoxicity (Figure 4B).

Is UPS Involved in Ab Neurotoxicity?

Cognitive impairment, the main characteristic of AD, is known to be generated by neuronal loss induced by Ab. Up to date, much research has focused upon determining the mechanisms underlying the toxicity associated with Ab for the prevention of neuronal loss in AD. It has been reported that the genes, such as prostate apoptosis response-4 (Par-4), tau protein kinase 1/GSK-3b, and calsenilin may play important roles in neuronal degeneration, and the blockade of their functions using anti-sense approaches suppressed the neuronal death by Ab (Takashima et al., 1993; Guo et al., 1998; Jo et al., 2001). Though several regulatory genes mediating Ab neurotoxicity have been isolated, there is a missing link in the signaling cascade directly connecting accumulation and aggregation of ubiquitin adducts and pathogenic molecules in AD patients to the neuronal death triggered by Ab. Recently, Song et al. (2003) reported that Ab_{1-42} induced the accumulation of ubiquitin conjugates and inhibition of proteasome activity in neurons, which were mediated by an E2 ubiquitin-conjugating enzyme, E2-25K/Hip-2. Furthermore E2-25K/Hip-2 was proposed to be a proximal mediator of Ab neurotoxicity as an upstream of ASKI/JNK neurotoxic pathway. Whether the resulting aggregation caused by cellular indigestion contributes to the AD pathogenesis remains to be focused on in the future. Also, identification of such genes modulating PDD will be required to understand aggregation-associated pathogenesis and for therapeutic advances.

References

Auluck, P.K., Chan, H.Y.E., Trojanowski, J.Q., Lee, V.M.-Y., and Bonni, N.M. (2002). Science *295*, 865-868.

Bence, N.F., Sampat, R.M., and Kopito, R.R. (2001). Science *292*,1552–1555.

Bennett, M.C., Bishop, J.F., Leng, Y., Chock, P.B., Chase, T.N., and Mouradian, M.M. (1999). J. Biol. Chem. *274*, 33855–33858.

Chen, Q., Kimura, H., and Schubert, D. (2002). J. Cell. Biol. *158*, 79–89.

Chung, K.K., Zhang, Y., Lim, K.L., Tanaka, Y., Huang, H., Gao, J., Ross, C.A., Dawson, V.L., and Dawson, T.M. (2001). Nat. Med. *10*, 1108–1109.

Cuppers, P., Orlans, I., Craessaerts,K., Annaert, W., and De Strooper, B. (2001). J. Neurochem. *78*, 1168–1178.

Da Costa, C.A., Ancolio, K., and Checler, F. (1999). Mol. Med. *5*, 160–168.

David, H.S., Mok, S.S., and Bornstein, J.C. (2001). Nat. Rev. Neurosci. *2*, 595–598.

Ding, Q., Lewis, J.J., Strum, K.M., Dimayuga, E., Bruce-Keller, A.J., Dunn, J.C., and Keller, J.N. (2002). J. Biol. Chem. *277*, 13935–13942.

Feany, M.B., and Bender, W.W. (2000). Nature *404*, 394–398.

Fraser, P.E., Levesque, G., Yu, G., Mills, L.R., Thirlwell, J., Frantseva, M., Gandy, S.E., Seeger, M., Carlen, P.L., and St George-Hyslop, P. (1998). Neurobiol. Aging *19*, S19–S21.

Giasson, B.I., Duda, J.E., Murray, Ian. V.J., Chen, Q., Souza, J.M., Hurtig, H.I., Ischiropoulos, H., Trojanowski, J.Q., and Lee, V. M.-Y. (2000). Science *290*, 985–988.

Guo, Q., Fu, W., Xie, J., Luo, H., Sells, S.F., Geddes, J.W., Bondada, V., Rangnekar, V.M., and Mattson, M.P. (1998). Nat. Med. *4*, 957–962.
Gregori, L., Hainfeld, J.F., Simon, M.N., and Goldgaber, D. (1997). J. Biol. Chem. *272*, 58–62.
Hershko, A., and Ciechanover, A. (1998). Annu. Rev. Biochem. *67*, 425–479.
Imai, Y., Soda, M., Inoue, H., Hatto, N., Mizuno, Y., and Takahashi, R. (2001). Cell *105*, 891–902.
Jesenberger, V., and Jentsch, S. (2002). Nat. Rev. Mol. Cell Biol. *3*, 112–121.
Jin, T., Gu, Y., Zanusso, G., Sy, M., Kumar, A., Cohen, M., Gambetti, P., and Singh, N. (2000). J. Biol. Chem. *275*, 38699–38704.
Jo, D.G., Kim, M.J., Choi, Y.H., Kim, I.K., Song, Y.H., Woo, H.N., Chung, C.W., and Jung, Y.K. (2001). FASEB J. *15*, 589–591.
Keller, J.N., Hanni, K.B., and Markesbery, W.R. (2000). J. Neurochem. *75*, 436–439.
Kopito, R.R., and Sitia, R. (2000). EMBO Rep. *3*, 225–231.
Lee, M.K., Stirling, W., Xu, Y., Xu, X., Qui, D., Mandir, A.S., Dawson, T.M., Copeland, N.G., Jenkins, N.A., and Price, D.L. (2002). Proc. Natl. Acad. Sci. USA *99*, 8968–8973.
Layfield, R., Alban, A., Mayer, R.J., and Lowe, J. (2001). Neuropathol. Appl. Neurobiol. *27*, 171–179.
Lindsten, K., de Vrij, F.M., Verhoef, L.G., Fischer, D.F., van Leeuwen, F.W., Hol, E.M., Masucci, M.G., and Dantuma, N.P. (2002). J. Cell Biol. *157*, 417–427.
Lopez Salon, M., Morelli, L., Castano, E.M., Soto, E.F., and Pasquini, J.M. (2000). J. Neurosci. Res. *62*, 302–310.
Lorick, K.L., Jensen, J.P., Fang, S., Ong, A.M., Hatakeyama, S., and Weissman, A.M. (1999). Proc. Natl. Acad. Sci. USA *96*, 11364–11369.
Lucking, C.B., Durr, A., Bonifati, V., Vaughan, J., Michele, G.D., Gasser, T., Harhangi, B.S., Meco, G., Denefle, P., Wood, N.W., et al. (2000). New Eng. J. Med. *342*, 1560–1567.
Masliah, E., Rockenstein, E., Veinbergs, I., Mallory, M., Hashimoto, M., Takeda, A., Sagara, Y., Sisk, A., and Mucke, L. (2000). Science *287*, 1265–1268.
McNaught, K.S., Bjorklund, L.M., Belizaire, R., Isacson, O., Jenner, P., and Olanow, C.W. (2002). Neuroreport *13*, 1437–1441.
Morishima-Kawashima, M., Hasegawa, M., and Fang, P. (1993). Neuron *10*, 1151–1160.
Nishitoh, H., Matsuzawa, A., Tobiume, K., Saegusa, K., Takeda, K., Inoue, K., Hori, S., Kakizuka, A., and Ichijo, H. (2002). Genes Dev. *16*, 1345–5135.
Perry, G., Friedman, R., Shaw, G., and Chau, V. (1987). Proc. Natl. Acad. Sci. USA *84*, 3033–3036.
Polymeropoulos, M.H., Lavedan, C., Leroy, E., Ide, S.E., Dehejia, A., Dutra, A., Pike, B., Root, H., Rubenstein, J., Boter, R., Strenroos, E.D., Chandrasekharappa, S., Athanassiadou, A., Papapetrropoulous, T., et al. (1997). Science *276*, 2045–2047.
Schwartz, A.L., and Ciechanover, A. (1999). Annu. Rev. Med. *50*, 57–74.
Song, S., Kim, S.Y., Hong, Y.M., Jo, D.G., Lee, J.Y., Shim, S.M., Chung, C.W., Seo, S.J., Yoo, Y.J., Koh, J.Y., Lee, M.C., Yates, A.J., Ichijo, H., and Jung, Y.K. (2003). Mol. Cell *12*, *In press*.
Takashima, A., Noguchi, K., Sato, K., Hoshino, T., and Imahori, K. (1993). Proc. Natl. Acad. Sci. USA *90*, 7789–7793.
Tanaka, Y., Engelender, S., Igarashi, S., Rao, R.K., Wanner, T., Tanzi, R.E., Sawa, A., L. Dawson, V., Dawson, T.M., and Ross, C.A. (2001). Hum. Mol. Genet. *10*, 919–926.
Taylor, J.P., Hardy, J., and Fischbeck, K.H. (2002). Science *296*, 1991–1995.
Van Leeuwen, F.W., de Kleijn, D.P., van den Hurk, H.H., Neubauer, A., Sonnemans, M.A., Sluijs, J.A., Koycu, S., Ramdjielal, R.D., Salehi, A., Martens, G.J., et al. (1998). Science *279*, 242–7.
Wilkinson, K.D. (1997). FASEB J. *11*, 1245–1256.
Wyttenbach, A., Carmichael, J., Swartz, J., Furlong, R.A., Narain, Y., Rankin, J., and Rubinsztein, D.C. (2000). Proc. Natl. Acad. Sci. USA *97*, 2898–2903.
Yankner, B.A. (1996). Neuron *16*, 921–932.
Zhou, H., Li, S.H., and Li, X.J. (2001). J. Biol. Chem. *276*, 48417–48424.
Zhang, Y., Chung, K.K., Huang, H., Dawson, V., and Dawson, T.D. (2000). Proc. Natl. Acad. Sci. USA. *97*, 13354–13359.

Chapter 17

The Mechanism of Apoptosis Regulation by IAP Antagonist Smac/DIABLO

JUN JIN, JIANXIN DAI, JIAN ZHAO, AND YAJUN GUO*

ABSTRACT: Caspases are central components of the machinery responsible for cell apoptosis. The inhibitor of apoptosis proteins (IAPs) could efficiently block the caspases activation. Recently, a novel cell death regulator Smac/DIABLO (second mitochondria derived activator of caspases / Direct IAP Binding protein with low PI) was identified. Smac/DIABLO performs a critical function in apoptosis by eliminating the inhibition of IAPs. This article will firstly review the role of caspases and IAPs in apoptosis and then focus on the mechanism of apoptosis regulation by IAP antagonist Smac. We will discuss what is currently known about Smac/DIABLO such as the structure and function of Smac/DIABLO in apoptosis; its relation with Bcl-2 family proteins as well as its potenial application in cancer therapy.

Key Words: Apoptosis, Smac/DIABLO, Caspases; IAP

Introduction

Apoptosis or programmed cell death is an essential physiological process that plays a critical role in development and tissue homeostasis. However, apoptosis is also involved in a wide range of pathological conditions. Apoptotic cells may be characterized by specifically morphological and biochemical changes, including cell shrinkage, chromatin condensation, and internucleosomal cleavage of genomic DNA. At the molecular level, apoptosis is tightly regulated and is mainly orchestrated by the activation of the aspartate-specific cysteine protease (caspase) cascade. Caspases are expressed in cells as inactive precursors, which are activated by proteolytic processing. Two classes of caspases, initiators and effectors, are involved in mammalian cell apoptosis. Activated initiator caspases, such as caspase 8 and caspase 9, cleave the precursor forms of effector caspases, such as caspases 3, 6, and 7. Activated effector caspases in turn cleave a specific set of cellular substrates resulting in the biochemical and morphological changes associated with the apoptotic phenotype. The activation of initiator caspases is thought to irreversibly trigger the caspase cascade; necessitating that caspase activation is tightly regulated by layered control mechanisms. Among the growing number of cellular proteins that have been shown to regulate caspase

*Shanghai International Joint Cancer Institute, The Second Military Medical University, Shanghai, 200433 China, and Eppley Cancer Institute, Omaha, NE, 68198 USA. *Correspondence:* Dr. Yajun Guo, International Joint Cancer Institute, 800 Xiang Yin Road, Shanghai 200433, People's Republic of China. Tel: 0086-21-25070241; Fax: 0086-21-25074349; E-mail: yguo@unmc.edu activation and activity are the IAPs, including c-IAP1, c-IAP2, XIAP, and survivin. These proteins have been reported to block both death receptor- and mitochondrially-mediated apoptotic pathways by directly inhibiting initiator and effector caspases (Ellis, et al., 1991; Salvesen, G. S., and Dixit, V. M. 1997; Thornberry, N. A., and Lazebnik, Y. 1998; Verhagen, A. M. et al., 2001).

Smac/DIABLO, a mitochondrial protein released into the cytosol in response to apoptotic stimuli, was recently found to promote caspase activation by eliminating IAP function. Smac binds to most known human IAP family members and relieves their inhibition of caspase activity. The 20 N-terminal amino acids of the mature Smac protein are crucial for Smac-IAP interaction, and removal of this region completely abrogates its ability to bind to XIAP. Since Smac blocks IAP activity, it has been proposed that Smac is a mammalian functional homologue of the *Drosophila* proapoptotic proteins Reaper, Grim, and Hid. (Du, C. et al., 2000; Verhagen, A. M. et al., 2000; Goyal, L. et al., 2000). This review will firstly review the role of caspases and IAPs in apoptosis and then focus on the mechanism of apoptosis regulation by IAP antagonist Smac.

Caspases and Apoptosis

The critical involvement of a cysteine protease CED-3 in apoptosis was first discovered in the nematode *Caenorhabditis elegans* 9 years ago (Yuan et al., 1993). Since then, compelling evidence has demonstrated that apoptosis is executed by a family of cysteine proteases, named as caspases. Caspase stands for cysteine-dependent aspartate specific protease (Alnemri, E.S., et al., 1996), and is a term coined to define proteases that belong to the family C14 in the Barrett and Rawlings classification (Rawlings, N.D., 1998). Their enzymatic properties are governed by a dominant specificity for substrates containing Asp, and by the use of a Cys side chain for catalyzing peptide bond cleavage. Although the first mammalian caspase, caspase-1 or ICE (interleukin1b-converting enzyme), was identified to be an important regulator of inflammatory response, at least 8 of 14 caspases play important roles during apoptosis (Budihardjo et al., 1999; Earnshaw et al., 1999; Fesik and Shi, 2001; Salvesen and Dixit, 1997; Thornberry and Lazebnik, 1998).

The Characteristic Structure of Caspases

Caspases are synthesized as inactive proenzymes, which are activated by cleavage at specific Asp residues to active enzymes containing both large (p20) and small (p10) subunits. In some cases these subunits are separated by a linker region, which may be involved in regulation of the activation of the caspase. All caspases contain an active-site pentapeptide of general structure QACXG (where X is R, Q or G). The amino acids Cys-285 and His-237 involved in catalysis, and those involved in forming the P1 carboxylate binding pocket in caspase-1 (Arg-179, Gln-283, Arg-341 and Ser-347), are conserved in all other caspases, except for caspase-8. However, the residues that form the P_2-P_4 binding pocket are not well conserved, suggesting that they may determine the substrate specificities of different caspases.

The N-peptide of caspases can vary in length from 22 amino acid residues in caspase 6 to over 200 in caspases 8 and 10. Embedded within the longer N-peptides of the caspases are distinct domains that are required for the activation of the caspases. There are two different types of elements known as the death effector domain (DED domain) and the caspase recruitment domain (CARD domain). The DED domains are found exclusively in caspases 8 and 10, which are associated with death receptor initiated cell death and essential for the recruitment of these caspases to the receptor via specific adapter molecules. The role of the CARD domain is more promiscuous since it is found in caspases involved in cytokine activation as well as those that trigger cell death. Thus, caspase 2 and 9 as well as caspases 1, 4, 5 and 13 all contain CARD domains. Both the DED and the CARD domains are 6-helix bundles with primarily hydrophobic (DED) or hydrophilic/charged (CARD) side chains positioned to accept complementary surfaces on the respective partners. Thus, caspase 8 DED is thought to bind to a complementary surface on the DED of its adapter FADD (Fas-associated death domain), and caspase 9 CARD binds to a complementary surface on the Apaf-1(apoptotic protease activating factor-1) CARD (Qin, H. et al., 1999; Eberstadt, B. et al., 1998). These specific protein recruitment events deliver the specific singals required to drive the extrinsic and intrinsic activation processes.

The Function of the Caspases in Apoptosis

When the caspases are activated they cleave a wide variety of proteins, including certain key substrates in the cell, and kill the cell via apoptosis. The function of some of the caspases is called "execution," and the caspases that do this are the executioner caspases. Caspases reside in cells as inactive proforms, and in most cases can be activated by proteolytic cleavage at two sites; one between the large and small subunits of what will be the mature enzyme and one that removes the prodomain. This occurs in two ways. The first is through the action of another proteinase (usually a caspase) to mediate the cleavage. This is the major way of activating executioner caspases. The second is through the action of adaptor proteins that bind protein interaction motifs in the prodomains of "initiator" caspases. This adaptor binding promotes the activation of the initiator caspases, which then go on to cleave and activate the executioner caspases. These adaptor molecules link signaling events to caspase activation, and define two basic pathways for apoptosis—the mitochondrial pathway and the death receptor pathway.

Caspases involved in apoptosis are generally divided into two categories, the initiator caspases, which include caspase-2, -8, -9, and -10, and the effector caspases, which include caspase-3, -6, and -7. An initiator caspase is characterized by an extended N-terminal prodomain (>90 amino acids) important for its function, whereas an effector caspase contains 20–30 residues in its prodomain sequence. All caspases are produced in cells as catalytically inactive zymogens and must undergo proteolytic activation during apoptosis. The activation of an effector caspase (such as caspase-3 or -7) is performed by an initiator caspase (such as caspase-9) through cleavage at specific internal Asp residues that separate the large and small subunits. The initiator caspases are activated with the assembly of a multicomponent complex under apoptotic conditions. For example, the activation of procaspase-9 is facilitated by Apaf-1 and cytochrome c (Cyt.-c) (Li, et al., 1997; Li, F. et al., 1998). After that, a cascade of downstream caspases activation is tiggered.

Once activated, the effector caspases are responsible for the proteolytic cleavage of a broad spectrum of cellular targets, leading ultimately to cell death. The known cellular substrates include structural components (such as actin and nuclear lamin), regulatory proteins (such as DNA-dependent protein kinase), inhibitors of deoxyribonuclease (such as DFF45 or ICAD), and other proapoptotic proteins and caspases. Cleavage of the

DFF45 (DNA fragmentation factor 45) leads to removal of its inhibition of DFF40 or CAD (caspase-activated deoxyribonuclease), which degrades the chromosomes into nucleosomal fragments during apoptosis (Enari, et al., 1998; Liu, et al., 1997, 1998; Liu, X. et al., 1998).

Currently there exist three recognized points at which apical proteases are activated to initiate apoptosis, namely caspase-8, 9 and 10. Caspase-8, also know as FLICE, MACH, and Mch5, is an intrcellular cysteine protease that exists as a proenzyme. It is regarded as a typical initiator caspases associated with CD95- or TNF-induced death receptor apoptosis. CD95 (a cell death receptor) uses FADD (an adapter molecule) to interact physically with caspase-8 (a cytosolic protease) and initiate the apoptotic cascade. The precise mechanism by which recruitment of caspase-8 results in its activation is not known. It has been suggested that, in the latent state, the two DEDs of caspase-8 bind to each other, so preventing activation. The binding of FADD following the triggering of apoptosis by CD95 or TNF may cause a conformational change in the DED of FADD, so facilitating its binding to one of the DEDs of caspase-8, thereby disrupting the association of the two DEDs of caspase-8 and allowing its caspase domain to undergo autocatalytic activation (Boldin, M.P. et al., 1996; Muzio, M. et al., 1996).

Caspase 9, know as ICE-LAP6 or Mch6, is an important upstream proenzyme in the cascade of enzymatic reactions required to induce cellular apoptosis. It mainly takes part in the mtrochodrial apoptosis pathway. After the release of mitrochodrial cytochrome C, cappase-9 is actived following its association with the protein complex of Apaf1 and cytochrome C. Active caspase-9 in turn activates pro-caspase-3 promoting the manifestation of some of more classical features of apoptosis. Activation of caspase-9 can be regulated through protein phosphorylation events. The optimal cleavage recognition sequance for caspase is LEHD. The ability of caspase-9 to associate with proapoptotic and anti-apoptotic proteins as well as its regulatory function during embryonic development suggests that caspase-9 is a key regulator of apoptosis in vivo. (li, P. et al., 1997; Cardone, M.H. et al., 1998; Thornberry, N.A. et al., 1997).

Caspase-10 is an intracellular cysteine protease that exists as a proenzyme activated during the cascade of events associated with apoptosis. Four isoforms of caspase-10 have been described: caspase-10/a (Mch-4), caspase-10/b (FLICE2), caspase-10/c, and caspase-10/d. Caspase-10 is a group III caspase highly homologous to caspase-8. Caspase-10 contains two death effector domains and can be recruited along with caspase-8 into apoptosis signaling complexes associated with TNFR-like death receptors including Fas (Fernandes, A.T. et al., 1996; Vincenz, C. and Dixit, V.M. 1997). Caspse 10 is an apical initiator caspase in DR signaling but with potentially different substrate specificiatics or function. Purified recombinant caspase-10 processes all caspases, including pro-caspases-3, -7 and -10, whereas neither caspase-3 nor caspase-7 cleaves pro-caspase-10, suggesting that the latter is upstream of both caspase-3 and caspase-7 and lies at or near the apex of a cascade of proteases. There is one interesting unusual cascade that mainly runs through caspase 10 with little effect of caspase-8 and 3. (Sprick, M.R. et al., 2002; Millet, A. et al., 2002). This function of caspase 10 is also proved in our current research about melanoma specific monoclonal antibody SM5-1 induced apoptosis.

The death signal transduction route may be divided into intrinsic and extrinsic pathways According to whether the apoptotic signal originates from the inside or the outside of the cell. The extrinsic pathways include delivery of granzyme B to the cells as well as receptor ligation. Following TNFR-1 or Fas receptor ligation, the initiator caspase 8 or 10 is activated by adapter-mediated recruitment to the receptor's cytosolic face (Muzio, M. et al., 1998;

Yang, X. et al., 1998; Peter, M.E. et al., 1998). Alternatively in the intrinsic pathway, the initiator caspase 9 is activated following release of mitochondrial components to form the Apaf complex (Li, P. et al., 1997; Zou, H. et al., 1997; Liu, X. et al., 1996). Both activated initiators converge on the proteolytic activation of caspase 3.

The most prevalent caspase in the cell is caspase-3. It is the one ultimately responsible for the majority of the apoptotic effects, although it is supported by two other caspases (caspase-6 and 7). Together, these three executioner caspases might cause the apoptotic phenotype by cleavage or degradation of several important substrates. For example, DNA fragmentation is caused by the action of caspase-3 on a complex of caspase-activated DNase (CAD)/DNA fragmentation factor (DFF) 40 and iCAD/DFF45 (Enari et al., 1998; Liu et al., 1997). In non-apoptotic cells, CAD (caspase activated deoxyribonuclease) is present as an inactive complex with iCAD. During apoptosis, caspase-3 cleaves the inhibitor, allowing the nuclease to cut the chromatin. Meanwhile, gelsolin and fodrin are cleaved and activated, and the plasma membrane is dissociated from the cytoskeleton, which causes blebbing (Kothakota ct al., 1997; Martin et al., 1995).

In conclusion, caspases play a central role in the execution of apoptosis. The key components of the pathway of caspase activations have been elucidated recently. Regulation of caspase activity can take a great effect on cell cycle and programmed cell death.

The Inhibitors of Apoptosis Proteins (IAP) Family

The IAP gene was first identified in baculovirus because it protected infected cells from death and enhanced virus replication (Crook, N.E. et al., 1993). IAPs are now known to be a wildly-expressed family of apoptotic inhibitors. To date, three IAPs have been discovered in virus (CpIAP, OpIAP and AcIAP), two in Drosophila (DIAP1 and DIAP2), and six in human (XIAP, cIAP1, cIAP2, NAIP, Survivin and BRUCE). Most of them have conserved domains and have associated biological activities of apoptosis suppression by binding and inhibiting caspases.

The Characteristic Structure of IAPs

IAPs family proteins are characterized by two defining motifs. The first of these is the BIR (Baculoviral Inhibitor of apoptosis Repeat) domain. The approximately 70 amino acid BIR domain, which is characterized by the consensus sequences: RX2TF, DX3CX2C and HX6C, contains four alpha helices linked by several loops. In the core of this BIR, there are extensive interactions between the large number of hydrophobic residues and a zinc atom coordinated by three cysteines and a histidine. On the surface of the BIR domain, a number of hydrophobic regions and conserved charged amino acids have been identified, which may participate in interactions of IAPs with other proteins. Deletion mutation confirms that the conserved BIR domains are necessary for inhibition of caspase activity and apoptosis (Sun, C. et al., 1999). The second defining motif found in the IAPs is RING finger zinc-binding domain C terminal to the BIR repeats. It contains a $CX_2CX_NCXHX_{2-3}CX_2CXMCX_2C$ motif, which forms a spherical cross brace structure that can coordinate two Zn atoms (Borden, K.L. 2000). The RING Zn finger has been known to be involved in DNA and protein interactions (Borden and Freemont, 1996). The presence of the RING Zn finger appears to be critical to the baculoviral IAP anti-apoptotic function.

Among the mammalian IAPs, c-IAP1, cIAP2, and XIAP have three BIRs in the N-terminal portion of the molecule and a RING finger at the C-terminus. NAIP contains three BIRs without RING. However survivin and BRUCE have just one BIR for each.

Mechanism of Caspase Inhibition by IAPs

Over-expression of IAPs renders the cell resistant to a wide varity of apoptotic stimuli. Structure-function studies reveal that BIR domains are required for their protective activity. Mammalian XIAP, c-IAP1, and c-IAP2 contain three BIR domains. The second BIR is able to bind and inhibit caspase-3 and -7 with the constant of 2–5 n*M* (Takahashi, R. et al., 1998). Additionally, it has been shown that several residues in the linker region N-terminal to the BIR2 domain of XIAP are critical for inhibiting caspase-3, even though they do not affect the structure of the BIR. A chimerical molecule containing this linker region and XIAP BIR1 is able to inhibit caspase-3, although the linker region or BIR1 alone cannot inhibit the caspase. The XIAP BIR2 domain interacts with both pro-caspase-3 and activated caspase-3, whereas the linker region only interacts with activated caspase. Since caspase-3 acts on consensus sequence DEVD, it has been suggested that the DxxD sequence in this linker region directly interacts with the active center of caspase-3 and blocks its activity (Sun, C. et al., 1999). Further studies reveal the structure of caspase-3 or 7 in complex with BIR2 domain and flanking sequences (Chai, J. et al., 2001; Huang, Y. et al., 2001; Riedl, S. et al., 2001). The XIAP BIR2/caspase complex consists of two heterodimers of an active caspase and two BIR2 domains. The catalytic site of each active caspase heterodimer is occupied by one BIR2 molecule. The N-terminal linker region of XIAP BIR2 contributes to the critical hydrogen bonds and hydrophobic interactions responsible for binding to the active site. The linker fills the substrate-binding site in an elongated manner and interacts with the catalytic groove. Residues 141, 147, 148, and 149 of XIAP are key determinants in the structure and form specific hydrogen bonds and hydrophobic interactions which is essential for the inhibition of binding.

In contrast, BIR3-RING fragment, but not BIR1-BIR2 fragment of XIAP, binds and inhibits caspase-9 (Deveraux, Q.L. et al., 1999). This suggests that the mechanism of caspase-9 inhibition by the BIR3 domain of XIAP is distinct from caspase-3 and -7 inhibition by XIAP BIR2 (Sun, C.H. et al., 2000). Recent study demonstrates that the initial proteolytic processing of caspase-9 at Asp315 exposes a peptide at the N-terminus of the caspase. The availability of this peptide allows free XIAP to bind cleaved caspase-9, thereby obstructing substrate to entry into the active site of the enzyme (Srinivasula, S. et al., 2001). This four amino acid peptide (ATPF), also presented in Smac/DIABLO (AVPI), determines the ability of this protein to bind IAPs. Normally XIAP BIR3, which is specific for caspase-9, is unable to inhibit caspase-3. However, engineering the four amino acid peptide into the N terminus of the caspase-3 small subunit, allowed binding and inhibition by BIR3, suggesting that this four amino acid IAP binding motif is required for inhibitory functioning by BIR3. Furthermore, the mutation of a single residue (E314) in the BIR3 domain that is important for Smac/DIABLO–IAP interaction abrogates the binding of the small subunit of caspase-9 to XIAP (Liu, Z. et al., 2000). This provides additional evidence that binding and inhibition of caspase-9 requires an intact BIR3 domain.

Caspase activation plays a key role in the execution of apoptosis. Mostly, two major pathways of caspase activation are involved, the cell surface death receptor pathway and the mitochondria initiated pathway respectively. IAPs' function can affect both major pathways. In the cell surface death receptor pathway, although IAPs do not bind or inhibit caspase-8, they do bind to and inhibit its substrate caspase-3 and 7. Thus they can arrest the cascade of proteolysis and provide protection from fas/caspase-8-induced apoptosis (Deveraux, Q.L. et al., 1997). In the mitochondrial-initiated pathway, IAPs can block the death signal by different mechanisms. They inhibit the initiator caspase pro-caspase-9 and prevent its processing and activation (Deveraux, Q.L. et al., 1998). More importantly, they bind directly to the effector caspase-3 and 7(Li, P. et al., 1997). IAPs family proteins can suppress apoptosis efficiently.

IAP Family Gene Expression and Cancer

IAPs expression can be detected in many normal tissues. *XIAP* mRNA was observed in all adult and fetal tissues except peripheral blood leukocytes, indicating that it is a ubiquitously expressed member of the family (Liston, P. et al., 1996). Highest expression of c-IAP1 and c-IAP2 is observed in the kidney, small intestine, liver, and lung. In contrast, the central nervous system has a very weak expression. *NAIP* mRNA appears to be only in adult liver and in placenta (Roy, N. et al., 1995). Survivin exhibits the most restricted expression of an IAP family member in adult tissues. Survivin mRNA is found only in occasional normal adult human or mouse tissues (Ambrosini, G. et al., 1997). Cell-cycle regulation and programmed cell death are associated with the function of these IAPs expressed.

In tumor, the most well studied member of IAP family is survivin. Unlike other IAP family members, survivin is expressed in most cancers cells (lung, colon, breast, prostate, pancreas, high-grade lymphomas, neuroblastomas, gastric). Further studies indicate that expression of survivin correlates with advanced worsening stage or unfavourable histology for neuroblastoma (Adida, C. et al., 1998.), and also correlated with decreased apoptotic index in gastric cancer (Lu, C.D. et al., 1998) or colon cancer (Kawasaki, H. et al., 1998). Over-expressed IAPs promotes tumor development and resistance to drug and radiation therapy. Recent studies also found that over expression of survivin is contributed to shortened 5-year disease survival rates (a 10% decrease compared to survivin negative patients). In addition, Survivin expression correlates with decreased cell apoptosis index and bcl-2 expression in colon cancer. In these patients, 5-year disease survival rates are 20% lower than the patients with a low apoptotic index (statistically significant). Suppressions of Survivin expression by using antisense cause apoptosis and sensitization to anticancer drugs. These studies suggest that the IAPs apparently provide a safeguard mechanism against minimal activation of the apoptosis and set up an endogenous threshold level for caspase activation. The cellular level of IAPs may determine the difference in sensitivities to apoptosis-inducing stimuli in various cell types. For these reasons, the regulation of IAPs levels becomes an important issue in apoptosis.

The Mammalian IAP Antagonist Diablo/Smac

Diablo/Smac was identified using two different approaches. Du and colleagues noted the presence of a factor in mitochondrial extracts that, together with cytochrome-c and dATP, was able to enhance the activation of caspases in cytosolic extracts, hence the name second mitochondria derived activator of caspases (Smac) (Du, C.Y. et al., 2000). This protein was purified from solubilised membrane extracts and sequenced by Edman degradation.

Diablo (Direct IAP Binding protein with low PI) was independently identified as a protein co-immunoprecipitating with mammalian IAP homologue A/X-linked IAP (MIHA/XIAP) and was shown to antagonise inhibition of UV induced apoptosis by MIHA (Verhagen, A.M. et al., 2000) These studies demonstrated that Diablo/Smac promotes cell death by interacting with IAPs and preventing their inhibition of caspases.

The Structural Basis for Binding of Smac to IAPs

The full length of smac cDNA has 1358bp encoding a 239-amino-acid precursor peptide. The mature form of Smac contains 184 amino acids and behaves as a monomer in solution. A helical wheel analysis revealed that N-terminal region of Smac resembles a typical mitochondrial targeting signal sequence: an amphipathic a helix with positively charged amino acid sidechains on one side (Arg-10, Arg-17, Arg-19, Lys-31, Lys-32, Arg-33, and Arg-40) (Schatz and Dobberstein, 1996). Direct sequencing analysis identified that the N terminus of mature Smac is amino acid 56. Since the 25kDa recombinant Smac expressed in Sf-21 cells is fully active, amino acids 1–55 likely represent mitochondrial targeting signal peptide that is subsequently cleaved after mitochondrial import. Indeed, this polypeptide was sufficient to target green fluorescent protein (GFP) to mitochondria when fused to the N terminus of GFP.

All members of the IAP family contain at least one BIR (baculoviral IAP repeat) motif and many contain three. Different BIR domains may exhibit distinct functions. The second BIR domain (BIR2) of XIAP is a potent inhibitor for caspase-3, whereas the third BIR domain of XIAP (XIAP-BIR3) primarily targets the active caspase-9. The crystal structure of Smac/DIABLO reveals that it homodimerizes through an extensive hydrophobic interface. The wild-type Smac protein (residues 1–184) forms a dimer in solution and interacts with both the BIR2 and BIR3 domains of XIAP. Point mutations at the dimeric interface, such as Phe33 → Asp, completely inactivate dimer formation. Such monomeric mutants can no longer interact with XIAP-BIR2, but they retain binding to the BIR3 domain. A missense mutation of the N-terminal Ala residue to Met (Ala1 → Met) in wild-type Smac abolishes binding to both the BIR2 and BIR3 domains of XIAP and results in complete loss of Smac function. In support of its critical role, a short 7-residue peptide derived from the Smac N-terminus was able to promote the activation of procaspase-3, raising the possibility of using small peptides or peptide mimics to treat cancer cells.

The amino acids of the BIR3 domain are critical for binding to the Smac peptide by testing mutant proteins. As expected, proteins containing alanine mutations of residues distant from the Smac binding site (D315, E318, H343) retain wild-type binding affinity. In contrast, when the hydrophobic amino acids that define the A1' binding pocket (L307, W310) or the P3' interaction site (W323) were mutated, a significant loss in binding to the Smac peptide was observed. Furthermore, the E314S mutant BIR3 domain lost almost all of binding abilites, which is consistent with an electrostatic interaction between E314 and the N-terminal amine of Smac. Despite its close proximity to the Smac peptide, mutation of Q319 caused only a slight loss in binding affinity. Another mutant, D296A, binds with a ten-fold low affinity to the Smac peptide and yet is not directly involved in binding to the Smac peptide. This can be explained by the importance of the D296 carboxyl group in stabilizing the structure of the binding site through interactions with the backbone NH of D309.

On the basis of primary sequence alignments, Smac interacts with IAPs in the same manner as the XIAP BIR3/Smac peptide complex. Not only are the residues in the Smac binding site conserved (for example, L307, W310, E314, W323), but the three-dimensional structures in this region of the BIR3 domain of XIAP—and of the other structurally characterized IAPs that bind to Smac—are also similar. Some of these structural differences may be due to conformational changes upon complexation. There are also a few differences in the amino acids that form the Smac binding site in these proteins that probably confer binding specificity. For example, the XIAP BIR2 domain contains bulkier amino acid residues compared to the XIAP BIR3 domain (K206 versus G306, and K208 versus T308) in the region that interacts with P3 and P4. This may explain the differences in binding affinities observed for the P3A and P4A peptides to the XIAP BIR2 and BIR3 domains. Functional homologues of Smac (such as Reaper, Hid, Grim), found in lower organisms, may also bind to IAPs in a similar fashion, on the basis of their sequence homology at the N terminus which is the critical portion of these proteins for interacting with IAPs. Although a Hid peptide with an N-terminal methionine did not bind to the XIAP BIR2 or BIR3 domain, tight binding of a Hid peptide was observed when the N-terminal methionine was removed which probably occurs to the native protein during normal processing in *Drosophila*. (Chai, J. et al., 2000; Liu, Z. et al., 2000; Wu, G. et al., 2000).

Smac Functions in Cell Apoptosis

The Smac/DIABLO function in cell apoptosis is briefly showed in figure1. Smac/DIABLO antagonizes the function of IAPs both of the initiator Apaf-1-caspase-9 and the effector: caspase-3 and 7. At the initiator level, the exposed BIR3 can bind to the IAP binding motif of caspase-9 and hinder the entry of substrate to the active site of the enzyme. Smac/DIABLO competes with caspase-9 for binding to the same pocket of the BIR domain of XIAP via the common IAP binding motif (Srinivasula et al., 2001) so that the cell undergoes apoptosis. By contrast, at the effector level, the linker region N-terminal to the BIR2 domain of XIAP is a major structural determinant in binding to the active form of caspase-3 and -7. This complex may be destabilized by the interaction of Smac/DIABLO with BIR2 due to a conformation change. Release of IAP complexed with Smac/DIABLO would relieve the inhibition on the caspase and the active caspase is then free to induce cell death.

The biological functions of Smac/DIABLO in apoptosis have been proved by many studies. For instance, Srinivasula S.M. reported the expression of a cytosolic Smac/DIABLO in cells allowed TRAIL to bypass Bcl-xL inhibition of death receptor-induced apoptosis (Srinivasula, S.M. et al., 2000); A.M. Verhagen and D.L. Vaux found that Diablo/Smac was very efficient at reversing IAP inhibition of UV induced cell death. (Verhagen, A.M. and Vaux, D.L., 2002). In our lab, we also find that the Smac plays a critical role in neutralizing IAP inhibition of the effector caspases in TRAIL-induced Saos-2 cell apoptosis. Overexpressed Smac can improve the TRAIL-induced apoptosis while its decreased expression will inhibit the apoptosis.

Nevertheless, a compensation rout is implied in the Okada H's research, in which Smac-deficient (Smac(−/−)) mice were created by using homologous recombination in embryonic stem (ES) cells. Smac(−/−) mice were matured normally and did not show any histological abnormalities. Although the cleavage of procaspase-3 in vitro was inhibited

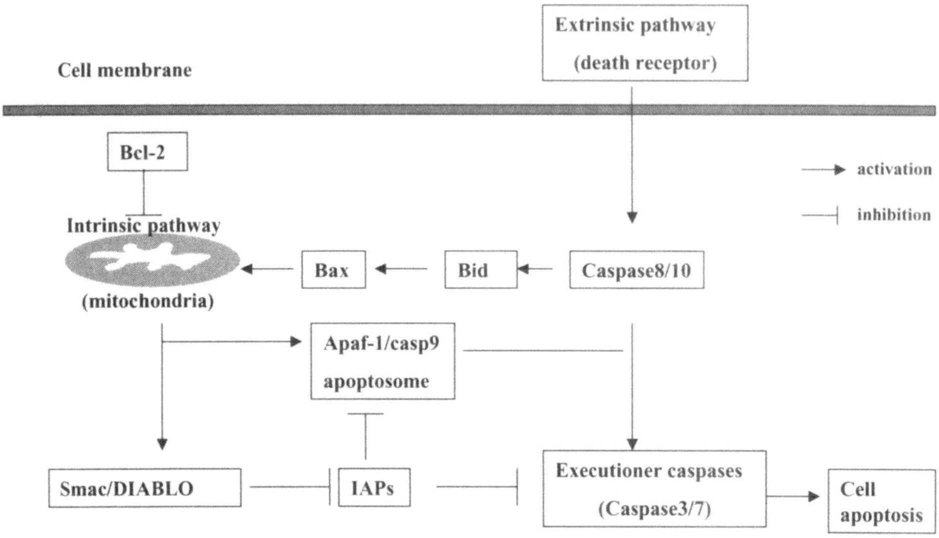

Figure 1. The role of Smac/DIABLO in cell apoptosis. The apoptosis pathway include intrinsic (mitochondria) and extrinsic pathway (death receptor) depending on whether the apoptotic signal originates from the inside or the outside of the cell. There is a cross talk between death receptor and mitochondria-mediated cell death pathway through Bid and Bax. The initiator caspases such as caspas-9, 8 and 10 are activated by different apoptotic signal and the activated initiators can directly process the executioner caspases (caspas-3, 6 and 7) activation. These ultimately lead the cell death. The IAPs can block the function of these initiator and executioners caspases (caspase 9, 3, and 7 et al.) by binding. During the apoptosis, the Smac/DIABLO was induced to release from mitochondrial to cytosol. Then cytosol Smac/DIABLO binds to the IAPs and eliminates the inhibition of IAPs.

in lysates of Smac(−/−) cells, all types of cultured Smac(−/−) cells tested responded normally to all apoptotic stimuli applied. There were also no detectable differences in Fas-mediated apoptosis in the liver in vivo. These suggest the existence of a redundant molecule or molecules capable of compensating for a loss of Smac function. (Okada, H. et al., 2002). This hypothesis had been proved currently. The mature serine protease Omi (also known as HtrA2) was discovered as another mitochondrial direct BIR3-binding protein and a caspase activator. Like mature Smac, matured Omi contains a conserved IAP-binding motif (AVPS) at its N terminus, which is exposed after processing of its N-terminal mitochondrial targeting sequence upon import into the mitochondria. The matured Omi is released together with mature Smac from the mitochondria into the cytosol during apoptosis. The matured Omi can induce apoptosis in human cells in a caspase-independent manner through its protease activity and in a caspase-dependent manner via its ability to disrupt caspase-IAP interaction. Over-expression of Omi/HtrA2 sensitizes cells to undergo apoptosis and its removal by RNA interference reduces cell death. Omi/HtrA2 thus extends the set of mammalian proteins with Reaper-like function that are released from the mitochondria during apoptosis. (Hegde, R. et al., 2002; Martins, L.M. et al., 2002).

Smac and the Bcl-2 Family Proteins

Bcl-2 family members are key regulators of mitochondria-mediated apoptosis by regulating the mitochondrial membrane permeability (Adams, J.M., Cory, S., 1998). They can be divided into two groups according to their functions, namely, anti-apoptotic proteins, like Bcl-2 and Bcl-xl, which promote cell survival, and pro-apoptotic ones like Bax and Bak, which induce cell death. The ratio between the two groups determines cell survival or death. Bax has been implicated in apoptosis induced by various stresses and plays a key role in mediating apoptosis induced by anti-cancer agents (Zhang, L. et al., 2000). Bax is present predominantly in cytosol and translocated to mitochondria when receiving death signals (Hsu, Y.T. et al., 1997; Wolter, K.G. et al., 1997). The translocation of Bax has been proved important for its pro-apoptosis function, and can be blocked by Bcl-2 (Nomura, M. et al., 1999; Murphy, K.M. et al., 2000). Bax-induced apoptosis is mediated at least in part facilitating cytochrome c release from mitochondria (Rsse, T. et al., 1998; Jurgensmeier, J.M. et al., 1998). It accomplishes so by forming pores on mitochondrial membrane, either through inserting into the outer mitochondrial membrane directly or by cooperating with the principle outer-membrane channel, the voltage-dependent anion channel (VDAC) (Antonsson, B. et al., 2000; Shimizu, S. et al., 1999).

More recently, the effects of Bcl-2 and Bax on regulating Smac/DIABLO release have been studied in our lab. We found that Bax induces the release of Smac during etopeside induced Saos-2 cell apoptosis. The mitochondrial Smac decreases as the mitochondrial Bax increases. This coincidence is blocked by Bcl-2 and has no effect on anti-sense smac expressed Soas-2 cells. This reveals Bax may be a key component regulating Smac release. The decrease of mitochondrial Bax in anti-sense smac expressed cells indicates certain factors inside mitochondria, like Smac, might regulate the distribution of Bax between cytosol and mitochondria.

Cell death induced through the p53 pathway is executed by the caspase proteinases, which, by cleaving their substrates, lead to the characteristic apoptotic phenotype. Caspase activation by p53 occurs through the release of apoptogenic factors from the mitochondria, including cytochrome c and Smac/DIABLO. Released Smac/DIABLO facilitates caspase activation through repression of the IAP caspase inhibitor proteins. The release of mitochondrial apoptogenic factors is regulated by the pro- and anti-apoptotic Bcl-2 family proteins, which either induce or prevent the permeabilization of the outer mitochondrial membrane (Schuler, M., and Green, D.R., 2001). In our research, we also proved that p53 induced cell apoptosis is partly through the release of mitochondrial Smac into the cytoplasm and can be blocked by Bcl-2 in our study. The human osteosarcoma cell line, Saos-2, was selected based on its $p53^{null}$ status and its ability to undergo cell death upon p53 expression. In the study, solely by expressing wt p53 in a $p53^{null}$ background, classical apoptotic cell death can be induced in Saos-2 cells, which might be prevented by either anti-sense smac or bcl-2 overexpression. Using the Saos-2 cell line with Bcl-2, overexpression we also demonstrated that transfection of Bcl-2 blocks the Smac release to the cytosol in etoposide-induced apoptosis (p53-independent pathway). It is conceivable that Bcl-2 prevents apoptosis partly throught regulating Smac release. This expends the newly established anti-apoptosis role of Bcl-2. To test if the Bcl-2 takes effect on the cytosolic Smac upstream apoptosis pathway or whether it can block the Smac releasing in cytosol, we successfully constructed expression vector Ad.smac for Smac expression. After Ad.smac is used to infect the Bcl-2 over expressing cell line, Soas-2, the effect of the Bcl-2 inhibiting apoptosis was not observed and the apoptosis was almost similar to what was observed in the control cells, which were absent of Bcl-2 (30%–40% apoptosis). These suggest that Bcl-2 has little effects on cytosolic Smac, or it may affect the Smac on the Smac upstream pathway. Also possible is that these cells are insensitive to apoptosis for lack of cytosolic Smac.

There are many research focused on the Smac function in apoptosis and relationship with the Bcl-2 family proteins. Adrain C reports that Smac/DIABLO export from mitochondria into the cytosol is provoked by cytotoxic drugs and DNA damage, as well as by ligation of the CD95 death receptor. Mitochondrial efflux of Smac/DIABLO, in response to a variety of pro-apoptotic agents, was profoundly inhibited in Bcl-2-overexpressing cells. Thus, in addition to modulating apoptosis-associated mitochondrial cytochrome c release, Bcl-2 also regulates Smac release. However, apoptosis-associated cytochrome c and Smac/DIABLO released from mitochondria do not occur via the same mechanism. Smac/DIABLO efflux from mitochondria is a caspase-catalysed event that occurs downstream of cytochrome c releasing (Adrain, C. et al., 2001). Sun, X.M. also proved that mitochondria and associated Bcl-2 and Bcl-xL proteins might play a functional role in death receptor-induced apoptosis by modulating the release of Smac. This suggests that the relative ratios of XIAP (and other inhibitor-of-apoptosis proteins) to active caspase-3 and Smac may dictate, in part, whether a cell exhibits a type I or type II phenotype. (Sun, X.M. et al., 2002) Deng, Y. suggest that release of Smac/DIABLO from mitochondria through the TRAIL-caspase-8-tBid-Bax cascade is required to remove the inhibitory effect of XIAP and allow apoptosis to process. Inhibition of caspase-9 activity has no effect on TRAIL-induced caspase-3 activation and cell death, whereas expression of the active form of Smac/DIABLO in the cytosol is sufficient to reconstitute TRAIL sensitivity in Bax-deficient cells. His study indicates that Bax-dependent release of Smac/DIABLO, not cytochrome c, from mitochondria mediates the contribution of the mitochondrial pathway to death receptor-mediated apoptosis. (Deng, Y. et al., 2002).

Application of Smac in Cancer Therapy

In most cell apoptosis, the activated caspases are needed. The activation of the effector caspases can be blocked by IAPs. The cytosolic Smac is able to neutralize all known IAPs. So, enough cytosolic Smac must be required in various apoptosis pathways. If this hypothesis is correct, cells not sensitive to apoptosis for lack of cytosol Smac might be re-sensitized to apoptosis by expressing Smac in cytosol. It has been demonstrated that Type II cells, which are not sensitive to death receptor-induced apoptosis, can be converted to sensitive to death receptor-induced apoptosis (Type I cells) by expression of a cytosolic Smac. These indicated that Smac might represent a biologically important and therapeutically promising cornerstone in the IAP realm specifically and apoptosis generally. The ablation or down regulation of a single IAP has modest impact. However, if Diablo/Smac does indeed interact and inhibit the majority of BIRs, its upregulation may mediate a far more general IAP inhibition with dramatic results, which may become a real possibility of cancer therapies. (Holcik, M. et al., 2001; Srinivasula, S.M. et al., 2001; Wu, G. et al., 2000; Du, C. et al., 2000; Srinivasula, S.M. et al., 2000; MacKenzie, A., and LaCasse, E., 2000).

Many experiments had demonstrated that transfection of smac gene into cells could increase their sensitivity to apoptotic stimuli such as UV, etoposide and TRAIL. Simone F. reported that transfection of smac gene or Smac peptides was able to sensitize various tumor cells *in vitro* and malignant glioma cells *in vivo* for death receptor- as well as cytotoxic drug-induced apoptosis. Expression of a cytosolic active Smac or cell-permeable Smac peptides bypassed the Bcl-2 block pathway, which prevented the release of Smac from mitochondria, and also sensitized neuroblastoma, melanoma cells and patient-derived primary neuroblastoma cells *ex vivo* which had been resistant to death receptor-initiated or cytotoxic drug-induced cell death. More importantly, Smac peptides strongly enhanced the antitumor activity of Apo-2L/tumor necrosis factor–related apoptosis-inducing ligand (TRAIL) in an intracranial malignant glioma xenograft model *in vivo*. Complete eradication of established tumors with long-term survival of one mouse was only achieved upon combined treatment with Smac peptides and Apo2L/TRAIL without detectable toxicity to normal brain tissues. Thus, Smac agonists may be promising candidates for potential cytotoxic cancer therapies (Fulda, S. et al., 2002). In our lab, the recombinant adenovirus Ad.samc with a high expression of smac in transfected cell line was constructed and had a significant anti-tumor potential in several models. The effects of Ad.smac on human hepatocarcinoma cell line were also examined both in vitro and in vivo. The result showed that Ad.smac strongly induced the apoptosis of HepG2, a hepatoma cell line, with a complete inhibition of xenograft hepatoma growth in nude mice. The treatmeant of Ad.smac plus 5-FU showed synthetic anti-tumor effect that eliminated the xenograft tumor in nude mice. The results suggests that Ad.smac gene therapy may provide a novel strategy of hepatocellular carcinoma treatment.

ACKNOWLEDGMENTS: This work is partly supported by the grants from National Science Foundation of China, Dept of Health of PLA, and Shanghai Science and Technology Commission.

References

Adams, J.M., and Cory, S. (1998). The Bcl-2 protein family: arbitries of cell survival. Science 281: 1322–1326.

Adida, C., Berrebi, D., Peuchmaur, M., Reyes-Mugica, M., and Altieri, D.C. (1998). Anti-apoptosis gene, survivin, and prognosis of neuroblastoma. Lancet 351(9106): 882–883.

Adrain, C., Creagh, E.M., and Martin, S.J. (2001). Apoptosis-associated release of Smac/DIABLO from mitochondria requires active caspases and is blocked by Bcl-2. EMBO J; 20(23): 6627–36.

Alnemri, E.S., Livingston, D.J., Nicholson, D.W., Salvesen, G., Thornberry, N.A., Wong, W.W., and Yuan, J. (1996). Human ICE/CED-3 protease nomenclature. Cell 87, p. 171.

Ambrosini, G., Adida, C., and Altieri, D. (1997). A novel anti-apoptosis gene, survivin, expressed in cancer and lymphoma. Nat. Med. 3: 917–921.

Antonsson, B., Montessuit, S., Lauper, S., Eskes, R., and Martinou, J.C. (2000). Bax oligomerization is required for channel-forming activity in liposomes and to trigger cytochrome c release from mitochondria. Biochem. J. 345: 271–278.

Boldin, M.P., Goncharov, T.M., Goltsev, Y.V., and Wallach, D. (1996). Involvement of MACH, a novel MORT1/FADD-interacting protease, in Fas/APO-1- and TNF receptor-induced cell death. Cell 14; 85(6): 803–15.

Borden, K.L., and Freemont, P.S. (1996). The RING finger domain: a recent example of a sequence-structure family. Curr. Opin. Struct. Biol. 6(3): 395–401.

Borden, K.L. (2000). RING domains: master builders of molecular scaffolds? J. Mol. Biol. 295: 1103–12.

Budihardjo, I., Oliver, H., Lutter, M., Luo, X., and Wang, X. (1999). Biochemical pathways of caspase activation during apoptosis. Annu. Rev. Cell. Dev. Biol. 15: 269–290.

Cardone, M.H., Roy, N., Stennicke, H.R., Salvesen, G.S., Franke, T.F., Stanbridge, E., Frisch, S., and Reed, J.C. (1998). Regulation of cell death protease caspase-9 by phosphorylation. Science 13; 282 (5392): 1318–21.

Chai, J., Du, C., Wu, J.W., Kyin, S., Wang, X., and Shi, Y. (2000). Structural and biochemical basis of apoptotic activation by Smac/DIABLO. Nature 24; 406(6798): 855–62.

Chai, J., Shiozaki, E., Srinivasula, S.M., Wu, Q., Datta, P., Alnemri, E.S., Shi, Y., and Dataa, P. (2001). Structural basis of caspase-7 inhibition by XIAP. Cell 104(5): 769–80.

Chaohong Sun, Mengli Cai, Robert, P., Meadows, Nan Xu, Angelo, H., Gunasekera, Julia Herrmann, Joe C. Wu, and Stephen W. Fesik (2000). NMR Structure and Mutagenesis of the Third Bir Domain of the Inhibitor of Apoptosis Protein XIAP J. Biol. Chem. 275: 33777–33781.

Crook, N.E., Clem, R.J., and Miller, L.K. (1993). An apoptosis-inhibiting baculovirus gene with a zinc finger-like motif. J. Virol. 67: 2168–74.

Deng, Y., Lin, Y., and Wu, X. (2002). TRAIL-induced apoptosis requires Bax-dependent mitochondrial release of Smac/DIABLO. Genes. Dev. 16(1): 33–45.

Deveraux, Q.L., Leo, E., Stennicke, H.R., Welsh, K., Salvesen, G.S., and Reed, J.C. (1999). Cleavage of human inhibitor of apoptosis protein XIAP results in fragments with distinct specificities for caspases. EMBO J. 18: 5242–51.

Deveraux, Q.L., N. Roy, H.R. Stennicke, T. Van Arsdale, Q. Zhou, M. Srinivasula, E.S. Alnemri, G.S. Salvesen, and J.C. Reed. (1998). IAPs block apoptotic events induced by caspase-8 and cytochrome c by direct inhibition of distinct caspases. EMBO J. 17: 2215–2223.

Deveraux, Q.L., R. Takahashi, G.S. Salvesen, and J.C. Reed. (1997). X-linked IAP is a direct inhibitor of cell death proteases. Nature 388: 300–303.

Du, C., M. Fang, Y. Li, L. Li, and Wang, X. (2000). Smac, a mitochondrial protein that promotes cytochrome c-dependent caspase activation by eliminating IAP inhibition. Cell 102: 33–42.

Earnshaw, W.C., Martins, L.M., and Kaufmann, S.H. (1999). Mammalian caspases: structure, activation, substrates, and functions during apoptosis. Annu. Rev. Biochem. 68: 383–424.

Ellis, R.E., J.Y. Yuan, and H.R. Horvitz. (1991). Mechanisms and functions of cell death. Annu. Rev. Cell. Biol. 7: 663–698.

Enari, M., Sakahira, H., Yokoyama, H., Okawa, K., Iwamatsu, A., and Nagata, S. (1998). A caspase-activated DNase that degrades DNA during apoptosis, and its inhibitor ICAD. Nature 391: 43–50.

Fernandes-Alnemri, T., Armstrong, R.C., Krebs, J., Srinivasula, S.M., Wang, L., Bullrich, F., Fritz, L.C., Trapani, J.A., Tomaselli, K.J., Litwack, G., and Alnemri, E.S. (1996). In vitro activation of CPP32 and Mch3 by Mch4, a novel human apoptotic cysteine protease containing two FADD-like domains. Proc. Natl. Acad. Sci. USA. 93(15): 7464–9.

Fesik, S.W., and Shi, Y. (2001). Controlling caspases. Science 294: 1477–1478.

Fulda, S., Wick, W., Weller, M., and Debatin, K.M. (2002). Smac agonists sensitize for Apo2L/TRAIL- or anticancer drug-induced apoptosis and induce regression of malignant glioma in vivo. Nat. Med. 8(8): 808–15.

Goyal, L., K. McCall, J. Agapite, E. Hartwieg, and Steller, H. (2000). Induction of apoptosis by Drosophila reaper, hid and grim through inhibition of IAP function. EMBO J. 19: 589–597.

Hegde, R., Srinivasula, S.M., Zhang, Z., Wassell, R., Mukattash, R., Cilenti, L., DuBois, G., Lazebnik, Y., Zervos, A.S., Fernandes-Alnemri, T., and Alnemri, E.S. (2002). Identification of Omi/HtrA2 as a mitochondrial apoptotic serine protease that disrupts inhibitor of apoptosis protein-caspase interaction. J. Biol. Chem. 277(1): 432–8.

Holcik, M., Gibson, H., and Korneluk, R.G. (2001). XIAP: apoptotic brake and promising therapeutic target. Apoptosis. 6(4): 253–61. Review.

Hsu, Y.T., Wolter, K.G., and Youle, R.J. (1997). Cytosol-to-membrane redistribution of Bax and Bcl-XL during apoptosis. Proc. Natl. Acad. Sci. USA 94: 3668–3672.

Huang, Y., Park, Y.C., Rich, R.L., Segal, D., Myszka, D.G., and Wu, H. (2001). Structural basis of caspase inhibition by XIAP: differential roles of the linker versus the BIR domain. Cell 104(5): 781–90.

Kawasaki, H., Altieri, D.C., Lu, C.D., Toyoda, M., Tenjo, T., and Tanigawa, N. (1998). Inhibition of apoptosis by survivin predicts shorter survival rates in colorectal cancer. Cancer Res. 58(22): 5071–4.

Kothakota, S., Azuma, T., Reinhard, C., Klippel, A., Tang, J., Chu, K., McGarry, T.J., Kirschner, M.W., Koths, K., Kwiatkowski, D.J., and Williams, L.T. (1997). Caspase-3-generated fragment of gelsolin: effector of morphological change in apoptosis. Science 278: 294–298.

Lee, N., MacDonald, H., Reinhard, C., Halenbeck, R., Roulston, A., Shi, T., and Williams, L.T. (1997). Activation of hPAK65 by caspase cleavage induces some of the morphological and biochemical changes of apoptosis. Proc. Natl. Acad. Sci. USA 94, 13642–13647.

Li, F., Ambrosini, G., Chu, E.Y., Plescia, J., Tognin, S., Marchisio, P.C., and Altieri, D.C. (1998). Control of apoptosis and mitotic spindle checkpoint by survivin. Nature 396: 580–584.

Li, P., D. Nijhawan, I. Budihardjo, S. Srinivasula, M. Ahmad, E. Alnemri, and Wang, X. (1997). Cytochrome c and dATP-dependent formation of Apaf-1/Caspase-9 complex initiates an apoptotic protease cascade. Cell 91: 479–489.

Liston, P., N. Roy, K. Tamai, C. Lefebvre, S. Baird, G. Cherton-Horvat, R. Farahani, M. McLean, J. Ikeda, A. MacKenzie, and R.G. Korneluk. (1996). Suppression of apoptosis in mammalian cells by NAIP and a related family of IAP genes. Nature 379: 349–353.
Liu, X., Li, P., Widlak, P., Zou, H., Luo, X., Garrard, W.T., and Wang, X. (1998). The 40-kDa subunit of DNA fragmentation factor induces DNA fragmentation and chromatin condensation during apoptosis. Proc. Natl. Acad. Sci. USA 95: 8461–8466.
Liu, X., Zou, H., Slaughter, C., and Wang, X. (1997). DFF, a heterodimeric protein that functions downstream of caspase-3 to trigger DNA fragmentation during apoptosis. Cell 89: 175–184.
Liu, Z., Sun, C., Olejniczak, E.T., Meadows, R.P., Betz, S.F., Oost, T., Herrmann, J., Wu, J.C., and Fesik, S.W. (2000). Structural basis for binding of Smac/DIABLO to the XIAP BIR3 domain. Nature 408(6815): 1004–8.
Lu, C.D., Altieri, D.C., and Tanigawa, N. (1998). Expression of a novel antiapoptosis gene, survivin, correlated with tumor cell apoptosis and p53 accumulation in gastric carcinomas. Cancer Res. 58(9): 1808–12.
Matthias, E., Baohua, H., Zehan, C., Robert P. Meadows, Shi-Chung Ng, Lixin Zheng, Michael J. Lenardo, and Stephen W. Fesik (1998). NMR structure and mutagenesis of the FADD (Mort1) death-effector domain: Nature 392 941–945.
MacKenzie, A., and Casse, E.L. (2000). Inhibition of IAP's protection by Diablo/Smac: new therapeutic opportunities? Cell Death and Differentiation 7: 866–867.
Martin, S.J., Reutelingsperger, C.P., McGahon, A.J., Rader, J.A., van Schie, R.C., LaFace, D.M., and Green, D.R. (1995). Early redistribution of plasma membrane phosphatidylserine is a general feature of apoptosis regardless of the initiating stimulus: inhibition by overexpression of Bcl-2 and Abl. J. Exp. Med. 182, 1545–1556.
Martins, L.M., Iaccarino, I., Tenev, T., Gschmeissner, S., Totty, N.F., Lemoine, N.R., Savopoulos, J., Gray, C.W., Creasy, C.L., Dingwall, C., Downward, J. (2002). The serine protease Omi/HtrA2 regulates apoptosis by binding XIAP through a reaper-like motif. J. Biol. Chem. 277(1): 439–44.
Millet, A., Bettaieb, A., Renaud, F., Prevotat, L., Hammann, A., Solary, E., Mignotte, B., and Jeannin, J.F. (2002). Influence of the nitric oxide donor glyceryl trinitrate on apoptotic pathways in human colon cancer cells. Gastroenterology 123(1): 235–46.
Murphy, K.M., Ranganathan, V., Farnsworth, M.L., Kavallaris, M., and Lock, R.B. (2000). Bcl-2 inhibits Bax translocation from cytosol to mitochondria during drug-induced apoptosis of human tumor cells. Cell. Death. Differ. 7: 102–111.
Muzio, M., Stockwell, B.R., Stennicke, H.R., Salvesen, G.S., and Dixit, V.M. (1998). An induced proximity model for caspase-8 activation. J. Biol. Chem. 273: 2926–2930.
N.D. Rawlings, in: A.J. Barrett, N.D. Rawlings, and J.F. Woessner (1998). Handbook of Proteolytic Enzymes, (Academic Press, San Diego, CA), Chapter 247.
Nomura, M., Shimizu, S., Ito, T., Narita, M., Matsuda, H., and Tsujimoto, Y. (1999). Apoptotic cytosol facilitates Bax translocation to mitochondria that involves cytosolic factor regulated by Bcl-2. Cancer Res. 59: 5542–5548.
Okada, H., Suh, W.K., Jin, J., Woo, M., Du, C., Elia, A., Duncan, G.S., Wakeham, A., Itie, A., Lowe, S.W., Wang, X., and Mak, T.W. (2002). Generation and characterization of Smac/DIABLO-deficient mice. Mol. Cell. Biol. 22(10): 3509–3517.
Peter, M.E., and Krammer, P.H. (1998). Mechanisms of CD95 (APO-1/Fas)-mediated apoptosis. Curr. Opin. Immunol. 10(5): 545–51.
Qin, H., Srinivasula, S.M., Wu, G., Fernandes-Alnemri, T., Alnemri, E.S., and Shi, Y. (1999). Structural basis of procaspase-9 recruitment by the apoptotic protease-activating factor 1. Nature 399(6736): 549–57.
Riedl, S.J., Renatus, M., Schwarzenbacher, R., Zhou, Q., Sun, C., Fesik, S.W., Liddington, R.C., and Salvesen, G.S. (2001). Structural basis for the inhibition of caspase-3 by XIAP. Cell 104(5): 791–800.
Rodriguez, J., and Lazebnik, Y. (1999). Caspase-9 and Apaf-1 form an active holoenzyme. Genes. Dev. 13: 3179–3184.
Roy, N., Mahadevan, M.S., McLean, M., Shutler, G., Yaraghi, R., Farahani, S., Baird, A., Besner-Johnson, Lefebvre, C., Kang, X., Salih, M., Aubry, H., Tamai, K., Guan, X., Ioannou, P., Crawford, T.O., Jong, P.J., Surh, L., Ikeda, J.E., Korneluk, R.G., and MacKenzie, A. (1995). The gene for neuronal apoptosis inhibitory protein is partially deleted in individuals with spinal muscular atrophy. Cell 80: 167–178.
Rudel, T., and Bokoch, G.M. (1997). Membrane and morphological changes in apoptotic cells regulated by caspase-mediated activation of PAK2. Science 276, pp. 1571–1574.

Saleh, A., Srinivasula, S.M., Acharya, S., Fishel, R., and Alnemri, E.S. (1999). Cytochrome c and dATP-mediated oligomerization of Apaf-1 is a prerequisite for procaspase-9 activation. J. Biol. Chem. 274: 17941–17945.

Salvesen, G.S., and Dixit, V.M. (1997). Caspases: intracellular signaling by proteolysis. Cell 91: 443–446.

Schuler, M., and Green, D.R. Mechanisms of p53-dependent apoptosis. Biochem. Soc. Trans. 2001 Nov; 29(Pt 6): 684–8.

Shimizu, S., narita, M., and Tsujimoto, Y. (1999). Bcl-2 family proteins regulate the release of apoptogenic cytochrome c by the mitochondrial channel VDAC. Nature 399: 483–487.

Sprick, M.R., Rieser, E., Stahl, H., Grosse, W.A., Weigand, M.A., and Walczak, H. (2002). Caspase-10 is recruited to and activated at the native TRAIL and CD95 death-inducing signalling complexes in a FADD-dependent manner but can not functionally substitute caspase-8. EMBO J. 21(17): 4520–30.

Srinivasula, S.M., Datta, P., Fan, X.J., Fernandes, A.T., Huang, Z., and Alnemri, E.S. (2000). Molecular determinants of the caspase-promoting activity of Smac/DIABLO and its role in the death receptor pathway. J. Biol. Chem. 275(46): 36152–7.

Srinivasula, S.M., Hegde, R., Saleh, A., Datta, P., Shiozaki, E., Chai, J., Lee, R.A., Robbins, P.D., Fernandes, A.T., Shi, Y., and Alnemri (2001). ESA conserved XIAP-interaction motif in caspase-9 and Smac/DIABLO regulates caspase activity and apoptosis. Nature 410(6824): 112–6.

Stennicke, H.R., Deveraux, Q.L., Humke, E.W., Reed, J.C., Dixit, V.M., and Salvesen, G.S. (1999). Caspase-9 can be activated without proteolytic processing. J. Biol. Chem. 274: 8359–8362.

Stennicke, H.R., Jurgensmeier, J.M., Shin, H., Deveraux, Q., Wolf, B.B., Yang, X., Zhou, Q., Ellerby, H.M., Ellerby, L.M., Bredesen, D., Green, D.R., Reed, J.C., Froelich, C.J., and Salvesen, G.S. (1998). Pro-caspase-3 is a major physiologic target of caspase-8. J. Biol. Chem. 273(42): 27084–90.

Stennicke, H.R., Jurgensmeier, J.M., Shin, Q.H., Deveraux, B.B., Wolf, X., Yang, Q., Zhou, H.M., Ellerby, L.M., Ellerby, D., Bredesen, D.R., Green, J.C., Reed, C.J., Froelich, G.S. and Salvesen (1998). Pro-caspase-3 Is a Major Physiologic Target of Caspase-8. J. Biol. Chem. 273 27084–27090.

Sun, C., Cai, M., Gunasekera, A.H., et al. (1999). NMR structure and mutagenesis of the inhibitor-of-apoptosis protein XIAP. Nature 401: 818–22.

Sun, X.M., Bratton, S.B., Butterworth, M., MacFarlane, M., and Cohen, G.M. (2002). Bcl-2 and Bcl-xL inhibit CD95-mediated apoptosis by preventing mitochondrial release of Smac/DIABLO and subsequent inactivation of X-linked inhibitor-of-apoptosis protein. J. Biol. Chem. 277(13): 11345–51.

Takahashi, R., Deveraux, Q., Tamm, I. et al. (1998). A single BIR domain of XIAP sufficient for inhibiting caspases. J. Biol. Chem. 273: 7787–90.

Thornberry, N.A., and Lazebnik, Y. (1998). Caspases: enemies within. Science 281: 1312–1316.

Thornberry, N.A., Rano, T.A., Peterson, E.P., Rasper, D.M., Timkey, T., Garcia-Calvo, M., Houtzager, V.M., Nordstrom, P.A., Roy, S., Vaillancourt, J.P., Chapman, K.T., and Nicholson, D.W. (1997). A combinatorial approach defines specificities of members of the caspase family and granzyme B. Functional relationships established for key mediators of apoptosis. J. Biol. Chem. 272, 17907–17911.

Tikoo, A., O'Reilly, L., Day, C.L., Verhagen, A.M., Pakusch, M., and Vaux, D.L. (2002). Tissue distribution of Diablo/Smac revealed by monoclonal antibodies. Cell Death Differ 9(7): 710–6.

Vanags, D.M., Porn-Ares, M.I., Coppola, S., Burgess, D.H., and Orrenius, S., (1996). Protease involvement in fodrin cleavage and phosphatidylserine exposure in apoptosis. J. Biol. Chem. 271, pp. 31075–31085.

Verhagen, A.M., Coulson, E.J., and Vaux, D.L. (2001). Inhibitor of apoptosis proteins and their relatives: IAPs and other BIRPs. Genome. Biol. 2: 3009.

Verhagen, A.M., Ekert, P.G., Pakusch, M., Silke, J., Connolly, L.M., Reid, G.E., Moritz, R.L., Simpson, R.J., and Vaux, D.L. (2000). Identification of DIABLO, a mammalian protein that promotes apoptosis by binding to and antagonizing IAP proteins. Cell 102: 43–53.

Verhagen, A.M., and Vaux, D.L. (2002). Cell death regulation by the mammalian IAP antagonist Diablo/Smac Apoptosis 7: 163–166.

Vincenz, C., and Dixit, V.M. (1997). Fas-associated death domain protein interleukin-1beta-converting enzyme 2 (FLICE2), an ICE/Ced-3 homologue, is proximally involved in CD95- and p55-mediated death signaling. J. Biol. Chem. 272(10): 6578–83.

Wolter, K.G., Hsu, Y.T., Smith, C.L., Nechushtan, A., Xi, X.G., and Youle, R.J. (1997). Movement of Bax from the cytosol to mitochondria during apoptosis. J. Cell. Biol. 139: 1281–1292.

Wu, G., Chai, J., Suber, T.L., Wu, J.W., Du, C., Wang, X., and Shi, Y. (2000). Structural basis of IAP recognition by Smac/DIABLO. Nature 408(6815): 1008–12.

Yang, X., Chang, H.Y., and Baltimore, D. (1998). Essential role of CED-4 oligomerization in CED-3 activation and apoptosis. Science 281: 1355–1357.

Yuan, J., Shaham, S., Ledoux, S., Ellis, H.M., and Horvitz, H.R. (1993). The C. elegans cell death gene ced-3 encodes a protein similar to mammalian interleukin-1 beta-converting enzyme. Cell 75, 641–652.

Zhang, L., Yu, J., Park, B.H., Kinzler, K.W., and Vogelstein, B. (2000). Role of BAX in the apoptotic response to anticancer agents. Science 290: 989–992.

Zhang, X.D., Zhang, X.Y., Gray, C.P., Nguyen, T., and Hersey, P. (2001). Tumor necrosis factor-related apoptosis-inducing ligand-induced apoptosis of human melanoma is regulated by smac/DIABLO release from mitochondria. Cancer Res. 61(19): 7339–48.

Zou, H., Li, Y., Liu, X., and Wang, X. (1999). An APAF-1-cytochrome c multimeric complex is a functional apoptosome that activates procaspase-9. J. Biol. Chem. 274: 11549–11556.

Chapter 18

Integration of TNF-a Signaling: Crosstalk between IKK, JNK, and Caspases

ANNING LIN*

ABSTRACT: The pro-inflammatory cytokine tumor necrosis factor-a (TNF-a) is a key regulator in immune responses, inflammation and programmed cell death (apoptosis). TNF-a exerts its biological activities through activation of multiple signaling effectors, including IKK, JNK and caspases. Activation of caspases is required for apoptosis, whereas IKK activation results in inhibition of apoptosis through the transcription factor NF-kB, which can induce expression of inhibitors of apoptosis. Activation of JNK stimulates the transcription factor c-Jun/AP-1, which regulates expression of many genes. In contrast to IKK and caspases, the role of JNK activation in apoptosis is highly controversial. Recent studies reveal that crosstalk between IKK, JNK and caspases determines the life or death of a cell in response to TNF-a. Here we propose a model that can explain how integration of TNF-a signaling affects the process of apoptosis.

Key Words: NF-kB; JNK; Apoptosis; TNF-a; Crosstalk

Introduction

The pro-inflammatory cytokine TNF-a exerts its biological activities by binding to its receptors (TNF-R1 ad TNF-R2) and activating multiple signaling pathways in many cell types (Tartaglia and Goeddel, 1992; Tracey and Cerami, 1993). The TNF-R1 signaling complex is composed of the trimerized receptor, TNF-R1-associated death domain protein (TRADD), FAS-associated death domain protein (FADD), TNF-receptor associated factor 2 (TRAF2) and receptor interacting protein (RIP) (Karin and Lin, 2002). FADD recruits and activates pro-Casp8, initiating the apoptotic pathways (Cryns and Yuan, 1998, Budihardjo et al., 1999; Green and Reed, 1998). In contrast, TRAF2 and RIP are involved in activation of c-Jun N-terminal kinase (JNK) and IkB kinase complex (IKK) (Karin and Lin, 2002), resulting in activation of transcription factors c-Jun and NF-kB, respectively.

IKK is a multiple protein complex, having two catalytic subunits, IKKa and IKKb (IKK-1 and IKK-2), and an essential regulatory subunit NEMO/IKKg /IKKAP1 (Ghosh and Karin, 2002). IKK is activated by a variety of extracellular stimuli, from pro-inflammatory cytokines to physical stress, through protein phosphorylation (Karin and Ben-Neriah, 2000; Ghosh and Karin, 2002). Activated IKK phosphorylates inhibitors of NF-kB (IkBs) on specific serines (Ser-32 and -36 in IkBa and Ser-19 and -21 in IkBb), triggering their

*Ben May Institute for Cancer Research, The University of Chicago, 5841 S. Maryland Avenue, MC 6027, Chicago, Il 60637. Tel. (773) 753-1408; Fax. (773) 702-6260; Email: alin@huggins.bsd.uchicago.edu

ubiquitination and subsequent degradation by the 26S proteosome. This degradation unmasks NF-kB's nuclear translocation signals, allowing NF-kB to translocate into the nucleus where it stimulates transcription of specific target genes that are involved in inflammation, immune responses, viral infection and cell survival (Karin and Ben-Neriah, 2000). Compelling genetic evidence shows that in many cell types NF-kB activation is required for cell survival (Karin and Lin, 2002).

c-Jun N-terminal protein kinase (JNK; also known as stress-activated protein kinase, SAPK) is a member of the mitogen-activated protein kinase (MAPK) superfamily (Lin 2003). JNK is activated by a variety of extracellular stimuli through a MAP kinase module, i.e., MAP3K→ MAP2K → MAPK. Two dual-specificity MAP2Ks (JNKK1/MKK4/SEK1 and JNKK2/MKK7) for JNK activation have been identified (Lin et al., 1995; Lu et al., 1997), whereas several MAP3Ks, including members of the MEKK family, ASK1, MLK, TAK1 and TPL-2, have been reported to act as MAP3Ks for JNK (Lin 2003). In addition, several scaffold proteins (JIP, b-arrestin and JSAP1) have been shown to regulate JNK activation (Davis 2000). Once activated, JNK phosphorylates c-Jun on its N-terminal transactivation domain at Ser63 and Ser73, resulting in the augmentation of c-Jun transcriptional activity and its ability to cooperate with *Ha*-Ras in transformation(Lin 2003). JNK also regulates transcription factors other than c-Jun, such as ATF2, Elk-1, p53 and c-Myc and non-transcription factors, such as members of the Bcl-2 family (Bcl-2 and Bcl-x_L) (Lin 2003). The role of JNK activation in TNF-a induced apoptosis is highly controversial, as being reported to have a pro-apoptotic, anti-apoptotic, or no role in this process (Lin 2003).

In this review, we will focus on the mechanisms underlying the crosstalk between IKK, JNK and casapases with respect to TNF-a -induced apoptosis. We propose a model to explain how integration of these signaling pathways determines the life or death of a cell in response to TNF-a .

Caspase-Mediated Proteolysis of IKKb Is Required for TNF-a Induced Apoptosis

Although TNF-a is able to activate caspases, it does not typically induce apoptosis in most cell types unless activation of NF-kB is inhibited (Karin and Lin, 2002). Under certain circumstances TNF-a does induce apoptosis, despite the concurrent activation of NF-kB (Chinnaiyan et al., 1995; Tang et al., 2001a). This is mainly due to inactivation of the NF-kB pathway by caspases, which overrides the protection conveyed by NF-kB during TNF-a -induced apoptosis (Karin and Lin, 2002).

It has been reported that caspases can cleave several key components of the NF-kB pathway. This list includes RIP, TRAF2, IKKb, IkBa, NF-kB and IAPs (Karin and Lin, 2002). Proteolysis of RIP by caspase 8 eliminates the ability of RIP to activate IKK (Lin et al., 1999). Similarly, TRAF2 can be proteolyzed or sequestered into a non-soluble cellular compartment upon recruitment to CD30, which is a member of the TNFR family (Duckett and Thompson, 1997). Caspase 3-mediated proteolysis of IkBa separates the regulatory N-terminal domain, which contains Ser32 and Ser36, from the body of IkBa (Barkett et al., 1997). The resultant N-terminal truncated IkBa mutant is resistant to Ubiquitination-mediated proteolysis of IkBa by the 26S proteosome and thereby functioning as a super-repressor-like IkBa to inhibit NF-kB activation (Fig. 1). Phosphorylation of IkBa by IKK or substitution of Ser32 and Ser36 with glutamates to mimic their phosphorylation blocks the

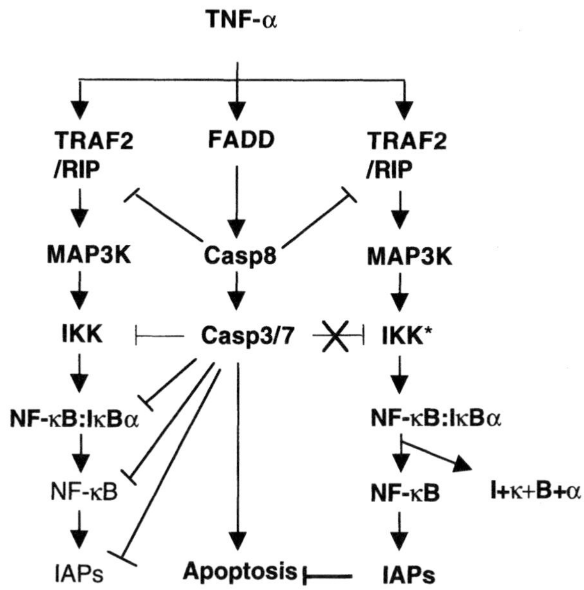

Figure 1. Schematic illustration of the role of IKKb proteolysis on TNF-a -induced apoptosis. See the text for details. IKK*, the IKK complex containing the caspase-resistant IKKb.

proteolysis of IkBa by caspases (Barkett et al., 1997). This suggests that IKK-dependent, proteosome-mediated proteolysis of IkBa can prevent caspase-mediated cleavage of IkBa (see discussion below).

Several caspases can cleave NF-kB/RelA, resulting in a C-terminal truncated RelA that can bind DNA but is unable to activate transcription (Karin and Lin, 2002). In this way, caspases converts NF-kB into its own inhibitor. Caspases also directly catalyzes the proteolysis of NF-kB-induced anti-apoptotic gene products, including cIAP1, xIAP and Bcl-X_L, thereby eliminating the anti-apoptotic effectors induced by NF-kB (Karin and Lin, 2002).

IKKb is essential for NF-kB activation by TNF-a (Karin and Ben-Neriah, 2000). Interestingly, IKKb, but not IKKa or IKKg, is proteolyzed by caspase 3-related caspases during TNF-a -induced apoptosis (Tang et al., 2001a). Proteolysis of IKKb eliminates its enzymatic activity and also generates its own inhibitor (Tang et al., 2001a). A caspase-resistant form of IKKb suppresses TNFa -induced apoptosis when overexpressed (Tang et al., 2001a). In cells expressing wild type IKKb, IKK was activated repeatedly in response to TNFa, but eventually its activation was diminished, as IKKb was proteolyzed (Tang et al., 2001a). By contrast, in cells expressing the caspase-resistant IKKb mutant, IKK could be activated repeatedly for a long time. As long as IKK is active, it phosphorylates IkBa, thereby preventing it to be proteolyzed by caspases (Fig. 1). The phosphorylation allows IkBa to be ubiquitinated and proteolyzed by the 26S proteosome, releasing NF-kB so it can translocate into the nucleus. This may enable NF-kB to escape from caspase-mediated proteolysis in the cytoplasm. Consequently, activation of NF-kB will lead to accumulation of IAPs. Although IAPs are substrates of caspases, accumulated IAPs may tip the balance toward inhibition of caspases by IAPs, thereby suppressing TNF-a induced apoptosis (Fig. 1). Thus, prolonged repeated IKK activation is critical in determining life or death of a cell in response to TNF-a.

Negative Regulation of JNK by NF-kB Contributes to Inhibition of TNF-a -Induced Apoptosis

The role of JNK activation in TNF-a -induced apoptosis is highly controversial. Several studies have implicated both pro-apoptotic and anti-apoptotic functions for JNK, whereas evidence also exists that JNK plays no role in the apoptotic process (Lin 2003). These inconsistencies have partly been attributed to different cellular context since the effect of JNK activation on apoptosis is influenced by other cellular activities. However, the molecular basis of this "cellular context" is obscure.

Recent studies reveal that the crosstalk between JNK and NF-kB affects the role of JNK activation in TNF-a -induced apoptosis. JNK activation by TNF-a was transient in wildtype fibroblasts, but was prolonged in cells deficient in either IKKb (Tang et al., 2001b), which is essential for TNF-a -induced NF-kB activation, or RelA (Tang et al., 2001b; De Smaele et al., 2001), which is a major transactivating subunit of NF-kB. Similar findings were reported using cells expressing a "super-repressor" IkBa mutant (Javelaud et al., 2001; De Smaele et al, 2001). Therefore, NF-kB activation prevents prolonged JNK activation by TNF-a (Tang et al., 2001b). The inhibition is likely mediated by NF-kB-induced inhibitors. Candidate NF-kB-induced JNK inhibitors include X-chromosome-linked IAP (XIAP) (Tang et al., 2001b) and GADD45b (De Smaele et al., 2001). Since genetic disruption of either gene has no significant effect on JNK activation in vivo, a compensation mechanism may exist and the inhibitory effect of NF-kB may be mediated by multiple proteins (Lin 2003). It is not clear at what point the JNK pathway is inhibited by NF-kB-induced inhibitors. It is likely that NF-kB induced inhibitor(s) may target the component(s) upstream of MAP2Ks in the TNF-a signaling pathway (Y. Minemoto and A. Lin, unpublished results). Note that NF-kB-mediated inhibition may target a unique component(s) in the TNF-a activated JNK pathway since interleukin (IL)-1 or UV-induced JNK activation is not affected (Tang et al., 2001b; Y. Minemoto, B.C. Dibling and A. Lin, unpublished results). Abrogation of prolonged JNK activation by expressing a dominant negative JNKK2(K149M) mutant (Tang et al., 2001b) or using a synthetic JNK inhibitor SP600125 (G. Tang, Y. Minemoto, F. Tang and A. Lin, unpublished results) suppressed TNF-a -induced apoptosis in *RelA*$^{-/-}$ or *Ikk*b$^{-/-}$ fibroblasts. Therefore, NF-kB inhibits TNF-a -induced apoptosis by blocking caspase activation and preventing prolonged JNK activation. Interestingly, it was found that JNK activation by TNF-a was transient in TNF-a resistant human breast carcinoma MCF-7-R cells, but was prolonged in TNF-a sensitive MCF-7 cells (Tang et al., 2002). Conversion of JNK activation from prolonged to transient in MCF-7 cells inhibited TNF-a -induced apoptosis (Tang et al., 2002). Conversely, inhibition of NF-kB resulted in prolonged JNK activation by TNF-a in MCF-7-R cells and sensitized cells to apoptosis (Tang et al., 2002). Restoration of transient JNK activation by SP600125 inhibited TNF-a -induced apoptosis in MCF-7-R cells (Tang et al., 2002). These observations further support the notion that prolonged, but not transient, JNK activation contributes to TNF-a -induced apoptosis (Lin 2003). It is important to note that prolonged JNK activation alone is not sufficient to induce apoptosis. Expression of a constitutively active JNKK2-JNK1 fusion protein (Zheng et al., 1998), which produces prolonged JNK activity, does not induce apoptosis in *RelA*$^{-/-}$ fibroblasts (Tang 2001b) or other cells examined (G. Tang, Y. Minemoto, F. Tang and A. Lin, unpublished results). Thus, prolonged JNK activation may only promote apoptosis in cells where the apoptotic process has been initiated.

We envision that JNK acts as a modulator rather an intrinsic component of the apoptotic machinery. Therefore, JNK activation may promote but not induce apoptosis. We

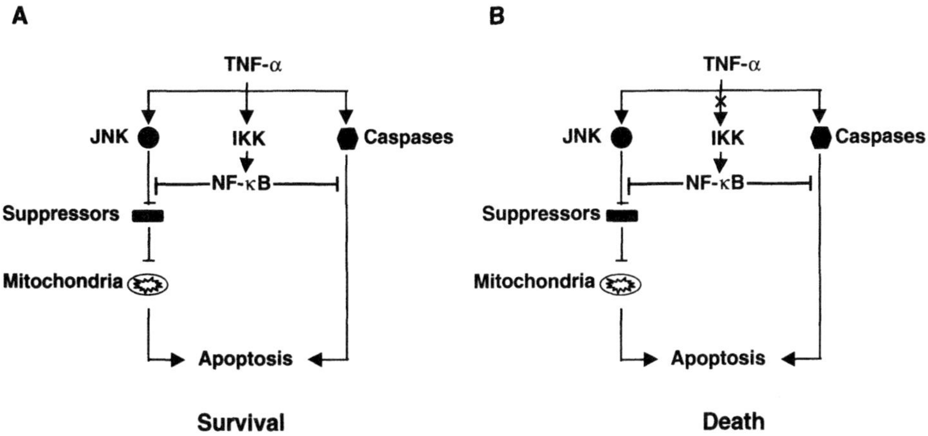

Figure 2. The "breaking the brake" on apoptosis model that explains the role of JNK activation in TNF-a induced apoptosis. (A) Activation of caspase activation initiates and executes apoptosis, whereas prolonged JNK activation promotes apoptosis by inactivating suppressors of the intrinsic death pathway, thereby breaking the brake on apoptosis. Activation of NF-kB by TNF-a blocks caspase activation and prevents prolonged JNK activation, resulting in inhibiting TNF-a -induced apoptosis. (B) Inactivation of NF-kB abolishes inhibition on both caspases and JNK. Therefore, TNF-a induces apoptosis.

hypothesize that JNK may inactivate some yet-to-be-identified suppressors of the apoptotic machinery, resulting in "breaking the brake" on apoptosis. Based on this model, activation of JNK contributes to apoptosis only if cells are undergoing apoptosis. This can explain how prolonged JNK activation alone does not kill, but is able to promote TNF-a induced apoptosis (Fig. 2). Since activation of NF-kB inhibits caspases and prevents prolonged JNK activation (Tang et al., 2001b; Tang et al., 2002; Lin 2003), TNF-a does not typically induce apoptosis in many cell types. However, in the absence of NF-kB, such as in *RelA−/−* fibroblasts, TNF-a induces apoptosis through activation of caspases, which initiate and execute apoptosis, and prolonged JNK activation, which releases inhibition on the apoptotic pathways (Fig. 2). This model can also explain how JNK does not induce apoptosis, but inhibition of JNK activation abolishes UV-induced apoptosis (Tournier et al., 2000). It is known that UV induces apoptosis by activating the mitochondrial-dependent death pathway (Tournier et al., 2000) and it also induces prolonged JNK activation by an unknown mechanism. Prolonged activation of JNK by UV may inactivate suppressors of the mitochondrial-dependent death pathway, thereby promoting apoptosis. Note that inhibition of the prolonged JNK activation only suppresses TNF-a induced death (Tang et al., 2001b), but completely abolishes UV-induced death (Tournier et al., 2000). Considering TNF-a activates both extrinsic (receptor-mediated) and intrinsic (mitochondrial-dependent) death pathways, while UV activates only the intrinsic death pathway, it is likely that prolonged JNK activation may inactivate suppressors involved in blocking the mitochondrial-dependent death pathway (Lin 2003). Therefore, the role of JNK activation in apoptosis can be thought of as being analogous to driving a car. If you only start the engine (activation of apoptotic pathways) or release the brake (prolonged activation of JNK), the car (apoptosis) will not move. The car (apoptosis) will move only when you start the engine (activation of apoptotic pathways) and release the brake (prolonged activation of JNK). It will be interesting to determine whether this "breaking the brake" on apoptosis model can also explain the role of JNK activation in other types of apoptosis.

Concluding Remarks

TNF-a regulates many cellular activities, such as immune responses, inflammation and apoptosis. The ultimate fate of a cell exposed to TNF-a largely depends on signal integration between TNF-a effectors, such as IKK, JNK and caspases. Although NF-kB activation by TNF-a can suppress apoptosis through induction of IAPs that inhibit caspases, which explains why TNF-a is not typically a killer in many cell types, proteolysis of the key components of the NF-kB pathway, such as IKKb, may tip the balance toward apoptosis. It will be of interesting to determine whether proteolysis of IKKb by caspases occurs under certain physiological or pathological conditions. NF-kB also negatively regulates TNF-a -induced JNK activation, which otherwise becomes prolonged and contributes to TNF-a -induced apoptosis. It has yet to be determined how NF-kB-induced inhibitors prevent prolonged JNK activation and how prolonged but not transient JNK activation promotes TNF-a -induced apoptosis. Since deregulation of IKK, JNK and caspases has been implicated in cancer and other diseases, such as Alzheimer's and Parkinson's disease, heart hypertrophy and ischemia (Lin 2003), investigation of signaling integration between IKK, JNK and caspases should provide important insight into the TNF-a biology and novel strategies to target these signaling pathways for prevention and treatment of human disease and cancer.

ACKNOWLEDGMENTS: I thank members of the Lin laboratory for discussion. I apologize to the authors for not being able to directly cite their work due to space constraints. This work is supported by grants from the National Institutes of Health (NIH).

References

1. Tartaglila, L.A., and Goeddel, D.V. (1992). Two TNF receptors. Immunol. Today *13*, 151–153.
2. Tracey, K.J., and Cerami, A. (1993). Tumor necrosis factor: an updeated review of its biology. Crit. Care Med. *21*, S415–S422.
3. Karin, M., and Lin, A. (2002). NF-kB at the crossroads of life and death. Nature Immunol. *3*, 221–227.
4. Cryns, V., and Yuan, J. (1998). Proteases to die for. Genes Dev. *12*, 1551–1570.
5. Bydihardjo, I., Oliver, H., Lutter, M., Luo, X., and Wang, X. (1999). Biochemical pathways of caspase activation during apoptosis. Annu. Rev. Cell Dev. Biol. *15*, 269–290.
6. Green, D.R., and Reed, J. C. (1998). Mitochondria and apoptosis. Science *281*, 1309–1313.
7. Ghosh, S., and Karin, M. (2002) Missing pieces in the NF-kB puzzle. Cell *109*, Suppl: S81–96.
8. Karin, M., and Ben-Neriah, Y. (2000). Phosphorylation meets ubiquitination: the control of NF-kB activity. Annu Rev Immunol. *18*, 621–663.
9. Lin, A. (2003). Activation of the JNK signaling pathway: breaking the brake on apoptosis. BioEssays *25*, 17–24.
10. Lin, A., Minden, A., Martinetto, H., Claret, F.X., Lange-Carter, C., Mercurio, F., Johnson, G.L., and Karin, M. (1995). Identification of a dual specificity kinase that activates the Jun kinases and p38-Mpk2. Science *268*, 286–290.
11. Lu, X., Nemoto, S., and Lin, A. (1997). Identification of c-Jun NH2-terminal protein kinase (JNK)-activating kinase 2 as an activator of JNK but not p38. J. Biol. Chem. *272*, 24751–24754.
12. Davis, R.J. (2000). Signal transduction by the JNK group of MAP kinases. Cell *103*, 239–252.
13. Chinnaiyan, A. M., O'Rourke, K., Tewari, M., and Dixit, V.M. (1995). FADD, a novel death domain-containing protein, interacts with the death domain of Fas and initiates apoptosis. Cell *81*, 505–512.
14. Tang, G., Yang, J., Minemoto, Y., Lin, A. (2001a). Blocking caspase-3-mediated proteolysis of IKKb suppresses TNF-a -induced apoptosis. Mol. Cell *8*, 1005–1016.
15. Lin, Y., Devin, A., Rodriguez, Y., Liu, Z.G. (1999). Cleavage of the death domain kinase RIP by caspase-8 prompts TNF-induced apoptosis. Genes Dev *13*, 2514–2526.

16. Duckett, C.S., and Thompson, C.B. (1997). CD30-dependent degradation of TRAF2: implications for negative regulation of TRAF signaling and the control of cell survival. Genes Dev *11*, 2810–2821.
17. Barkett, M., Xue, D., Horvitz, H.R., Gilmore, T.D. (1997). Phosphorylation of IkBa inhibits its cleavage by caspase CPP32 in vitro. J. Biol. Chem. *272*, 29419–29422.
18. Tang, G., Minemoto, Y., Dibling, B., Purcell, N.H., Li, Z., Karin, M., and Lin, A. (2001b). Inhibition of JNK activation through NF-kB target genes. Nature *414*, 313–317.
19. Javelaud, D., Besancon, F. (2001). NF-kB activation results in rapid inactivation of JNK in TNF-a -treated Ewing sarcoma cells: a mechanism for the anti-apoptotic effect of NF-kB. B. Oncogene *20*, 4365–4372.
20. De Smaele, E., Zazzeroni, F., Papa, S., Nguyen, D.U., Jin, R., Jones, J., Cong, R., Franzoso, G. (2001). Induction of gadd45b by NF-kB downregulates pro-apoptotic JNK signaling. Nature *414*, 308–311.
21. Tang, F., Tang, G., Xiang, J., Dai, Q., Rosner, M.R., and Lin, A. (2002). The absence of NF-kB-mediated inhibition of c-Jun N-terminal kinase activation contributes to tumor necrosis factor a -induced apoptosis. Mol. Cell. Biol. *22*, 8571–8579.
22. Zheng, C., Xiang, J., Hunter, T., and Lin, A. (1999). The JNKK2-JNK1 fusion protein acts as a constitutively active c-Jun kinase that stimulates c-Jun transcription activity. J. Biol. Chem. *274*, 28966–28971.
23. Tournier, C., Hess, P., Yang, D.D., Xu, J., Turner, T.K., Nimnual, A., Bar-Sagi, D., Jones, S.N., Flavell, R.A., and Davis, R.J. (2000). Requirement of JNK for stress-induced activation of the cytochrome c-mediated death pathway. Science *288*, 870–874.

Index

MIX
Papier aus verantwortungsvollen Quellen
Paper from responsible sources
FSC® C105338

If you have any concerns about our products,
you can contact us on
ProductSafety@springernature.com

In case Publisher is established outside the EU,
the EU authorized representative is:
Springer Nature Customer Service Center GmbH
Europaplatz 3, 69115 Heidelberg, Germany

Printed by Libri Plureos GmbH
in Hamburg, Germany